The Human Brain

Student's Self-test Coloring Book

The Human Brain Student's Self-test Coloring Book

Chief Consultants:
Joshua Gowin, Ph.D.
W. Wade Kothmann, Ph.D.

First edition for North America published in 2016
by Barron's Educational Series, Inc.

First published in 2016 by Global Book Publishing Pty Ltd
Part of The Quarto Group
Level 4, Sheridan House,
114 Western Road,
Hove, BN3 1DD, UK

Conceived, designed, and produced by Global Book Publishing Pty Ltd

Edited and designed by D & N Publishing, Baydon, Wiltshire, UK

Illustrations by Medical Artist Ltd (www.medical-artist.com)

Additional illustrations Mike Gorman, Thomson Digital, Glen Vause

CONTRIBUTORS

Zubair Ahmed, Ph.D.
Bethany Brookshire, Ph.D.
Shelly Xuelai Fan, Ph.D.
Jordan Gaines Lewis, Ph.D.
James Graham, Ph.D.
Ashley Lauren Juavinett, B.S.
W. Wade Kothmann, Ph.D.
Lauren Sakowski, B.S.
Nick Wan, B.S.

All inquiries should be addressed to:
Barron's Educational Series, Inc.
250 Wireless Boulevard
Hauppauge, NY 11788
www.barronseduc.com

ISBN: 978-1-4380-0870-7

Printed in China

9 8 7 6 5 4 3 2 1

Contents

Introduction 6

How to Use This Book 7

Overview of the Human Body 8

Nervous System Overview 12

Microscopic Structure of the Nervous System 19

Development and Aging of the Nervous System 31

Spinal Cord 40

Spinal Nerves and Innervation of Skin, Muscles, and Joints 52

Autonomic and Enteric Nervous System 65

Brainstem 70

Cranial Nerves 78

Cerebellum 99

Thalamus, Subthalamus, Epithalamus, and Pretectum 107

Hypothalamus and Pituitary Gland 114

Basal Ganglia 121

Topography of the Cerebral Hemispheres 126

Cerebral Cortex 136

Limbic System 146

Language, Sleep, Spatial Perception, Neural Plasticity, and Aging 151

Overview of Major Pathways 162

Overview of Neurotransmitter Systems 170

Blood Supply and Support of the Central Nervous System 174

Index 188

Introduction

At nearly 100 billion, the number of neurons in the human brain is about the same as the number of stars in the Milky Way galaxy. What's more, the average neuron is connected to 1,000 other neurons, meaning there are roughly 100 trillion synapses in each human brain. It is safe to say that this 3 lb (1.4 kg) organ is complex. Many scientists believe the brain is the next great frontier in human exploration. There are multitudes of unsolved questions about how it functions, what its cells do, and how all of the nervous system works together to make us each unique as individuals and yet also very similar collectively as human beings. Despite the many mysteries remaining, we have learned a great deal about the brain and the nervous system over the past century. In the pages that follow, we have tried to distill this knowledge into usable, engaging, entertaining, and informative text and illustrations.

Aside from being fascinating, learning about the brain is also practical for medical and nursing students, scientists, healthcare workers, educators, and many other professionals who require a working knowledge of neuroanatomy and physiology. For people who already have an understanding, it helps to brush up from time to time, and the visual aid provided here can also be an excellent resource for teaching others. The advantage of this book relative to other learning materials is that it provides an active learning experience. On each page, you can quiz yourself by filling in the names of the structures, and then you can color-code them to help store the knowledge. And while studying the brain may cause stress for some, coloring has been shown to reduce anxiety! This should not only make using this book more enjoyable, but also better for retaining the information presented herein. We hope you will find it a valuable resource as you dive into learning about the human brain.

How This Book Is Organized

Featuring more than 200 computer-rendered line drawings in a clean design, this book is divided into 20 comprehensive chapters. It covers the major divisions of the nervous system—from the peripheral to the central and from the microscopic to the large—and also includes an overview of how all the various parts work together. It covers both the development and aging of the brain, and provides an overview of some of the functions of the brain, such as how we store memories and navigate the world. This is an important facet in the study of anatomy—not only learning the names and locations of the parts of the body, but also discovering how these parts relate to each other and function together.

As you work your way through the book, you will gain both a clear understanding of anatomy and a deeper appreciation for the human brain—an amazingly complicated yet highly coordinated machine.

How to Use This Book

This book is designed to assist students and professionals to identify body parts and structures, and the colored leader lines aid the process by clearly pointing out each body part. The functions of coloring and labeling allow you to familiarize yourself with individual parts of the body and then check your knowledge.

Coloring is best done using either pencils or ballpoint pens (not felt-tip pens) in a variety of dark and light colors. Where possible, you should use the same color for like structures, so that all completed illustrations can be utilized later as visual references. According to anatomical convention, the color green is usually reserved for lymphatic structures, yellow for nerves, red for arteries, and blue for veins.

Labeling the colored leader lines that point to separate parts of the illustration enables you to test and then check your knowledge using the answers that are printed at the bottom of the page.

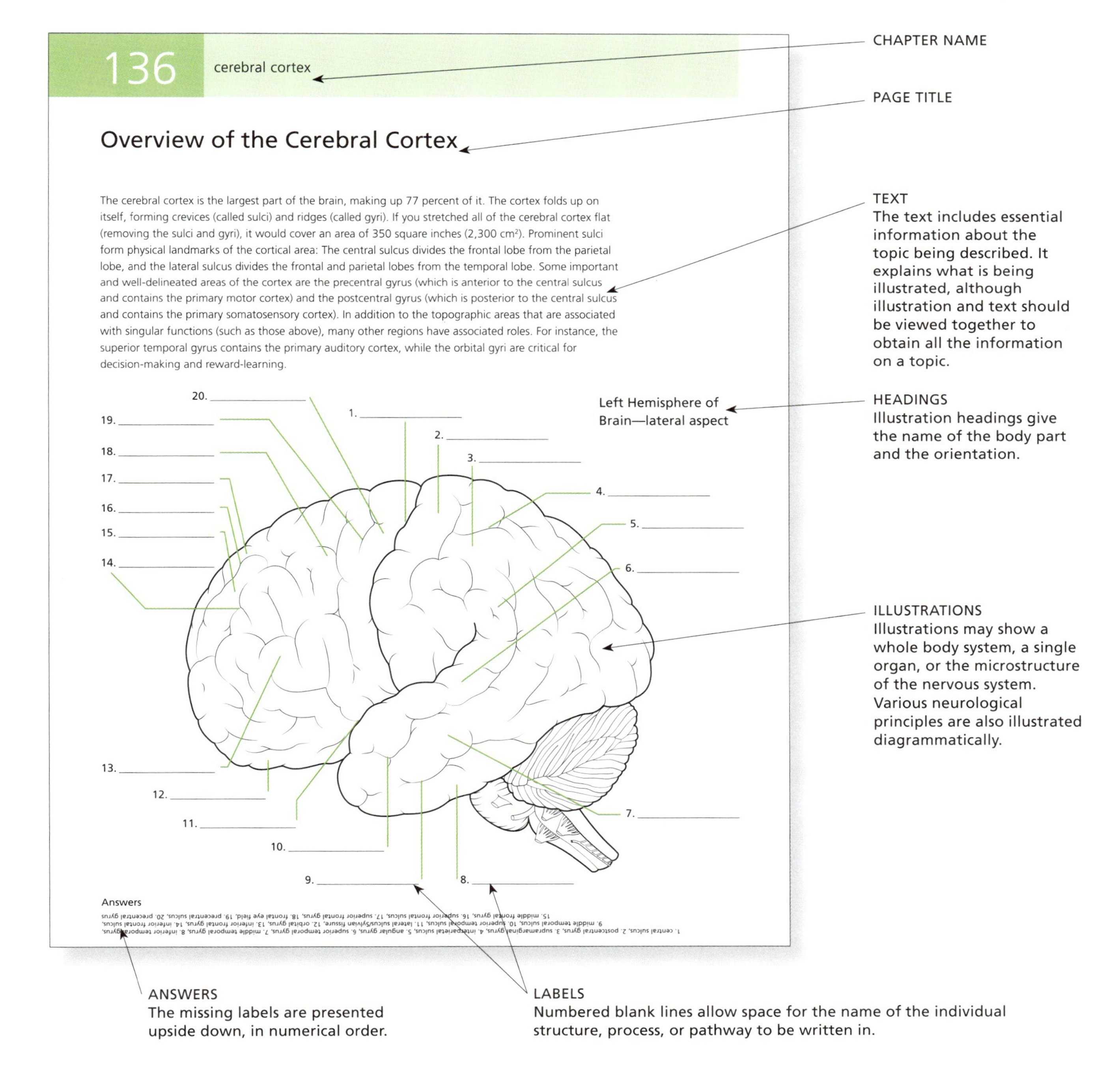
136
cerebral cortex
Overview of the Cerebral Cortex
The cerebral cortex is the largest part of the brain, making up 77 percent of it. The cortex folds up on itself, forming crevices (called sulci) and ridges (called gyri). If you stretched all of the cerebral cortex flat (removing the sulci and gyri), it would cover an area of 350 square inches (2,300 cm²). Prominent sulci form physical landmarks of the cortical area: The central sulcus divides the frontal lobe from the parietal lobe, and the lateral sulcus divides the frontal and parietal lobes from the temporal lobe. Some important and well-delineated areas of the cortex are the precentral gyrus (which is anterior to the central sulcus and contains the primary motor cortex) and the postcentral gyrus (which is posterior to the central sulcus and contains the primary somatosensory cortex). In addition to the topographic areas that are associated with singular functions (such as those above), many other regions have associated roles. For instance, the superior temporal gyrus contains the primary auditory cortex, while the orbital gyri are critical for decision-making and reward-learning.
Left Hemisphere of Brain—lateral aspect
1.
2.
3.
4.
5.
6.
7.
8.
9.
10.
11.
12.
13.
14.
15.
16.
17.
18.
19.
20.
Answers
1. central sulcus, 2. postcentral gyrus, 3. supramarginal gyrus, 4. intraparietal sulcus, 5. angular gyrus, 6. superior temporal gyrus, 7. middle temporal gyrus, 8. inferior temporal gyrus, 9. middle temporal sulcus, 10. superior temporal sulcus, 11. lateral sulcus/Sylvian fissure, 12. orbital gyrus, 13. inferior frontal gyrus, 14. inferior frontal sulcus, 15. middle frontal gyrus, 16. superior frontal gyrus, 17. superior frontal sulcus, 18. frontal eye field, 19. precentral sulcus, 20. precentral gyrus
CHAPTER NAME
PAGE TITLE
TEXT
The text includes essential information about the topic being described. It explains what is being illustrated, although illustration and text should be viewed together to obtain all the information on a topic.
HEADINGS
Illustration headings give the name of the body part and the orientation.
ILLUSTRATIONS
Illustrations may show a whole body system, a single organ, or the microstructure of the nervous system. Various neurological principles are also illustrated diagrammatically.
ANSWERS
The missing labels are presented upside down, in numerical order.
LABELS
Numbered blank lines allow space for the name of the individual structure, process, or pathway to be written in.

Overview of the Human Body

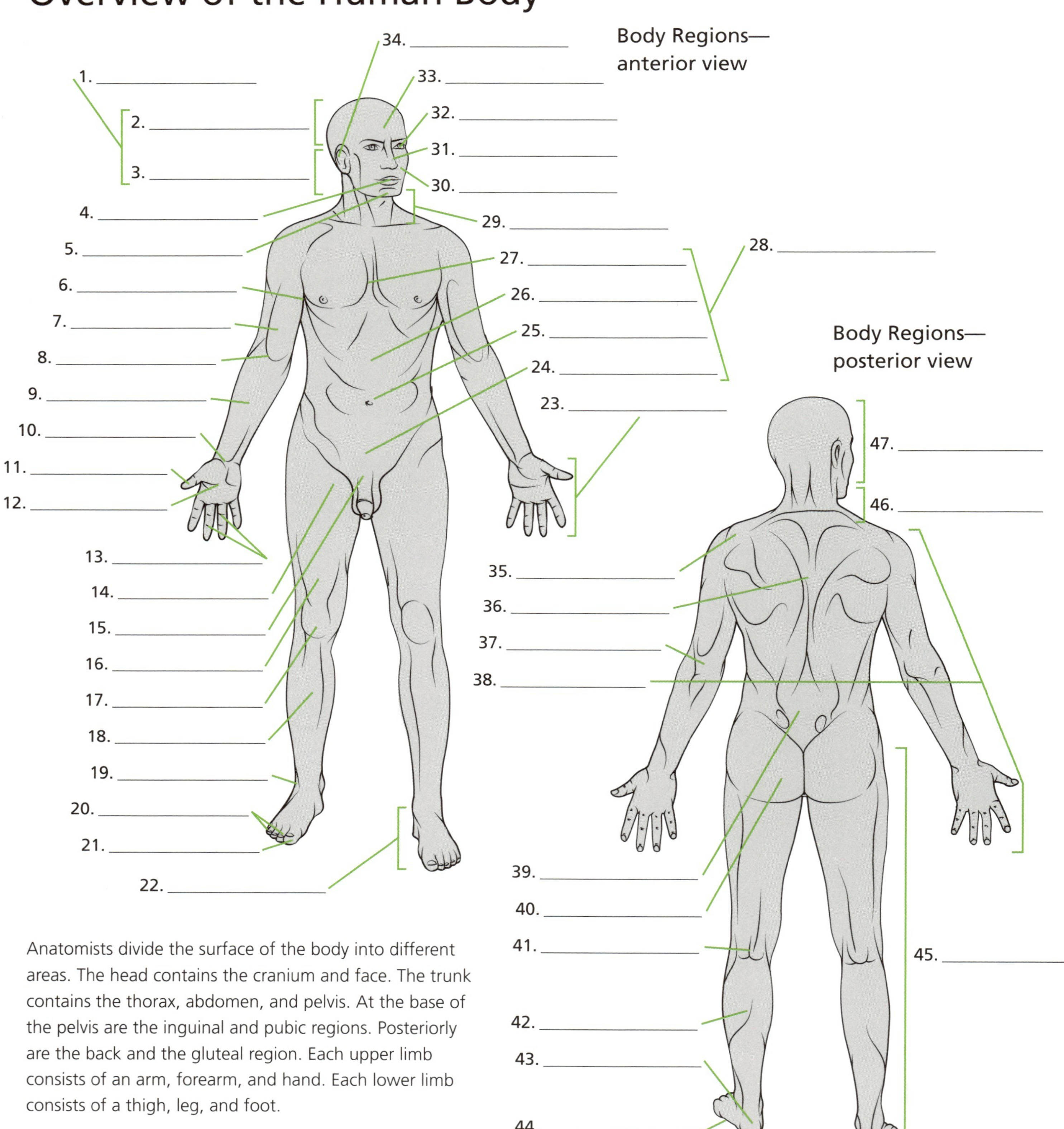

Anatomists divide the surface of the body into different areas. The head contains the cranium and face. The trunk contains the thorax, abdomen, and pelvis. At the base of the pelvis are the inguinal and pubic regions. Posteriorly are the back and the gluteal region. Each upper limb consists of an arm, forearm, and hand. Each lower limb consists of a thigh, leg, and foot.

Answers

1. head, 2. cranium (cranial), 3. face (facial), 4. mouth (oral), 5. chin (mental), 6. axilla (axillary), 7. brachium (brachial), 8. elbow (antecubital), 9. antebrachium (antebrachial), 10. wrist (carpal), 11. pollex (thumb), 12. palm (palmar), 13. digits (digital or phalangeal), 14. inguen (inguinal), 15. pubis (pubic), 16. femur (femoral), 17. patella (patellar), 18. crus (crural), 19. tarsus (tarsal), 20. digits (digital or phalangeal), 21. hallux (big toe), 22. pes (pedal), 23. hand, 24. pelvis (pelvic), 25. umbilicus (umbilical), 26. abdomen (abdominal), 27. thorax (thoracic), 28. trunk, 29. neck (cervical), 30. cheek (buccal), 31. nose (nasal), 32. eye (orbital or ocular), 33. forehead (frontal), 34. ear (otic), 35. shoulder (acromial), 36. back (dorsal), 37. olecranon (olecranal), 38. upper limb, 39. lower back (lumbar), 40. gluteus (gluteal), 41. popliteus (popliteal), 42. sura (sural), 43. calcaneus (calcaneal), 44. sole (plantar), 45. lower limb, 46. neck (cervical), 47. head

The body cavities are spaces containing the internal organs (viscera). The main cavities are the thoracic and abdominopelvic cavities in the torso and the cranial cavity in the head. The thoracic (or chest) cavity contains the heart, lungs, trachea, and esophagus. The abdominopelvic cavity is divided into the abdominal cavity, which contains most of the gastrointestinal tract, the kidneys, and the adrenal glands; below the abdominal cavity, the pelvic cavity contains the urogenital system and the rectum. The cranial cavity contains the brain and extends caudally as the spinal canal.

Body Cavities—sagittal view

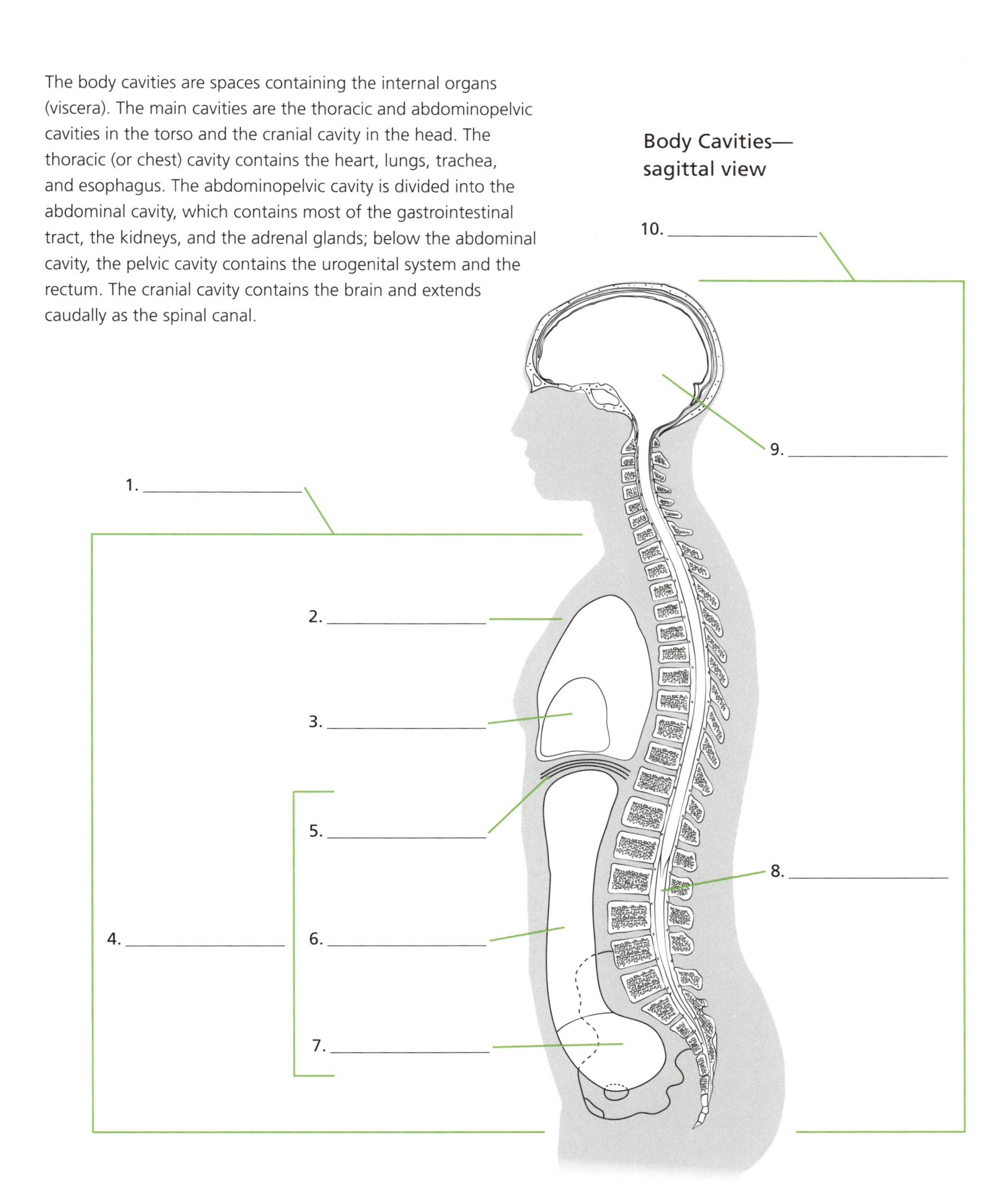

Answers

1. ventral cavity, 2. thoracic cavity, 3. pericardial cavity, 4. abdominopelvic cavity, 5. diaphragm, 6. abdominal cavity, 7. pelvic cavity, 8. spinal canal, 9. cranial cavity, 10. dorsal cavity

View Orientation and Anatomical Planes

Specific terms describe the orientation and relationships of the body and its parts. Sections of the body are described in terms of anatomical planes (flat surfaces). These are imaginary lines—vertical or horizontal—drawn through a body in the anatomical position (that is, with the body standing erect, feet together with toes pointed forward, and arms at the sides with palms facing forward). A transverse (axial or horizontal) plane cuts the body across from side to side, separating superior areas above from inferior areas below. A coronal (frontal) plane divides the body into dorsal (posterior, or back) and ventral (anterior, or front) pieces. A sagittal plane separates one side of the body from the other side (left from right). The midsagittal (median) plane is the sagittal plane exactly in the middle of the body. The relationships of one body part to another are identified by terms such as medial (toward the midline of the body) or lateral (away from the midline of the body); inferior (below, or lower) or superior (upper, or above); cranial (toward the head) or caudal (toward the tail); anterior (ventral, or toward the front) or posterior (dorsal, or toward the back); and proximal (closer to a reference point) or distal (farther from that reference point).

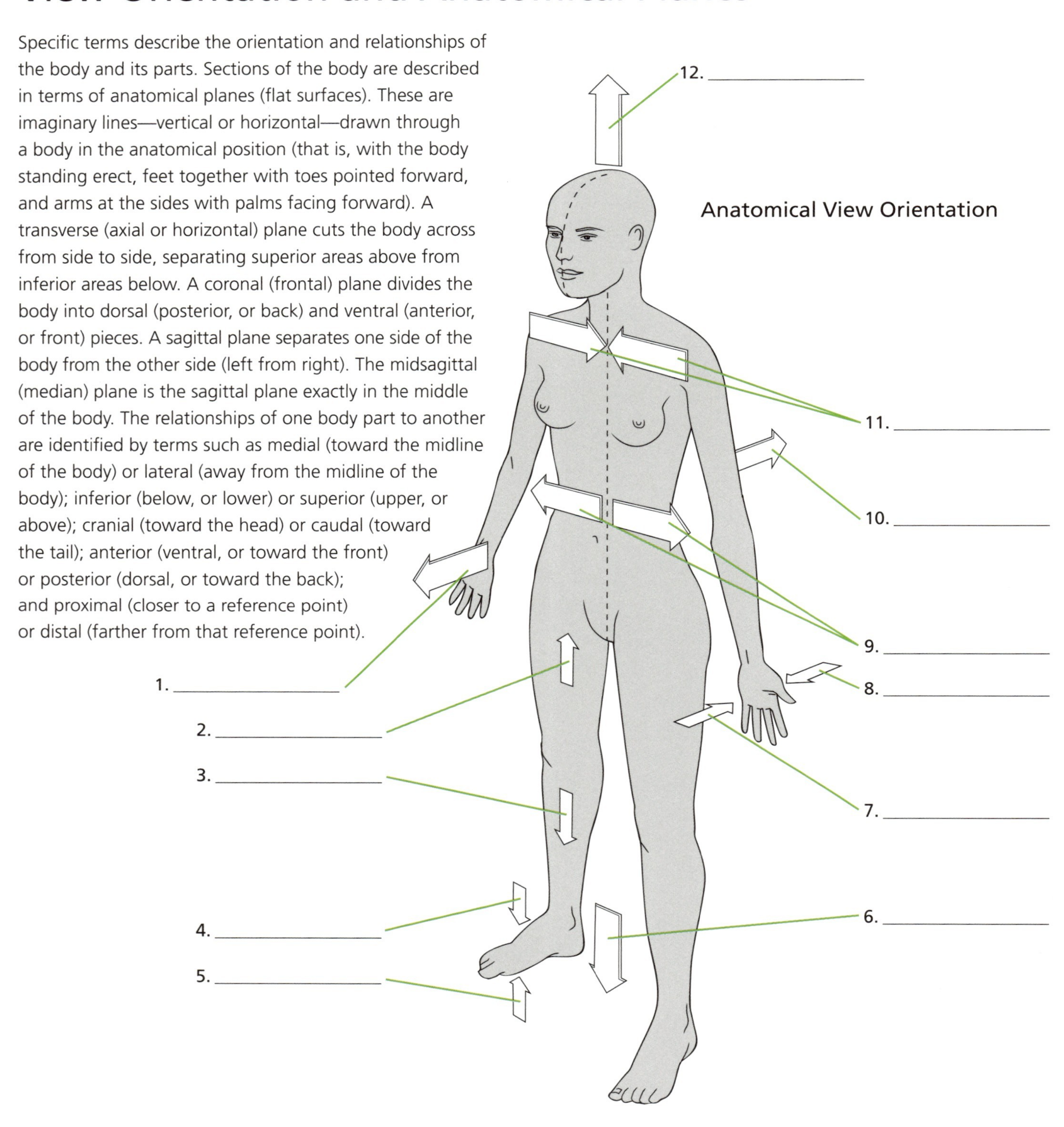

Answers

1. anterior, 2. proximal, 3. distal, 4. dorsal, 5. plantar, 6. inferior, 7. palmar, 8. dorsal, 9. lateral, 10. posterior, 11. medial, 12. superior

Anatomical Planes

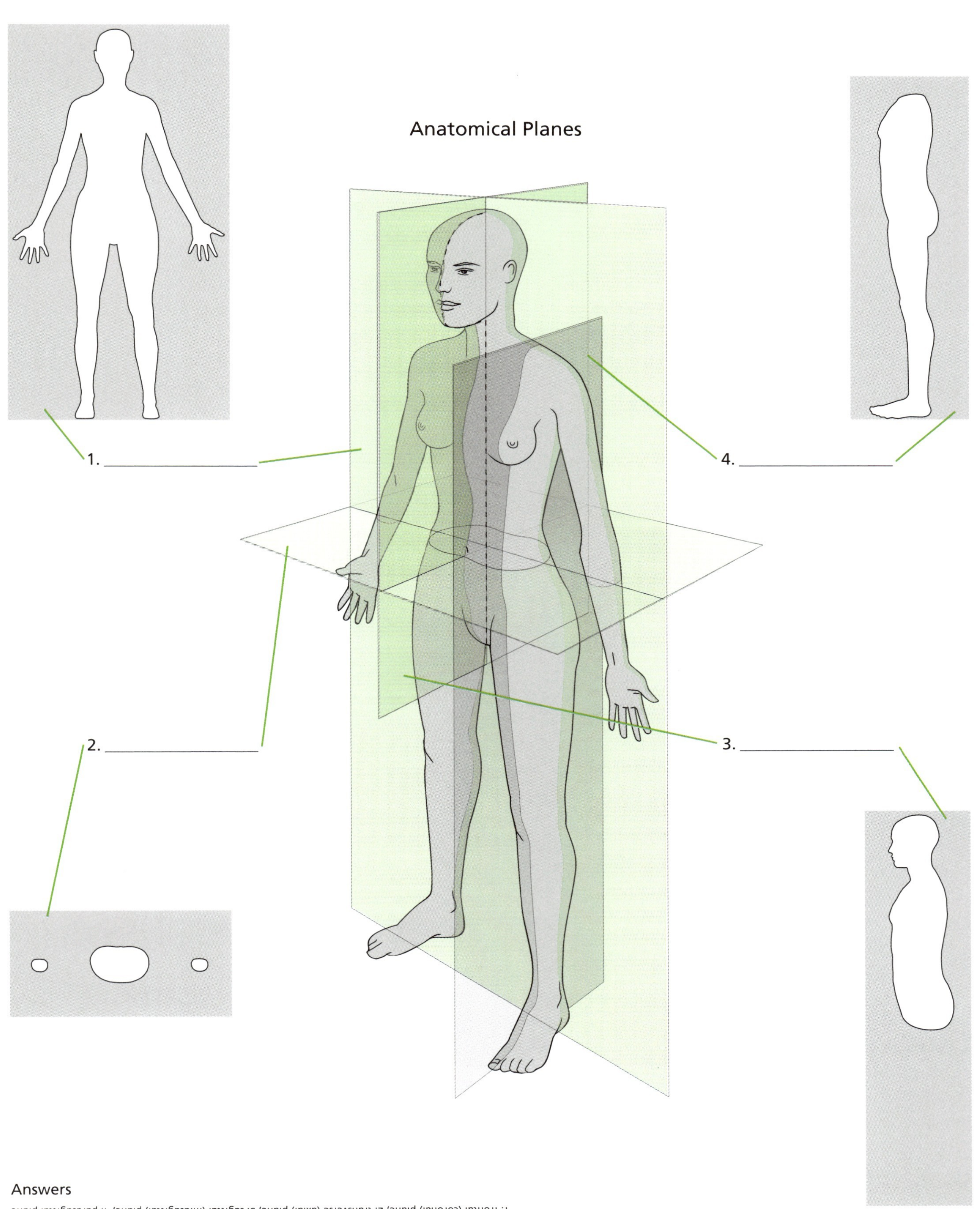

Answers

1. frontal (coronal) plane, 2. transverse (axial) plane, 3. sagittal (midsagittal) plane, 4. parasagittal plane

Central Versus Peripheral Nervous System

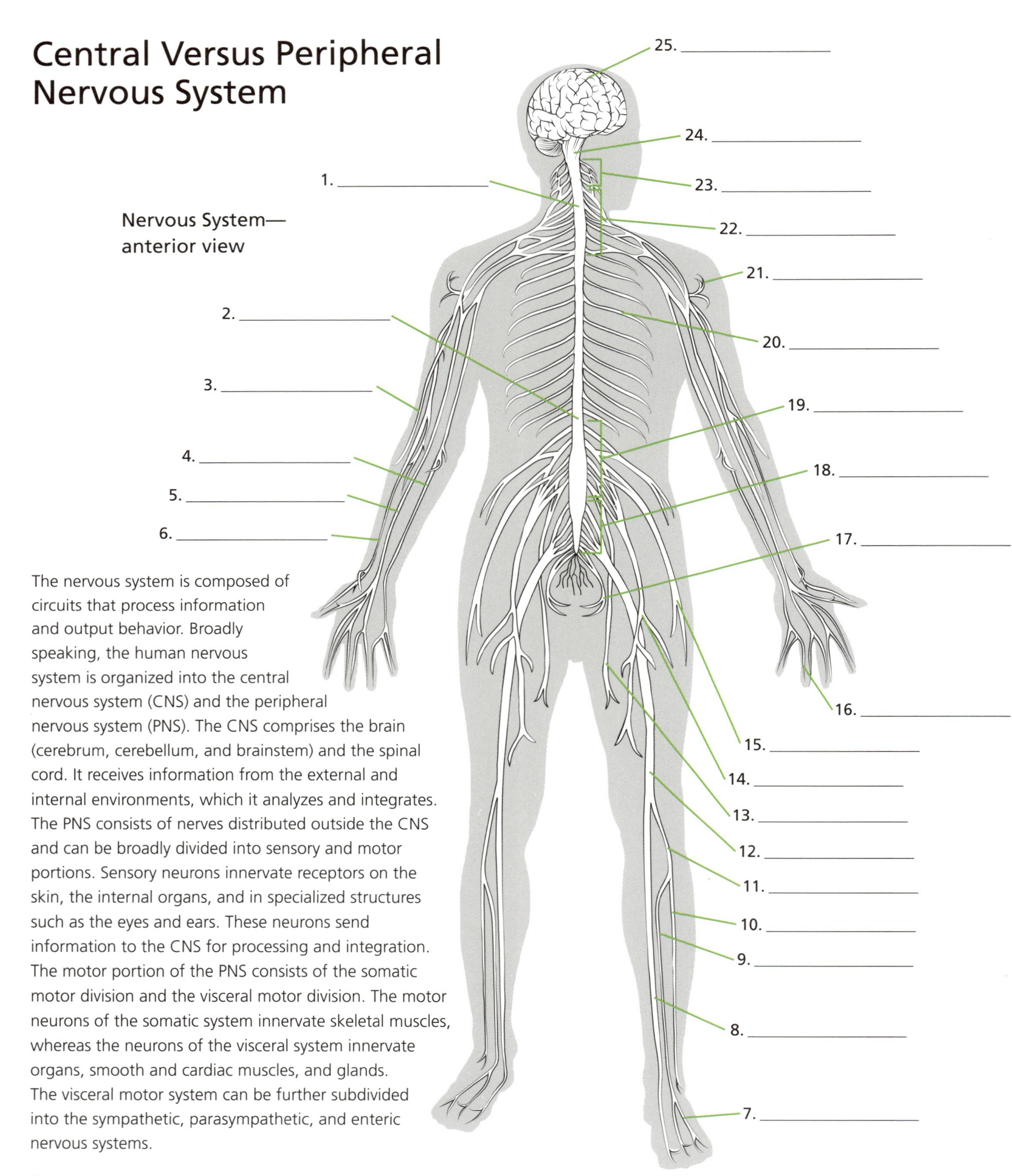

Nervous System—anterior view

The nervous system is composed of circuits that process information and output behavior. Broadly speaking, the human nervous system is organized into the central nervous system (CNS) and the peripheral nervous system (PNS). The CNS comprises the brain (cerebrum, cerebellum, and brainstem) and the spinal cord. It receives information from the external and internal environments, which it analyzes and integrates. The PNS consists of nerves distributed outside the CNS and can be broadly divided into sensory and motor portions. Sensory neurons innervate receptors on the skin, the internal organs, and in specialized structures such as the eyes and ears. These neurons send information to the CNS for processing and integration. The motor portion of the PNS consists of the somatic motor division and the visceral motor division. The motor neurons of the somatic system innervate skeletal muscles, whereas the neurons of the visceral system innervate organs, smooth and cardiac muscles, and glands. The visceral motor system can be further subdivided into the sympathetic, parasympathetic, and enteric nervous systems.

Answers

1. cervical enlargement of spinal cord, 2. lumbosacral enlargement of spinal cord, 3. musculocutaneous nerve, 4. ulnar nerve, 5. median nerve, 6. radial nerve, 7. plantar nerve, 8. posterior tibial nerve, 9. deep fibular nerve, 10. superficial fibular nerve, 11. common fibular nerve, 12. sciatic nerve, 13. obturator nerve, 14. femoral nerve, 15. lateral femoral cutaneous nerve, 16. digital nerve, 17. pudendal nerve, 18. sacral plexus, 19. lumbar plexus, 20. intercostal nerve, 21. axillary nerve, 22. brachial plexus, 23. cervical plexus, 24. medulla oblongata, 25. cerebral hemisphere

Overview of Nervous System Components

Neurons in the PNS are located in tight clusters called ganglia. In the CNS, neurons organize into two main structures: the brain and spinal cord. In these structures, functionally and structurally linked neurons organize into compact clusters called nuclei. Alternatively, neurons can also organize into layered sheets, the most prominent examples being in the cerebral cortex and the cerebellum. The cerebral cortex covers the outmost layer of the cerebrum; its sheet-like structure is folded into ridges called gyri (sing. gyrus) and furrows called sulci (sing. sulcus). This folding greatly increases the surface area of the cortex and allows the brain to pack more cortical space inside the finite dimensions of the skull.

Axons of neurons in the CNS are bundled into tracts. In the PNS these bundled axons are called nerves. On a gross histological level, nervous tissue is divided into "gray" and "white" matter. Gray matter refers to neural tissue that contains neuronal cell bodies, glia cells, and unmyelinated nerve fibers, whereas white matter consists largely of bundles of myelinated nerve fibers. Myelin significantly increases the speed of signal conduction.

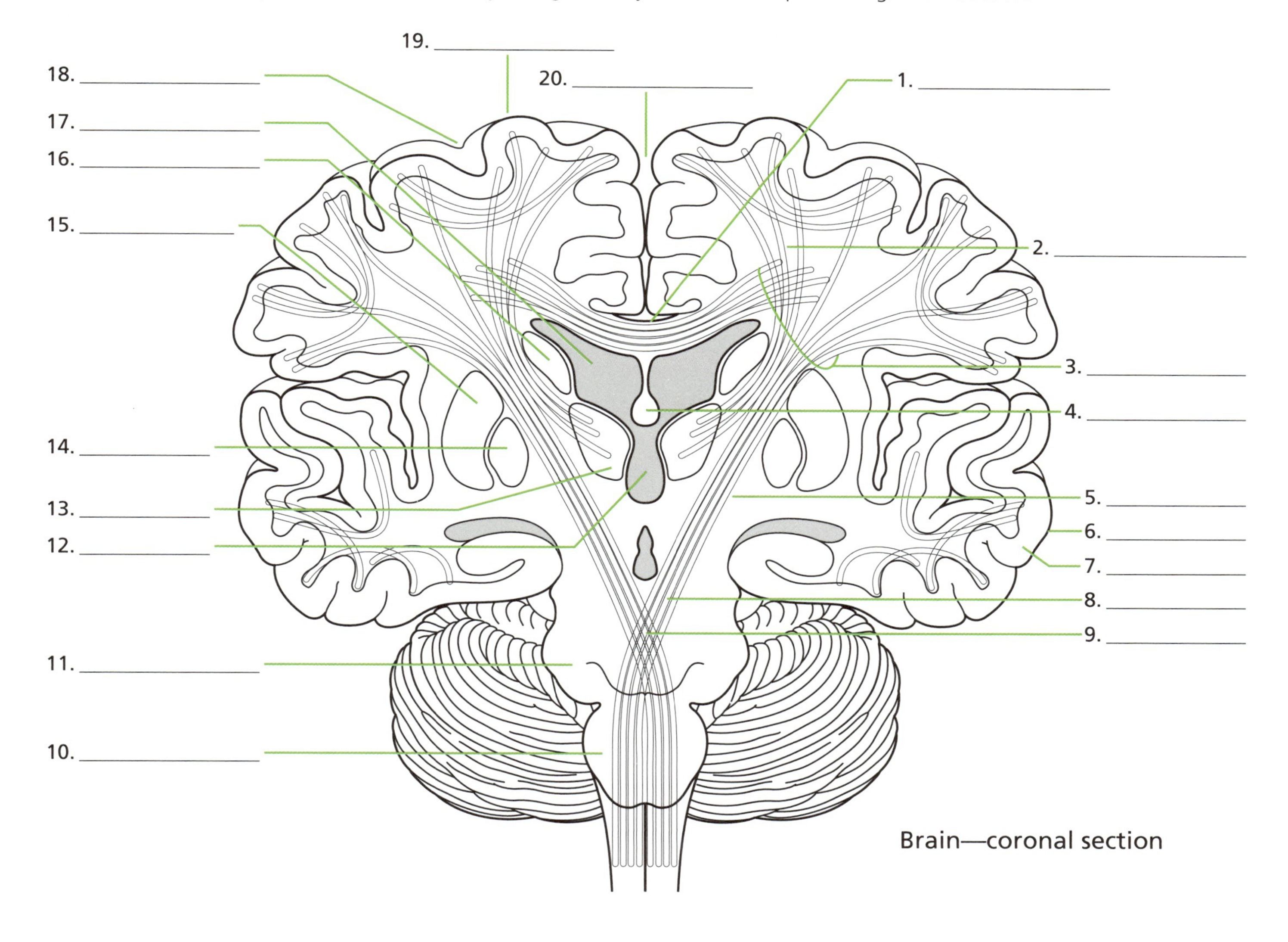

Brain—coronal section

Answers

1. corpus callosum, 2. association fibers, 3. corona radiata, 4. fornix, 5. internal capsule, 6. gray matter, 7. white matter, 8. projection fibers, 9. decussation of pyramids, 10. medulla oblongata, 11. pons, 12. third ventricle, 13. thalamus, 14. globus pallidus, 15. putamen, 16. caudate, 17. lateral ventricle, 18. sulci, 19. gyri, 20. longitudinal fissure

Terminology Used for Directions in the Nervous System

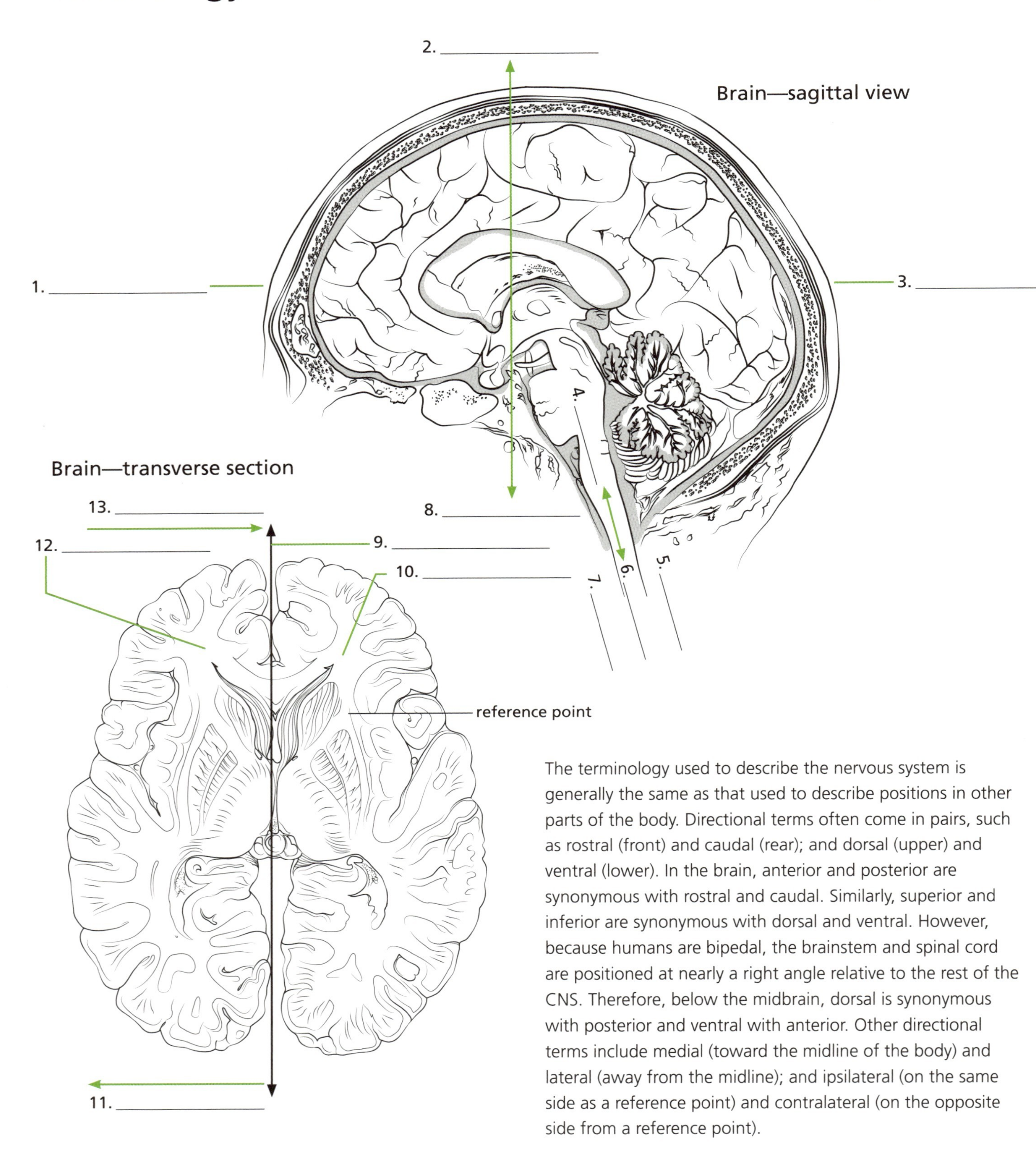

The terminology used to describe the nervous system is generally the same as that used to describe positions in other parts of the body. Directional terms often come in pairs, such as rostral (front) and caudal (rear); and dorsal (upper) and ventral (lower). In the brain, anterior and posterior are synonymous with rostral and caudal. Similarly, superior and inferior are synonymous with dorsal and ventral. However, because humans are bipedal, the brainstem and spinal cord are positioned at nearly a right angle relative to the rest of the CNS. Therefore, below the midbrain, dorsal is synonymous with posterior and ventral with anterior. Other directional terms include medial (toward the midline of the body) and lateral (away from the midline); and ipsilateral (on the same side as a reference point) and contralateral (on the opposite side from a reference point).

Answers

1. rostral (anterior), 2. dorsal (superior), 3. caudal (posterior), 4. rostral, 5. dorsal (posterior), 6. caudal, 7. ventral (anterior), 8. ventral (inferior), 9. midline, 10. ipsilateral, 11. lateral, 12. contralateral, 13. medial

Planes of Section

Standard planes are used for histological or topographical analysis of the internal anatomy of the brain. The planes correspond to the major axes of the brain and are at right angles with one another. The coronal (also called frontal) plane refers to sections parallel to the face and divides the brain into anterior and posterior sections. The horizontal (also called transverse or axial) plane runs parallel to the floor in a standing person and cuts the brain into superior and inferior portions. The sagittal plane divides the two hemispheres of the brain. The spinal cord can be sectioned along the same planes as the brain.

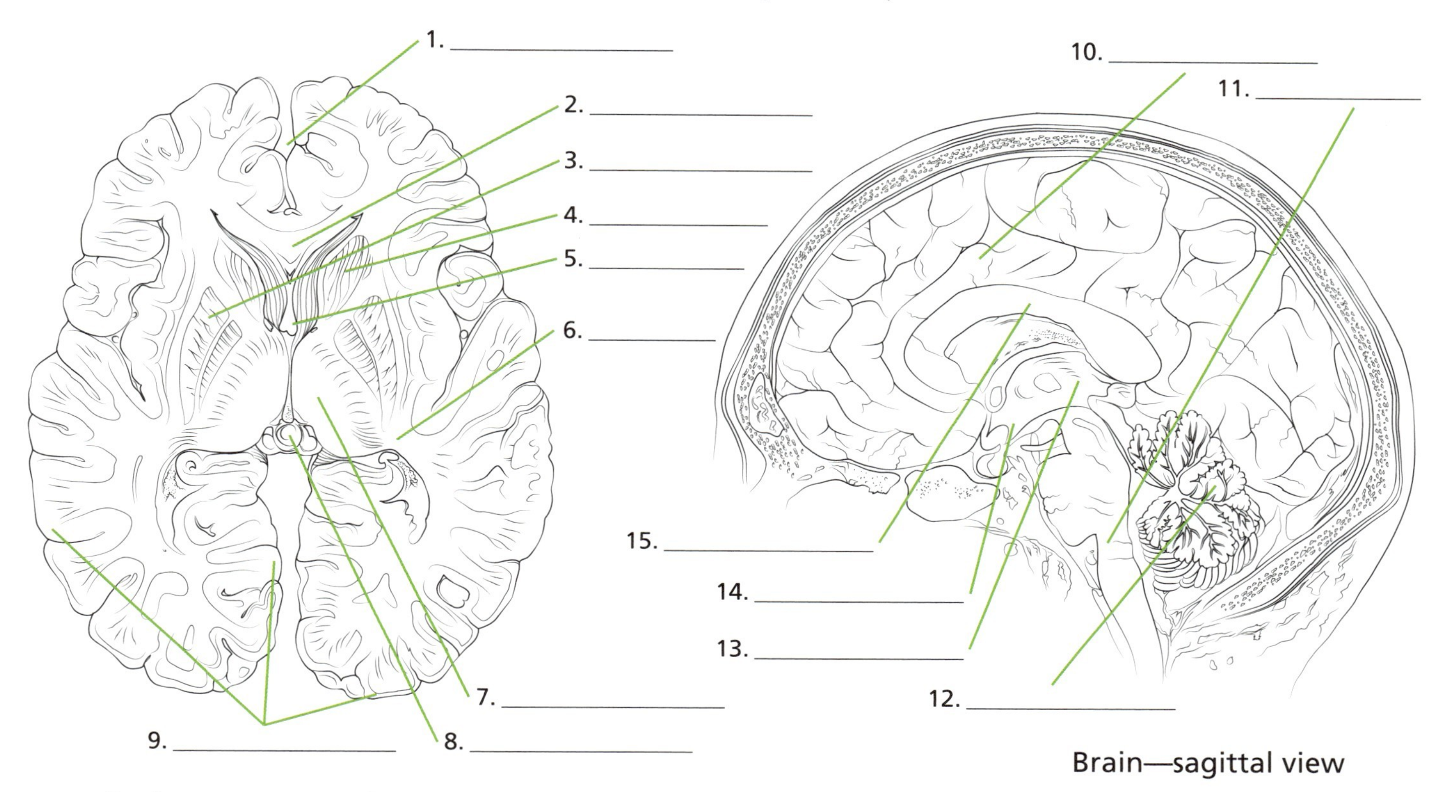

Brain—transverse view

Brain—sagittal view

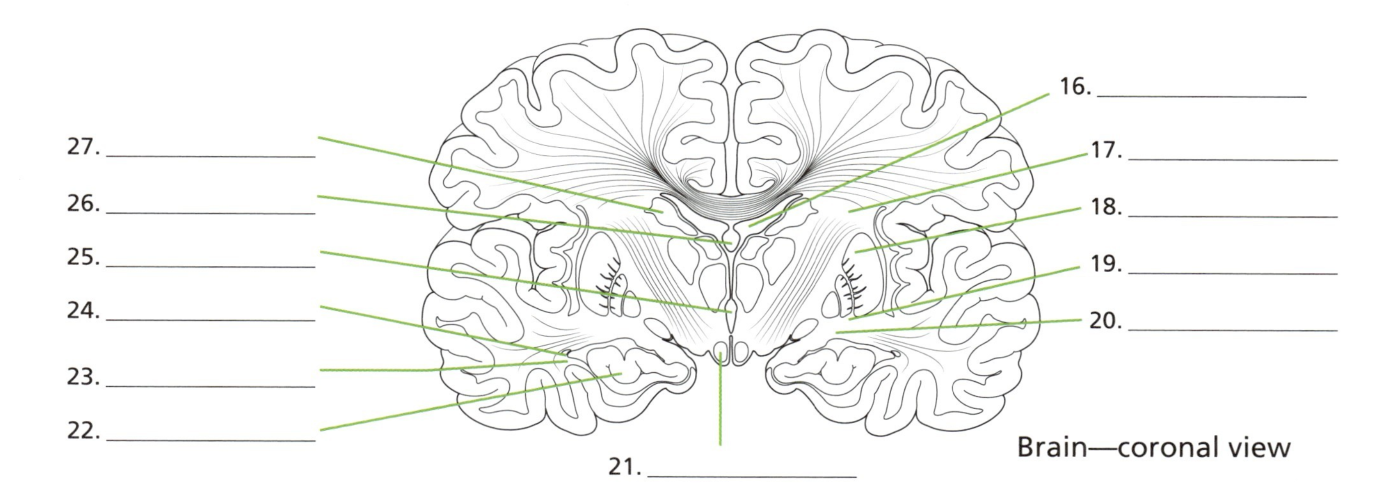

Brain—coronal view

Answers

1. sagittal fissure, 2. corpus callosum, 3. putamen, 4. caudate nucleus, 5. fornix, 6. globus pallidus, 7. thalamus, 8. pineal gland, 9. cerebral cortex, 10. cerebrum, 11. brainstem, 12. cerebellum, 13. thalamus, 14. hypothalamus, 15. corpus callosum, 16. corpus callosum, 17. caudate, 18. thalamus, 19. putamen, 20. globus pallidus, 21. mammillary body, 22. hippocampus, 23. lateral ventricle, 24. tail of the caudate nucleus, 25. third ventricle, 26. fornix, 27. lateral ventricle

Lobes of the Brain (Frontal, Occipital, Parietal, Temporal)

The cerebrum is the largest part of the brain and contains the cerebral cortex as well as subcortical structures. The cerebral cortex is divided into lobes that are associated with, but not limited to, certain functions. The frontal lobe is the most anterior; its posterior boundary is the central sulcus. It is important for higher-order cognitive functions. The parietal lobe is posterior to the central sulcus on the superior aspect of the brain; it is important for integrating sensory information and some higher-order visual processing. The temporal lobes sit inferior to the frontal and parietal lobes on both lateral aspects of the brain, separated from them by the lateral fissure. The temporal lobes are involved in the processing of auditory information, as well as some higher-order visual information. The insula, or insular cortex, is buried beneath the frontal and temporal lobes and is not visible from the surface of the brain. The insula is involved in many functions relating to awareness, cognition, and the regulation of homeostasis. The occipital lobe is the most posterior of the lobes and is separated from the parietal lobe somewhat arbitrarily by a line that runs from the parieto-occipital sulcus to the occipital notch. The occipital lobe is involved in many levels of visual processing.

Lobes of the Brain—lateral view

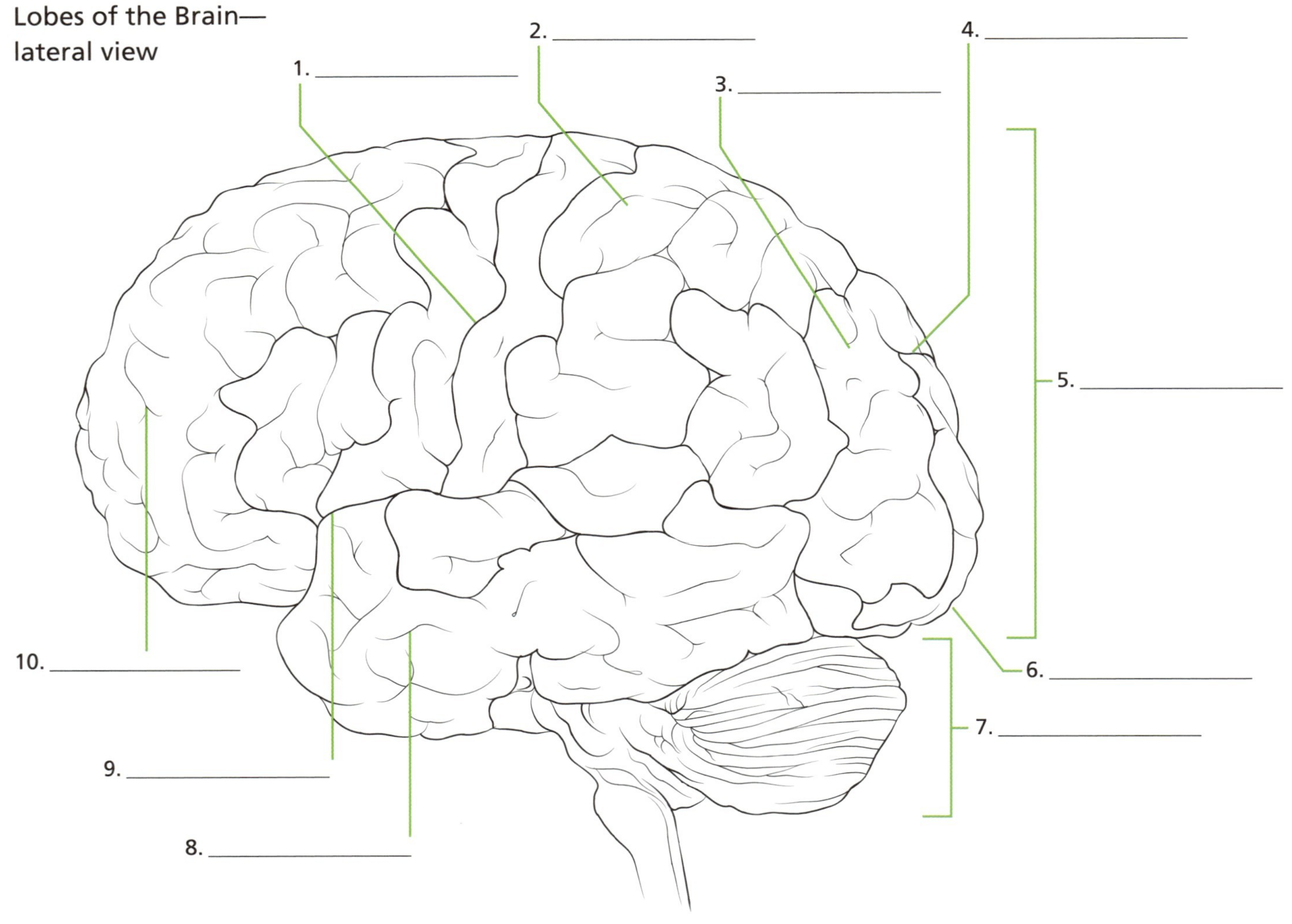

Answers

1. central sulcus, 2. parietal lobe, 3. occipital lobe, 4. parieto-occipital sulcus, 5. cerebrum, 6. occipital notch, 7. cerebellum, 8. temporal lobe, 9. lateral fissure, 10. frontal lobe

Hemispheres of the Brain (Left, Right)

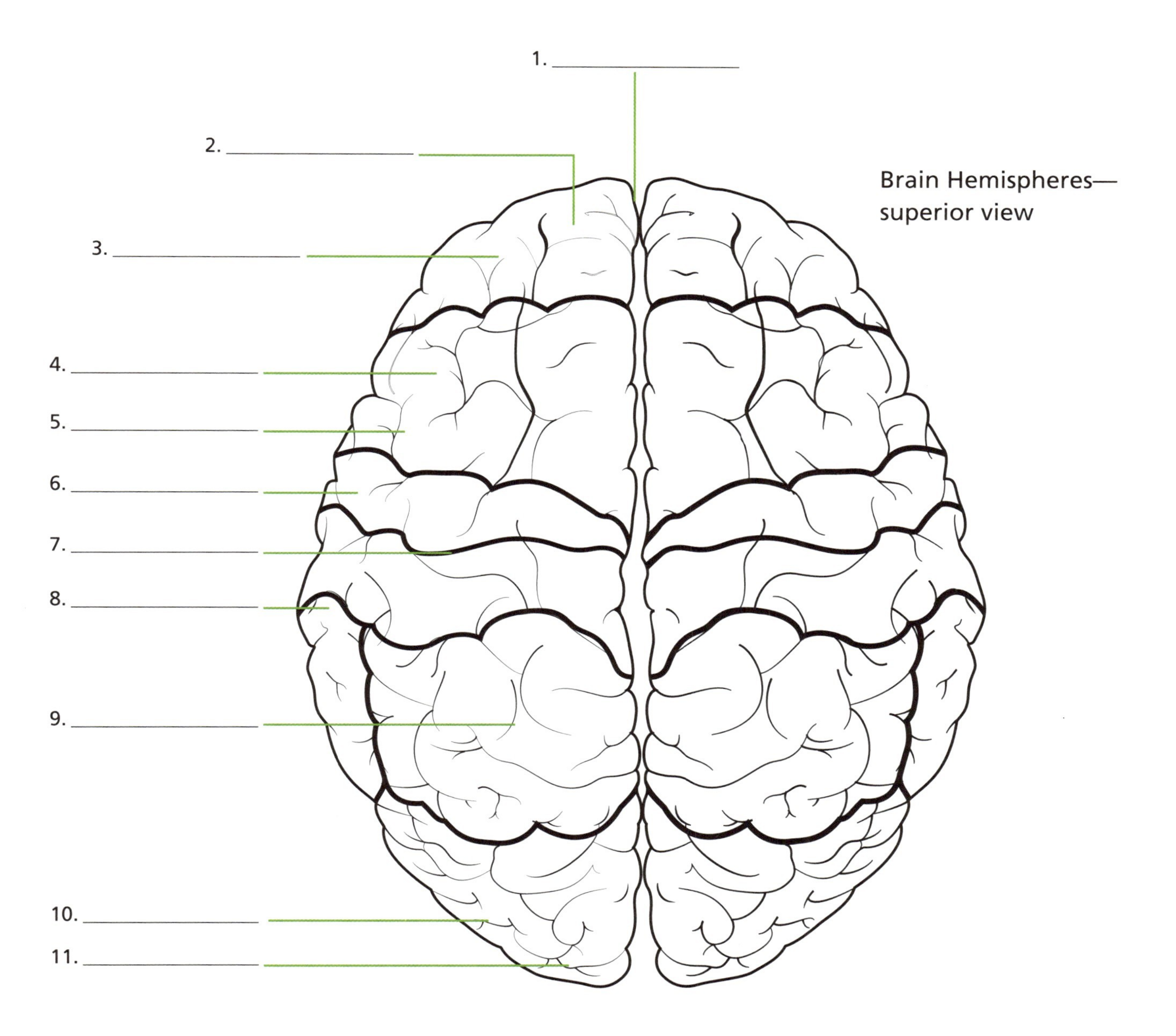

Brain Hemispheres—superior view

The human brain is separated into two cerebral hemispheres by a deep groove called the medial longitudinal fissure, which runs along the sagittal plane. The hemispheres are connected by a large white matter structure called the corpus callosum. The surface of each hemisphere is covered by gyri (ridges) and sulci (furrows) of the cerebral cortex. Overall, the macroscopic structure and organization of the hemispheres are relatively similar; however, notable differences do exist. For example, two cortical areas important for language (Broca's and Wernicke's areas) are typically located only on the left hemisphere, and indeed in most individuals the left hemisphere exhibits dominance for language processing and production. That said, most cognitive functions are performed bilaterally in both hemispheres.

Answers

1. medial longitudinal fissure, 2. frontal lobe, 3. superior frontal sulcus, 4. middle frontal gyrus, 5. precentral sulcus, 6. central sulcus, 7. postcentral gyrus, 8. postcentral sulcus, 9. parietal lobe, 10. lunate sulcus, 11. occipital lobe

Division of the Brain (Frontal, Mid, Hind)

The brain can be divided into forebrain (prosencephalon), midbrain (mesencephalon), and hindbrain (rhombencephalon) based on the embryonic development of the CNS. The forebrain is further separated into the diencephalon and the telencephalon. The most prominent feature of the diencephalon is the thalamus. Other crucial diencephalon structures include the hypothalamus and pituitary gland. The telencephalon develops into the cerebral cortex, the corpus callosum, and many subcortical structures, such as most of the basal ganglia and limbic system. The midbrain is caudal to the thalamus. The superior and inferior colliculi, together called the tectum, define its dorsal surface, while the cerebral peduncle comprises the rest. The hindbrain is further separated into the metencephalon and the myelencephalon. The myelencephalon forms the medulla oblongata, while the metencephalon forms the pons and cerebellum. The pons lies caudal to the midbrain, and the cerebellum is just inferior to the occipital lobe. The hindbrain and midbrain (minus the cerebellum) form the brainstem.

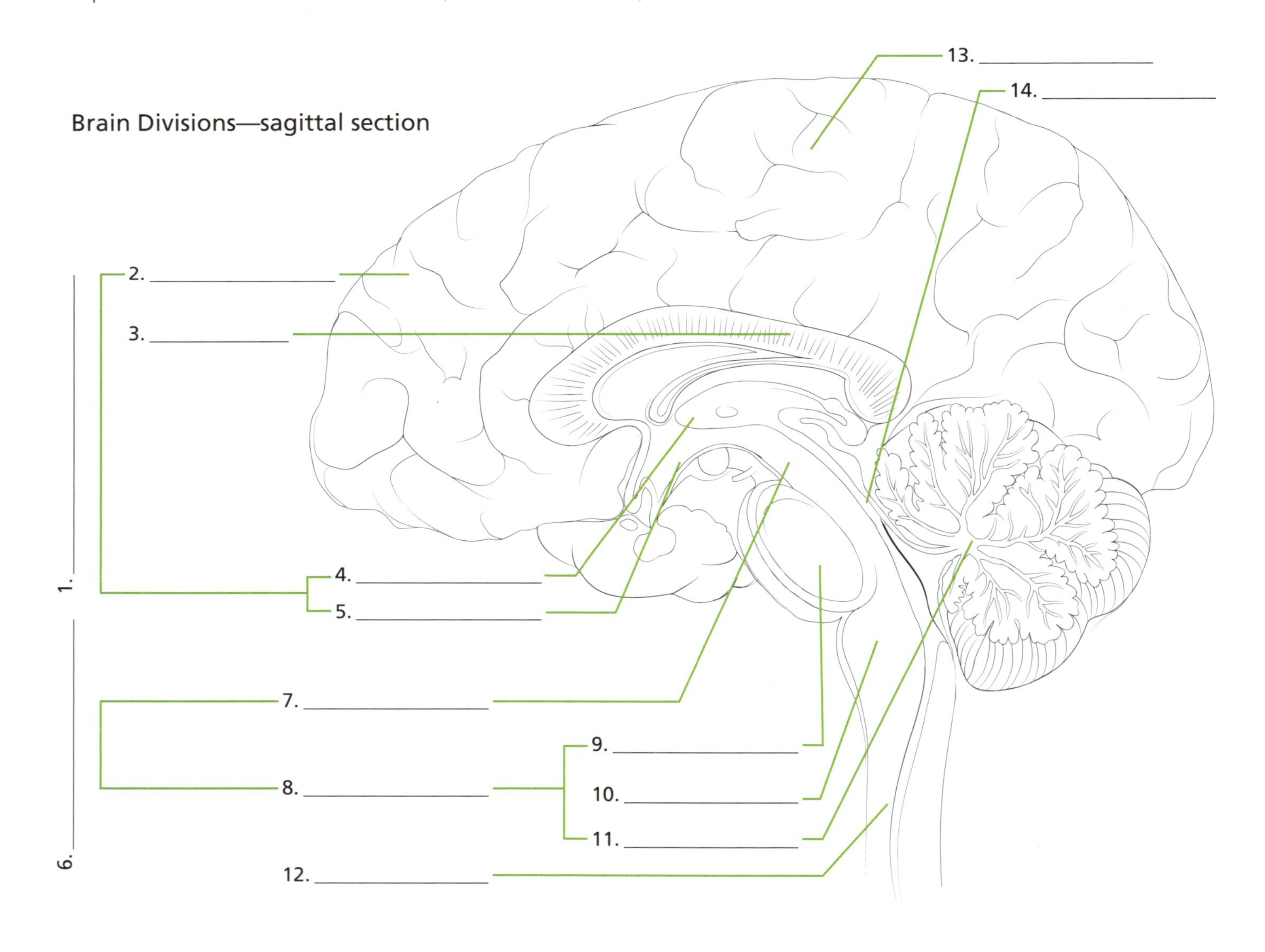

Answers

1. forebrain, 2. cerebrum, 3. corpus callosum, 4. thalamus, 5. hypothalamus, 6. brainstem, 7. midbrain, 8. hindbrain, 9. pons, 10. medulla oblongata, 11. cerebellum, 12. spinal cord, 13. cerebral hemisphere, 14. cerebral aqueduct

Neuron Overview

The neuron is the basic information-processing unit of the CNS. It consists of a cell body (soma) that contains various organelles, including the nucleus, as well as processes that extend from the soma. There are two kinds of processes: axons and dendrites.

Dendrites form extensive branches called dendritic trees, which range from tens of micrometers to several millimeters long. Dendrites receive incoming signals from other neurons at specialized junctions called synapses. Synapses are formed either directly onto dendritic membranes or onto small protrusions called dendritic spines.

Axons emerge from a protrusion on the soma called the axon hillock. Most neurons have a single axon that carries information away from the soma. Axons are sometimes encapsulated in a fatty insulating substance called myelin, which is produced by oligodendrocytes, a type of glia that wrap around the axons. Myelinated axons have gaps at evenly spaced intervals called nodes of Ranvier, which enable a rapid form of signal propagation called saltatory conduction. Axons end in terminal branches containing specialized domains called synaptic boutons. These synaptic terminals initiate information transfer from the axon of one neuron to the dendrites of other neurons.

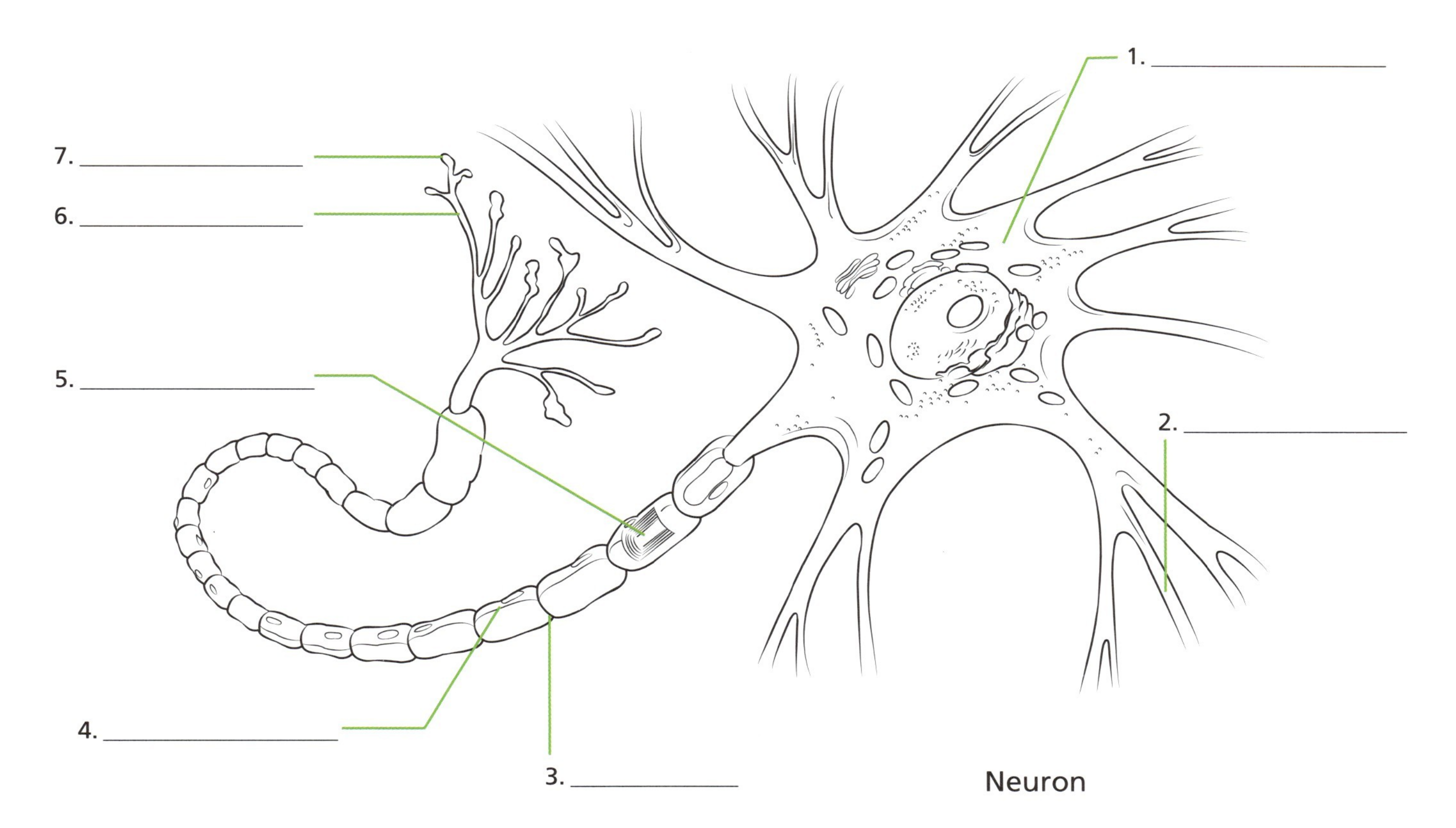

Neuron

Answers

1. soma, 2. dendrite, 3. node of Ranvier, 4. myelin sheath, 5. axon, 6. axon terminal, 7. synaptic bouton

Classification of Neurons

There are many different kinds of neurons in the nervous system, which vary in their structure, function, and gene expression. Broadly speaking, neurons can be classified by either their anatomy or their function.

Anatomical classification of neurons is done according to their appearance. Pseudounipolar neurons have two processes that fuse into one during development. The process extends from the cell body and forms central and peripheral branches. The central branch is associated with secretory functions, whereas the peripheral branch is associated with sensory receptors.

Bipolar neurons have two processes: a dendrite and an axon. The processes extend from opposite sides of the soma, forming a spindle shape. These neurons are usually sensory neurons found in select areas in the nervous system.

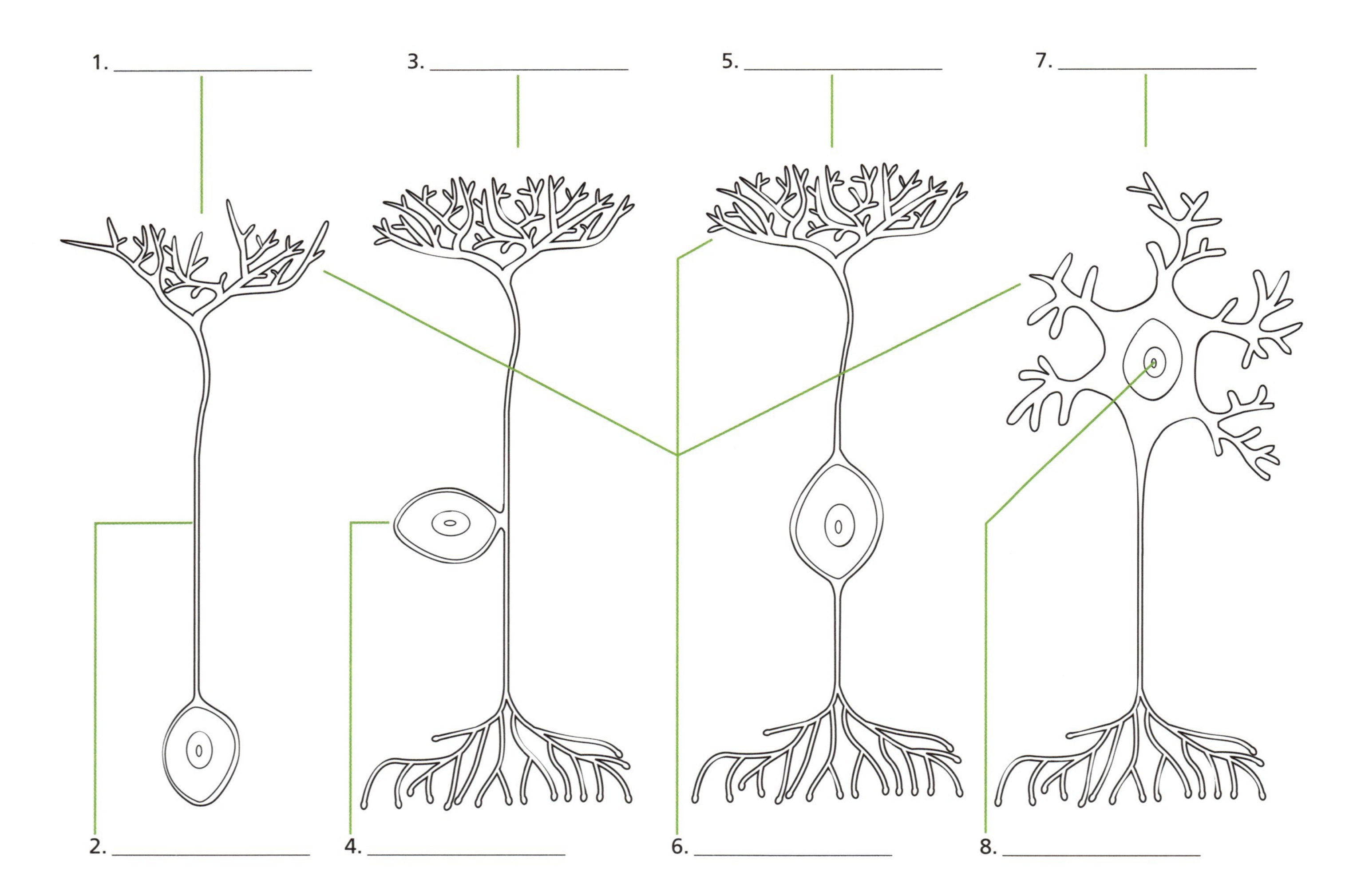

Neuron Types—structure

Answers

1. unipolar neuron, 2. axon, 3. pseudounipolar neuron, 4. cell body, 5. bipolar neuron, 6. dendrites, 7. multipolar neuron, 8. nucleus

Multipolar neurons are the most common type of neurons in the CNS. They have one axon and two or more dendrites emerging from the soma and are extremely diverse in morphology. Anatomical classification of neuronal subtypes is often made according to the appearance of their dendritic arbors or soma shape (such as the radially branching dendrites of starburst cells or the large triangular soma of pyramidal cells).

Neurons can also be classified by function, that is, according to which direction they transmit nerve impulses relative to the CNS. Sensory (afferent) neurons carry signals from sensory organs to the spinal cord and brain.

Motor (efferent) neurons conduct signals away from the CNS to effector muscles and organs.

Interneurons are found in the CNS and are by far the most abundant type of neuron. Interneurons integrate and transform incoming sensory information and output appropriate behavioral information to motor neurons. Functional classification of neuronal subtypes can also be made according to their diverse electrical responses to stimulation (such as the relative rapidity of response time and whether the response is transient or sustained).

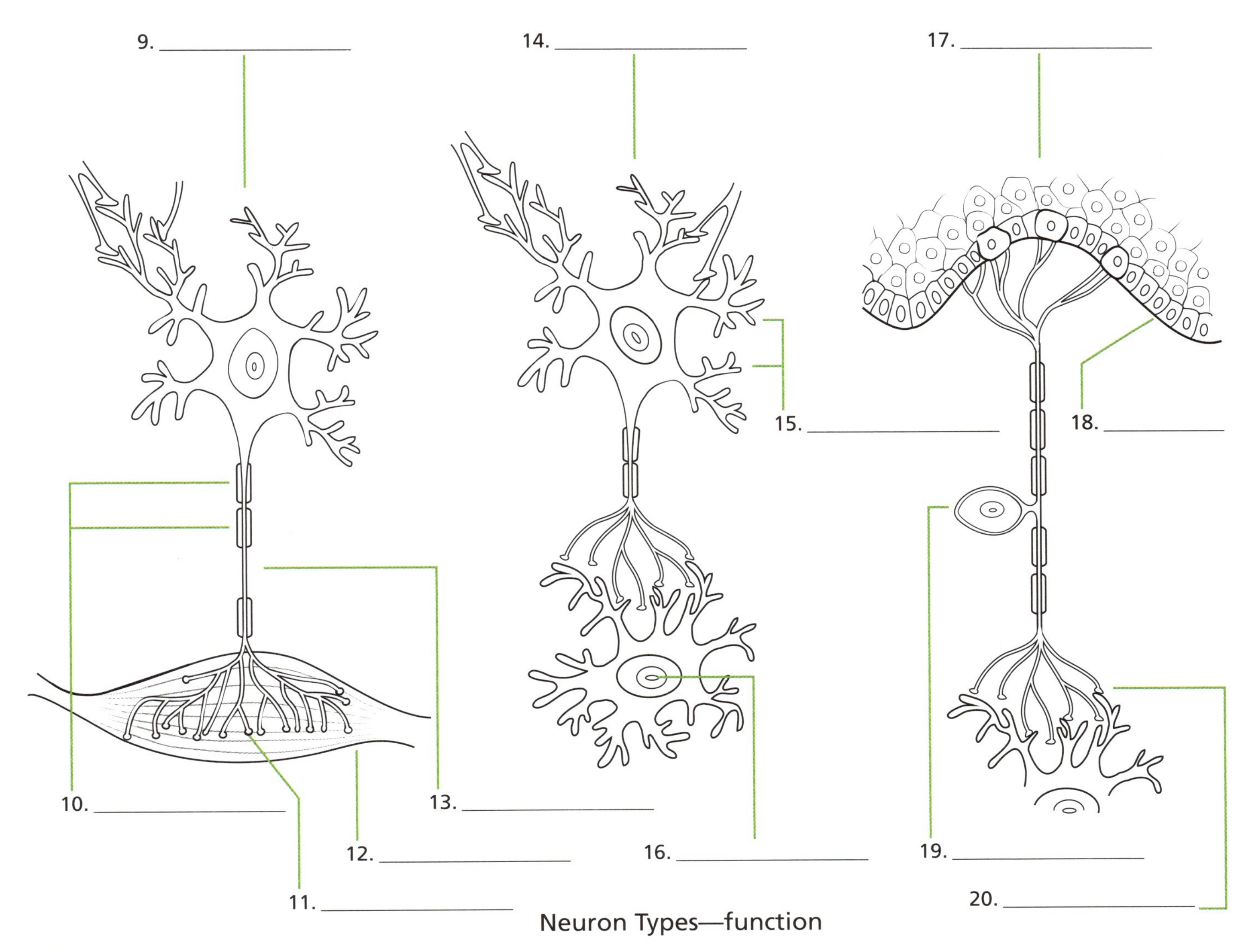

Neuron Types—function

Answers

9. motor neuron, 10. myelin sheath, 11. synaptic terminal, 12. muscle, 13. axon, 14. interneuron, 15. dendritic tree, 16. nucleus, 17. sensory neuron, 18. skin, 19. soma (or cell body), 20. synapses

Ultrastructure of a Typical Neuron

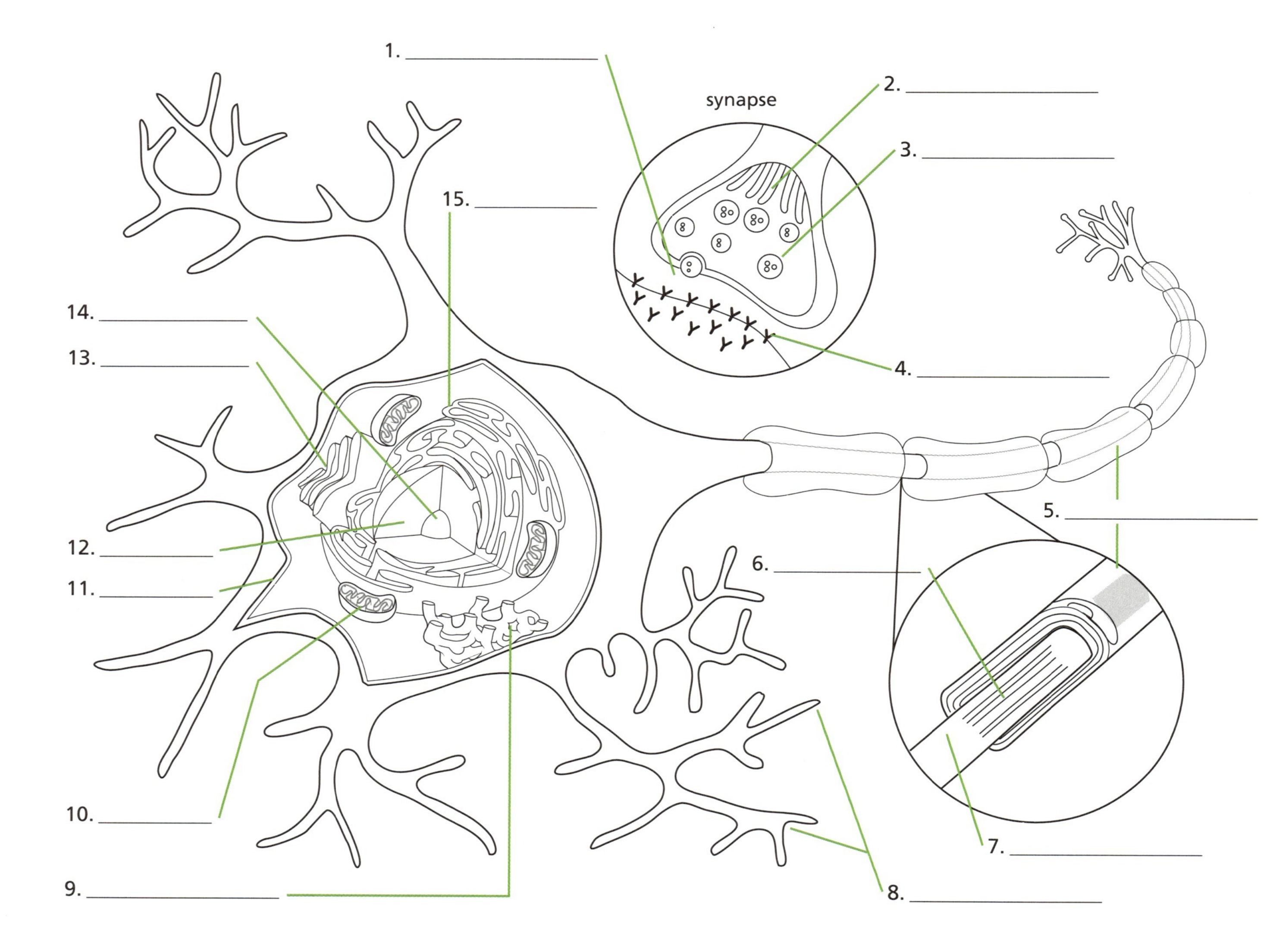

Neurons are surrounded by a semipermeable, double-layered phospholipid membrane that contains many proteins to detect and secrete neurotransmitters. The membrane acts as a barrier to ions, and other proteins regulate the influx and efflux of these ions dependent on the incoming and outgoing neuronal signals.

Mitochondria are usually concentrated in the most metabolically intensive areas of neurons, such as the synaptic terminals, to support the release of neurotransmitters from synaptic vesicles. This process is a prominent example of exocytosis, in which membrane-encapsulated vesicles fuse with the cellular membrane, thus releasing their contents into the cell exterior.

In general, the products of protein synthesis are transported down the axon and dendrite(s) on microtubules, which are embedded in a structural matrix made out of microfilaments. Some proteins, particularly those related to synaptic plasticity, can be synthesized locally in axons and synaptic terminals. Microtubules form part of the neuron's cytoskeleton.

Answers

1. synaptic cleft, 2. microtubule, 3. synaptic vesicle, 4. receptor, 5. myelin sheath, 6. microtubule, 7. axon, 8. dendrites, 9. smooth endoplasmic reticulum, 10. mitochondrion, 11. neuronal membrane, 12. nucleus, 13. Golgi apparatus, 14. nucleolus, 15. rough endoplasmic reticulum

Graded Potential

At rest, there is a difference in electrical potential (voltage) across the neuronal membrane called the resting potential. This potential is generated by an unequal distribution of ions between the cell exterior and interior, which creates an electrochemical gradient across the membrane. The major types of ions that contribute to the resting potential are sodium (Na^+), potassium (K^+), and chloride (Cl^-) ions, and they can move across the membrane only through special protein channels. Neurons expend energy to pump ions against their electrochemical gradient and maintain the resting potential. Na^+ is concentrated on the outside of the cell, and K^+ and Cl^- on the inside. The resting potential is around –70 mV (millivolts) in most neurons.

Neurotransmitters stimulate or inhibit neurons by causing protein channels that are permeable to one or more of these ions to open or close. This results in local, graded (analog) changes in the membrane potential. The magnitude of graded potentials directly correlates with the intensity of the stimulus. Graded potentials can depolarize the membrane (exciting the neuron) or hyperpolarize it (inhibiting the neuron). Graded potentials decrease with distance traveled laterally along a neuronal membrane.

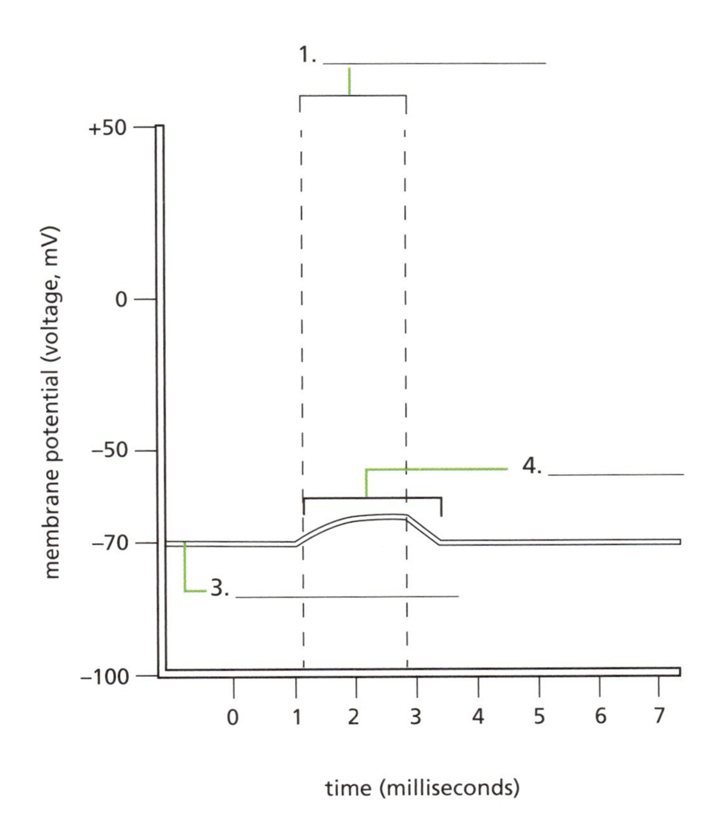

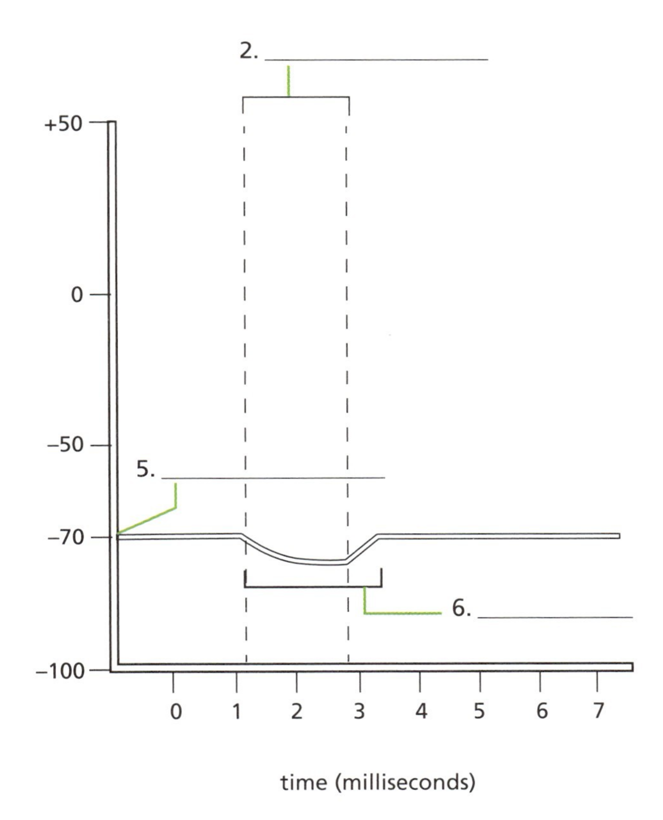

Voltage Traces of Graded Potential

Answers

1. depolarizing stimulus, 2. hyperpolarizing stimulus, 3. resting potential, 4. depolarization, 5. resting potential, 6. hyperpolarization

Action Potential

Action potentials, or spikes, are signals that actively propagate down a neuron's axon and carry the information that is integrated in the dendrites and soma. Local changes in membrane potential spread to the axon hillock, where they are spatiotemporally summed up. If the overall change results in sufficient depolarization and reaches a certain threshold, the neuron responds by opening voltage-gated ion channels and fires an action potential. In contrast to graded potentials, action potentials are all or none (digital) and propagate without decay in amplitude.

Action potentials are generated by sequential changes in the permeability of Na^+ and K^+ channels. Depolarization briefly opens Na^+ channels, which allows Na^+ to rush rapidly into the cell. This results in depolarization, which in turn opens K^+ channels. As K^+ exits the cell, the membrane potential reverts back toward its resting potential, causing repolarization. During this phase, the membrane potential overshoots briefly by becoming more negative than the resting potential, before returning back to its resting state. This is called after-hyperpolarization.

Following an action potential, it is more difficult to fire a second one. The absolute refractory period is when Na^+ channels are inactivated and cannot open in response to stimuli. The relative refractory period is when a stronger than usual stimulus is required to trigger an action potential.

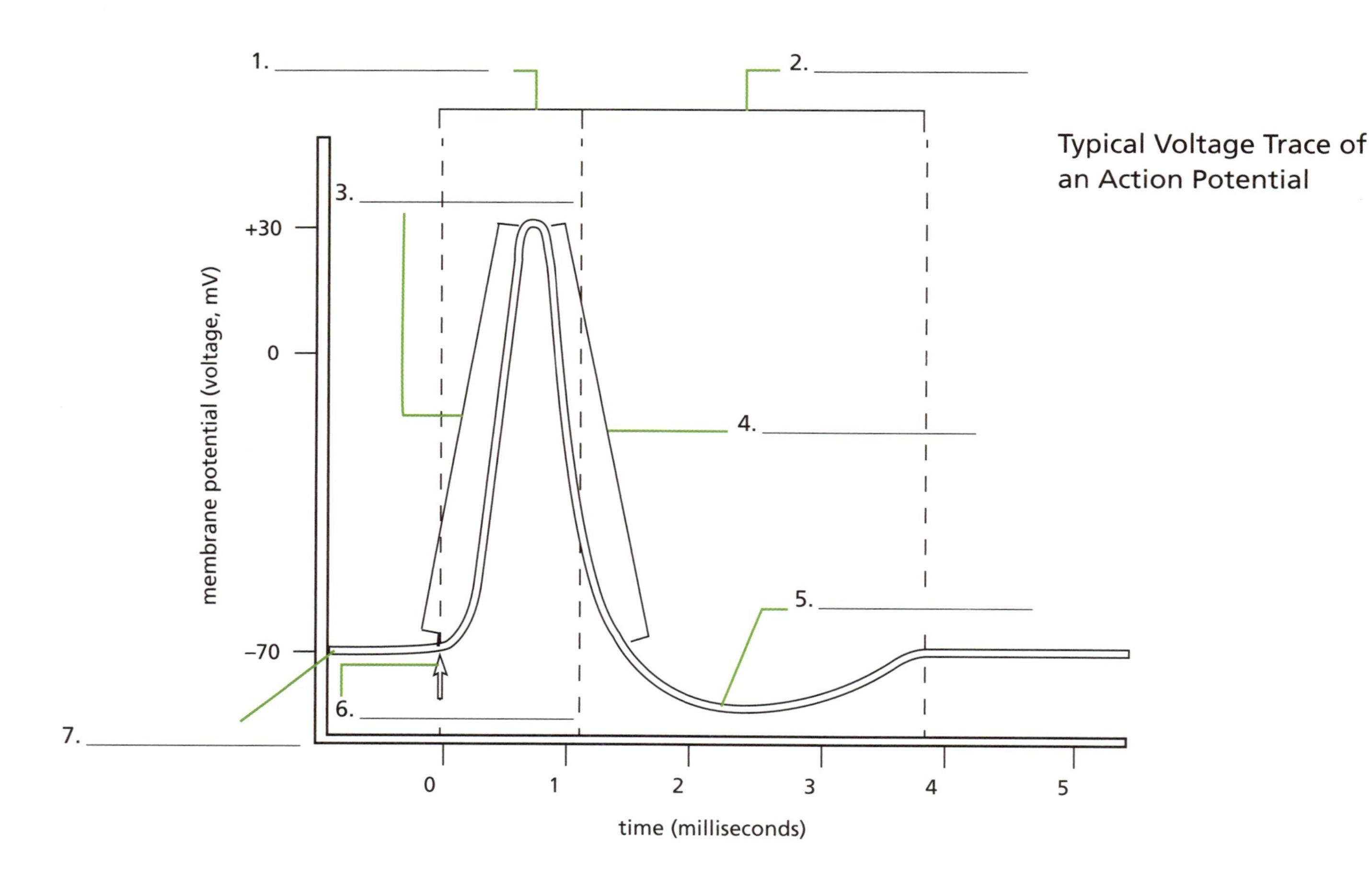

Typical Voltage Trace of an Action Potential

Answers

1. absolute refractory period, 2. relative refractory period, 3. depolarization (Na^+ enters), 4. repolarization (K^+ exits), 5. after-hyperpolarization, 6. stimulus, 7. resting potential

Glia (e.g., Astrocytes and Microglia)

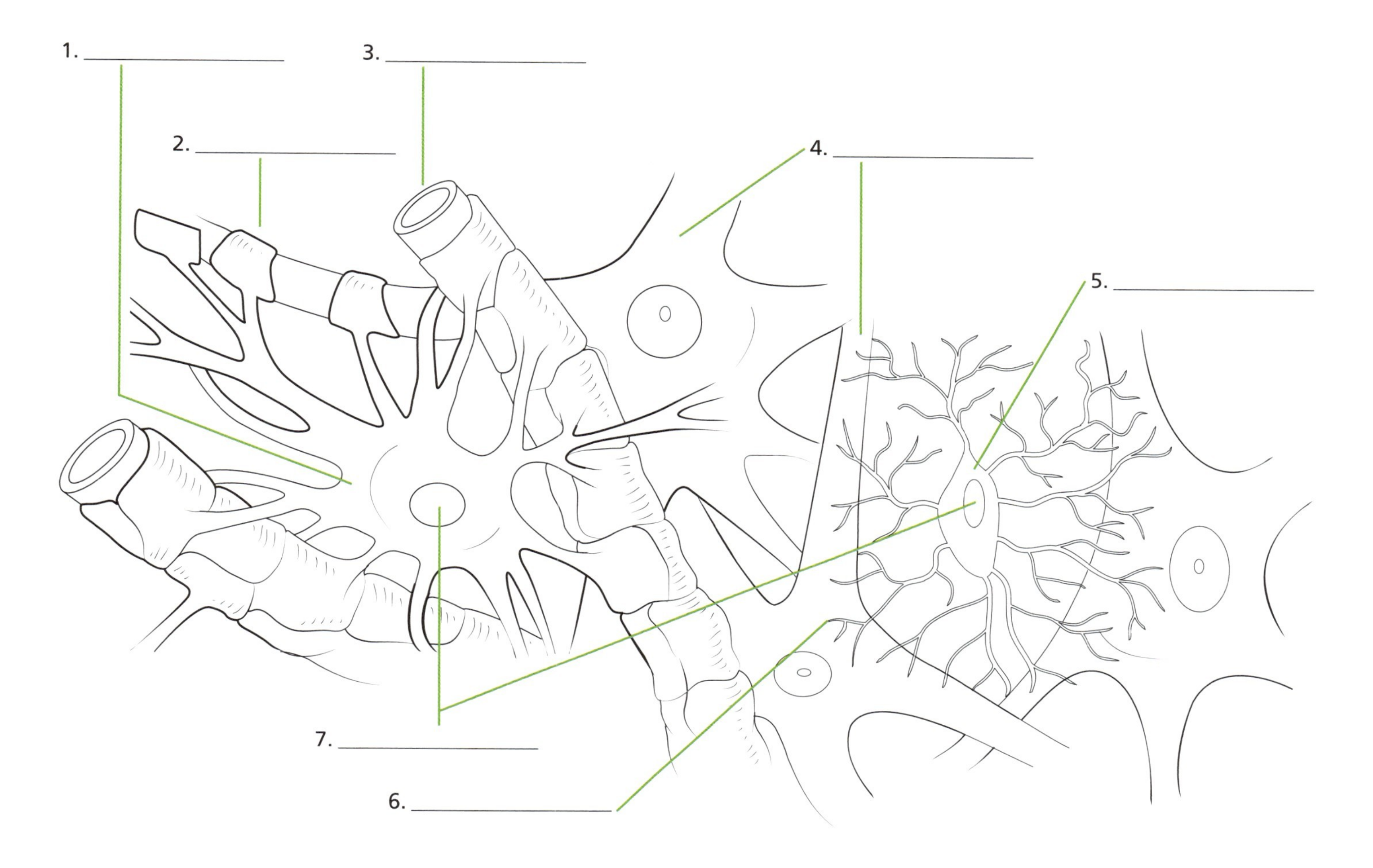

Glial cells are non-neuronal cells in the nervous system and have many structural and physiological functions. The two main types of glia in the PNS are Schwann cells and satellite cells. Schwann cells wrap around axons in the periphery and form the myelin sheath. They are also involved in clearing damaged and dying neurons in a process called phagocytosis. Satellite cells surround the somas of neurons in the PNS and control the chemical environment around them.

Astrocytes are the most common type of glia in the CNS. They have numerous processes with bulbous end-feet that wrap around neurons and blood vessels, and they facilitate chemical exchange between the neurons and blood. In addition, astrocytes maintain the blood–brain barrier, regulate extracellular ion balance, provide nutrients to neurons, regulate synaptic transmission, and play a role in the repair and scarring of nervous system tissue after damage.

The CNS has three other types of glia cells. Microglia are specialized immune cells with highly branched processes. They are the primary immune defense system against infections and contribute to the inflammatory response. Oligodendrocytes wrap around axons in the CNS and produce the myelin sheath. Ependymal cells line the ventricles of the brain and help circulate cerebrospinal fluid.

Answers

1. astrocyte, 2. astrocyte end-feet, 3. capillary, 4. neuron, 5. microglia, 6. processes, 7. nuclei

Myelination

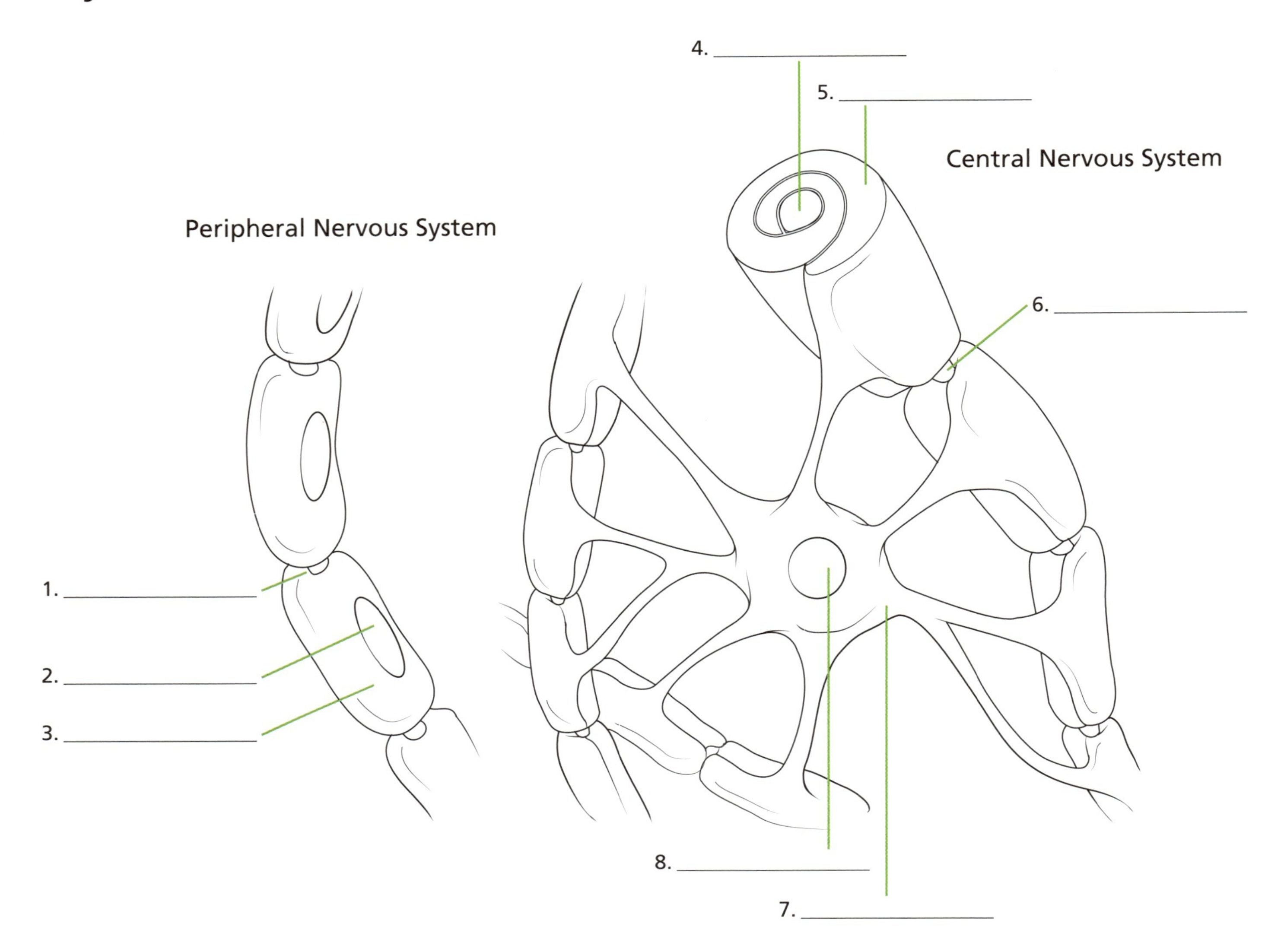

Myelination is the generation of a myelin sheath around axons in the nervous system. Myelin is a concentrically laminated structure made up of proteins and phospholipids. In the CNS, myelin is made by oligodendrocytes, which send their processes to myelinate multiple axons. In the PNS, myelin is produced by Schwann cells, which myelinate only one axon. The basic arrangement of the myelin sheath and its electrophysiological characteristics are relatively similar in the PNS and CNS.

Axons can be either unmyelinated or myelinated. Although myelinated axons are wrapped with myelin along their entire length, the sheath is not perfectly continuous. Between two adjacent myelin segments are gaps called nodes of Ranvier. The nodes have a high concentration of ion channels and allow ion flow across the axonal membrane.

Myelin effectively insulates segments of the axons and greatly enhances the speed of action potential propagation by means of saltatory conduction. Proper myelination is essential for normal nerve function, and demyelination can result in diseases such as multiple sclerosis.

Answers

1. axon, 2. nucleus, 3. Schwann cell, 4. nerve fiber, 5. myelin sheath, 6. node of Ranvier, 7. oligodendrocyte, 8. nucleus

Saltatory Conduction

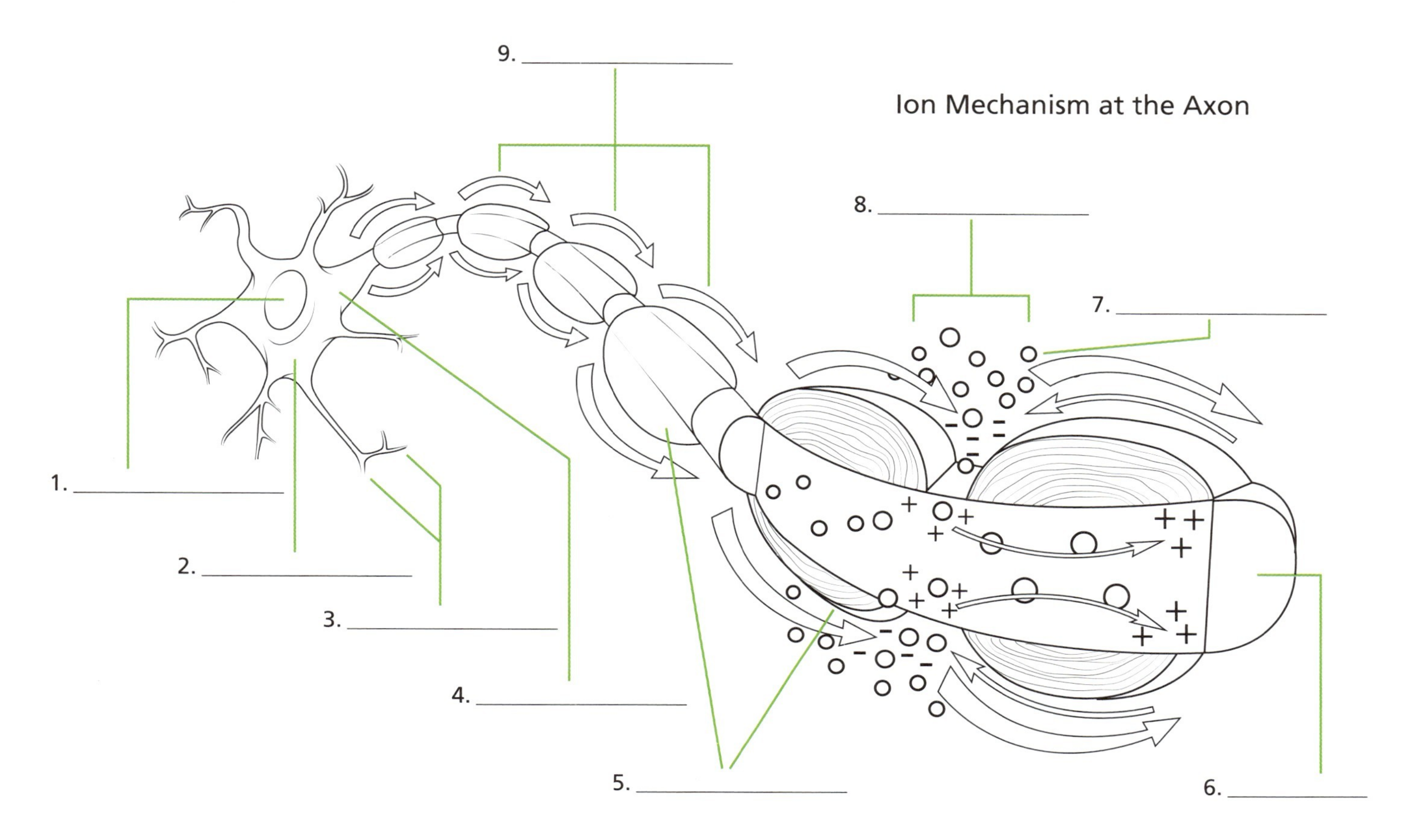

In myelinated axons, the myelin sheath acts as an electrical insulator, and transmembrane ion flow is possible only at the nodes of Ranvier. The nodes are highly enriched in voltage-gated Na^+ channels, which are responsible for the generation of action potentials. The presence of the insulating myelin sheath between the nodes prevents the ion current from leaking across internodal regions. Instead, the current flows to the next node, where the axon is exposed and regenerates the action potential. Thus, the action potential seems to jump from node to node. This type of propagation is called saltatory ("to dance") conduction.

Because the depolarization of the neuronal membrane needs to happen only at the nodes of Ranvier, the speed of action potential conduction in myelinated axons is much higher than that in unmyelinated axons. For example, unmyelinated axons generally conduct action potentials at around 0.5–10 m/s (meters per second), whereas myelinated axons can have propagation velocities up to 150 m/s.

Answers

1. nucleus, 2. soma, 3. dendrites, 4. axon hillock, 5. myelin sheath, 6. axon, 7. ions, 8. node of Ranvier, 9. saltatory conduction

Synapse Structure and Function

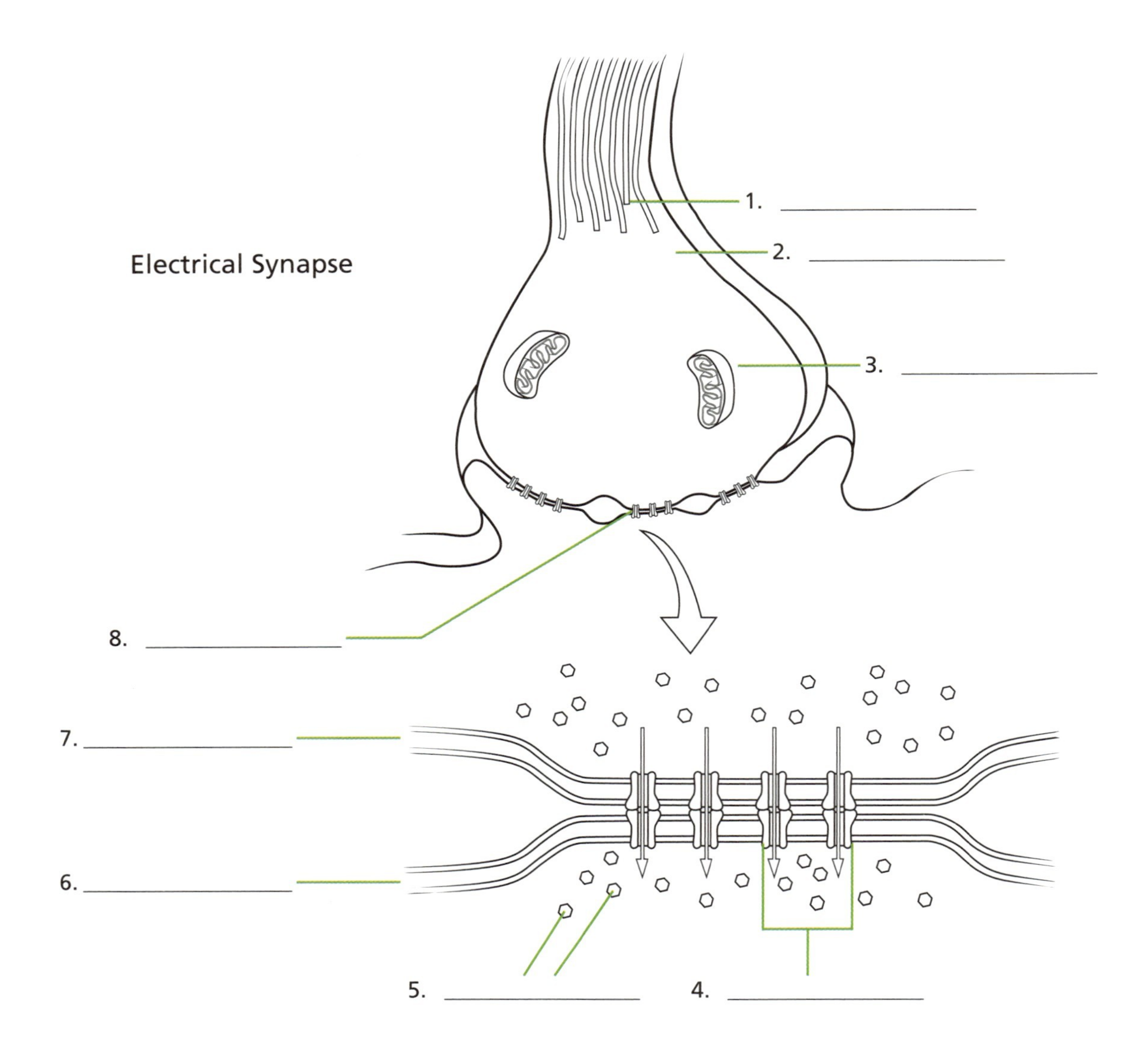

The synapse is a structure where one neuron communicates with another, typically from the axon terminal of one neuron to the dendrites of the other. Synapses permit information transfer between neurons and are the basis for the formation of complex neural circuits.

According to the transmission mechanism, synapses are classified into electrical and chemical. Electrical synapses are less common, and they directly connect the membranes of two communicating cells via intercellular protein channels called gap junctions. Gap junctions allow bidirectional signaling between neurons by facilitating simple diffusion of ions and small molecules between them. Because the diffusion of ions is passive through these channels, electrical synapses can be both excitatory and inhibitory. Furthermore, the signal amplitude always decays through electrical synapses due to their passive electrical properties.

Answers

1. microtubule, 2. cytoplasm, 3. mitochondrion, 4. gap junction channels, 5. ions, 6. postsynaptic membrane, 7. presynaptic terminal, 8. gap junction

Chemical Synapse

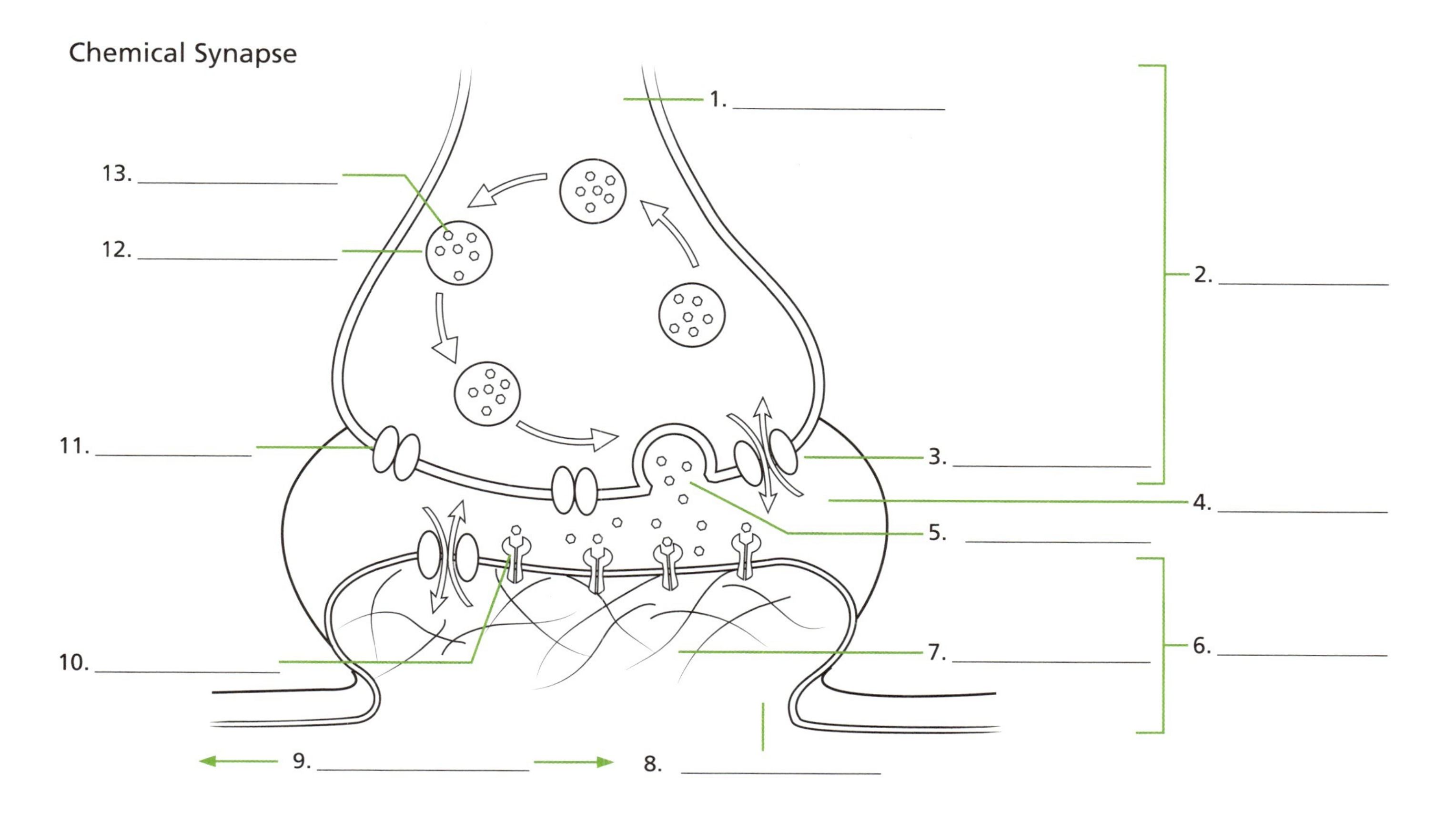

In contrast, chemical synapses (usually just called "synapses") transmit information between neurons via the release of chemical signals called neurotransmitters. Chemical synapses are by far the most common type of synapse in the nervous system and are classified by the identity of the neurotransmitter molecule that is released.

The structure of a synapse includes the presynaptic terminal, the synaptic cleft (the approximately 20 nanometer gap between two communicating neurons), and the postsynaptic membrane.

The presynaptic terminal is a specialized area on the axonal terminal of the transmitting neuron that contains many small spheres of phospholipid membrane called synaptic vesicles. These vesicles are loaded with neurotransmitter molecules, which then dock at specific sites on the neuron's plasma membrane called active zones.

Opposite the presynaptic terminal is the postsynaptic membrane, which contains neurotransmitter receptor proteins. Postsynaptic terminals can be found on axons, dendrites, and somas, though they typically occur on dendrites. Furthermore, some postsynaptic terminals (those that detect the neurotransmitter glutamate) are located on tiny bulbous projections of membrane called dendritic spines, which sprout from dendrites proper.

Neurotransmitter release is facilitated by a series of elaborate events. When the action potential reaches the presynaptic terminal, it depolarizes the membrane and rapidly increases intracellular Ca^{2+} levels by opening voltage-gated Ca^{2+} channels. Elevation of intracellular Ca^{2+} causes synaptic vesicles to fuse with the presynaptic membrane and release neurotransmitter molecules into the synaptic cleft. Following exocytosis—the release of neurotransmitters from the presynaptic terminal—the neurotransmitters diffuse across the synaptic cleft and bind to receptors on the postsynaptic membrane. Binding of neurotransmitters to postsynaptic receptors triggers excitation or inhibition of the postsynaptic cell. This in turn changes the probability that the postsynaptic neuron will fire an action potential. Neurotransmitters eventually dissociate from their receptors and are subsequently broken down by enzymes, diffuse out of the synaptic cleft, or are reabsorbed by the presynaptic cell.

Answers

1. axon terminal, 2. presynaptic terminal, 3. neurotransmitter transporter, 4. synaptic cleft, 5. active zone, 6. postsynaptic membrane, 7. postsynaptic density, 8. dendritic spine, 9. dendrite, 10. receptor, 11. Ca^{2+} channels, 12. synaptic vesicle, 13. neurotransmitter

Blood–Brain Barrier

The blood–brain barrier (BBB) is a highly selective barrier that limits and regulates molecular exchange between the brain and its blood supply.

The main components of the BBB are capillary endothelial cells, a basement membrane, and the end-feet of astrocytes. Tight junctions between adjacent endothelial cells form barriers that physically prevent molecules from diffusing into the brain through the intercellular spaces. The endothelial cell membrane contains many protein transporters that selectively allow the influx and efflux of molecules. Surrounding the endothelial cells is a layer of extracellular matrix called the basement membrane, which provides structural support and a further barrier to diffusion. Embedded in the basement membrane are contractile cells called pericytes, which wrap around endothelial cells and regulate capillary blood flow and the selective permeability of the BBB. The end-feet of astrocytes ensheath the capillaries and basement membrane and are critical for the induction and maintenance of the BBB, although they do not provide a physical barrier to diffusion. Instead, they provide a link between neurons and components of the BBB, and they also regulate capillary blood flow and BBB permeability. Finally, microglia patrol the BBB and act as the main form of active immune defense.

Although the BBB makes drug delivery into the CNS difficult, its integrity is essential for maintaining a healthy brain environment.

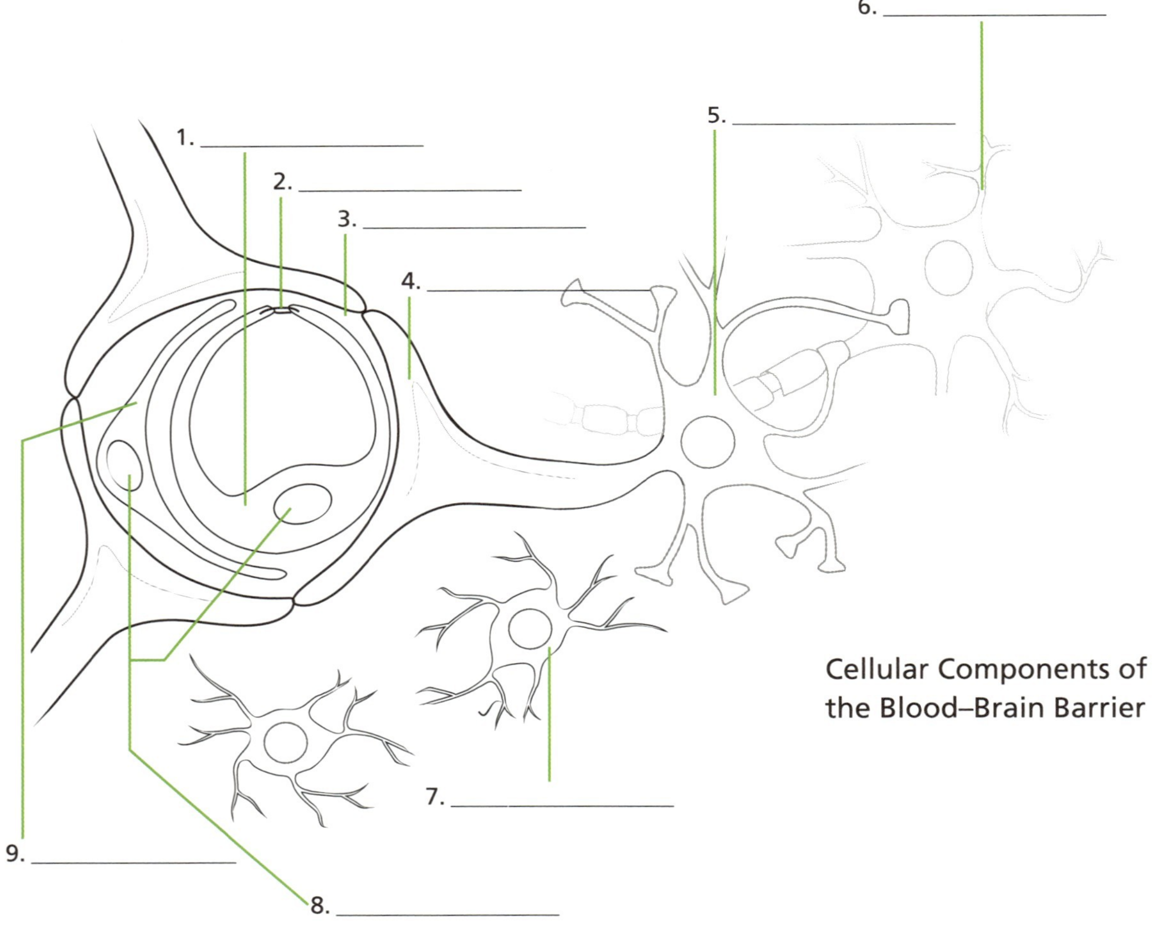

Cellular Components of the Blood–Brain Barrier

Answers

1. capillary endothelial cell, 2. tight junction, 3. basal lamina, 4. astrocyte end-feet, 5. astrocyte, 6. neuron, 7. microglia, 8. nucleus, 9. pericyte

Neural Plate and Neural Tube

The nervous system begins forming in the early stages of embryonic development with the neuroectoderm at three weeks. The neuroectoderm comes from the ectoderm, which is the outermost of the three layers of embryonic tissue; the inner two layers are the mesoderm and endoderm. Key players in the developing nervous system are the neural plate, the source of the majority of neurons and glia, and the neural tube, the precursor to the brain and spinal cord. The anterior and posterior ends of the neural plate become the brain and spinal cord, respectively. The notochord (similar to a spinal cord) is a flexible rod-shaped body present in all developing chordates. It is derived from the mesoderm and induces the development of the neural plate when the rate of cell proliferation is high. The neural plate thickens and extends, forming the neural groove. The neural groove is the longitudinal furrow of the neural plate that forms the neural tube. Cells on the margin of the neural groove migrate to make up the neural crest. The ends of the neural plate, known as the neural folds, fuse along the midline to form the neural tube. The neural tube closes and becomes a major component of brain and spinal cord development.

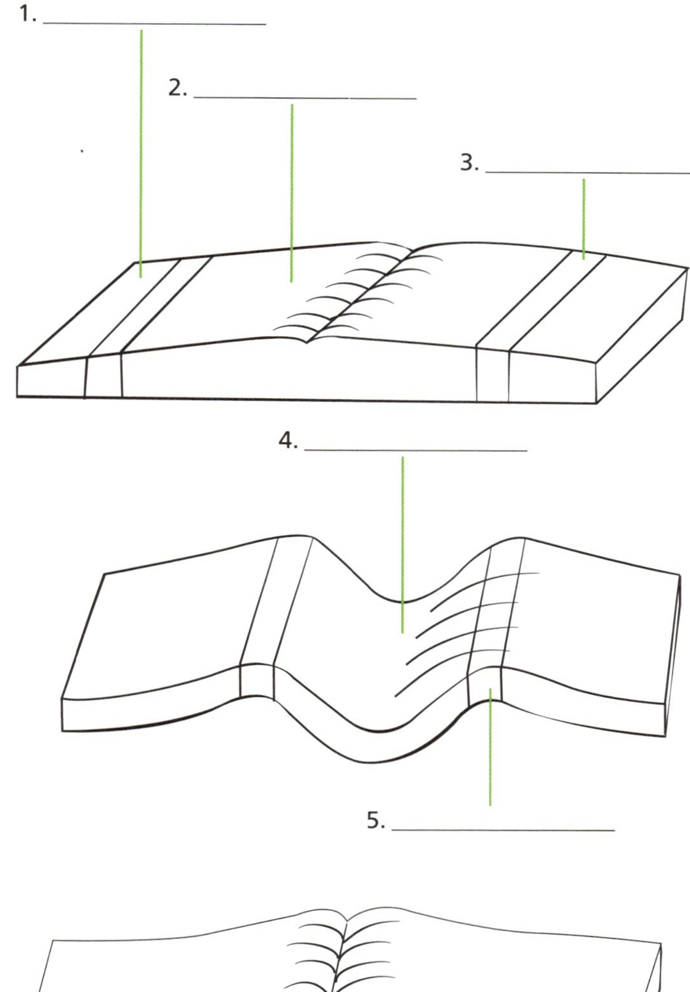

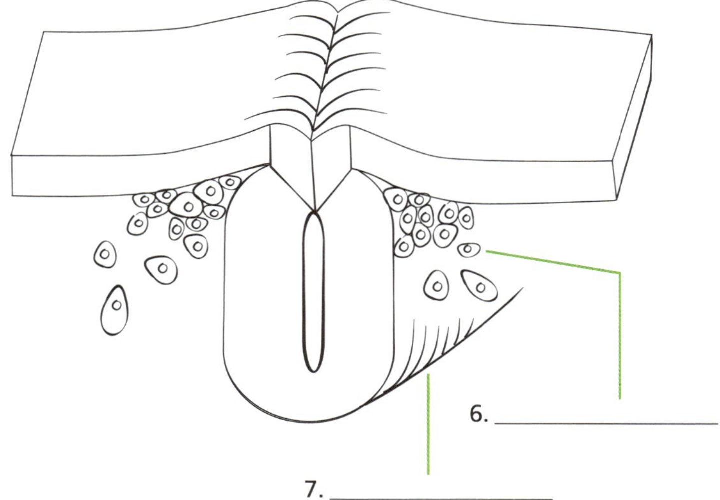

Development of the Neural Tube

Answers

1. ectoderm, 2. neural plate, 3. neural plate border, 4. neural groove, 5. neural fold, 6. neural crest cells, 7. neural tube

Genesis of Neurons and Glia

A developed human brain contains approximately 100 billion neurons and even more glial cells. A population of self-renewing cells known as stem cells gives rise to progenitor cells that will eventually become neurons and glia. Progenitor cells of the nervous system, including neurons, as well as astrocytes and oligodendrocytes—the glial cells—arise from specialized regions of the ectoderm in the late embryo known as the neuroectoderm. Progenitor cells have the potential to give rise to any of these cell types in the nervous system. Prior to differentiation, cells can revert to the previous stage, but they lack the self-renewing capacity once differentiated.

Differentiation from progenitor cell to neuron or glia occurs through a tightly controlled process involving chemical messengers such as hormones and growth factors. The process of neuron generation from neural stem cells and progenitor cells is known as neurogenesis. Neurogenesis occurs very early in mammalian embryonic development, whereas gliogenesis typically begins between weeks 15 and 20 after fertilization. Prenatally, neurogenesis occurs throughout the brain, including regions such as the dentate gyrus and nucleus accumbens. To date, adult mammals have two known regions of the brain that are capable of neurogenesis: the hippocampus and the subventricular zone.

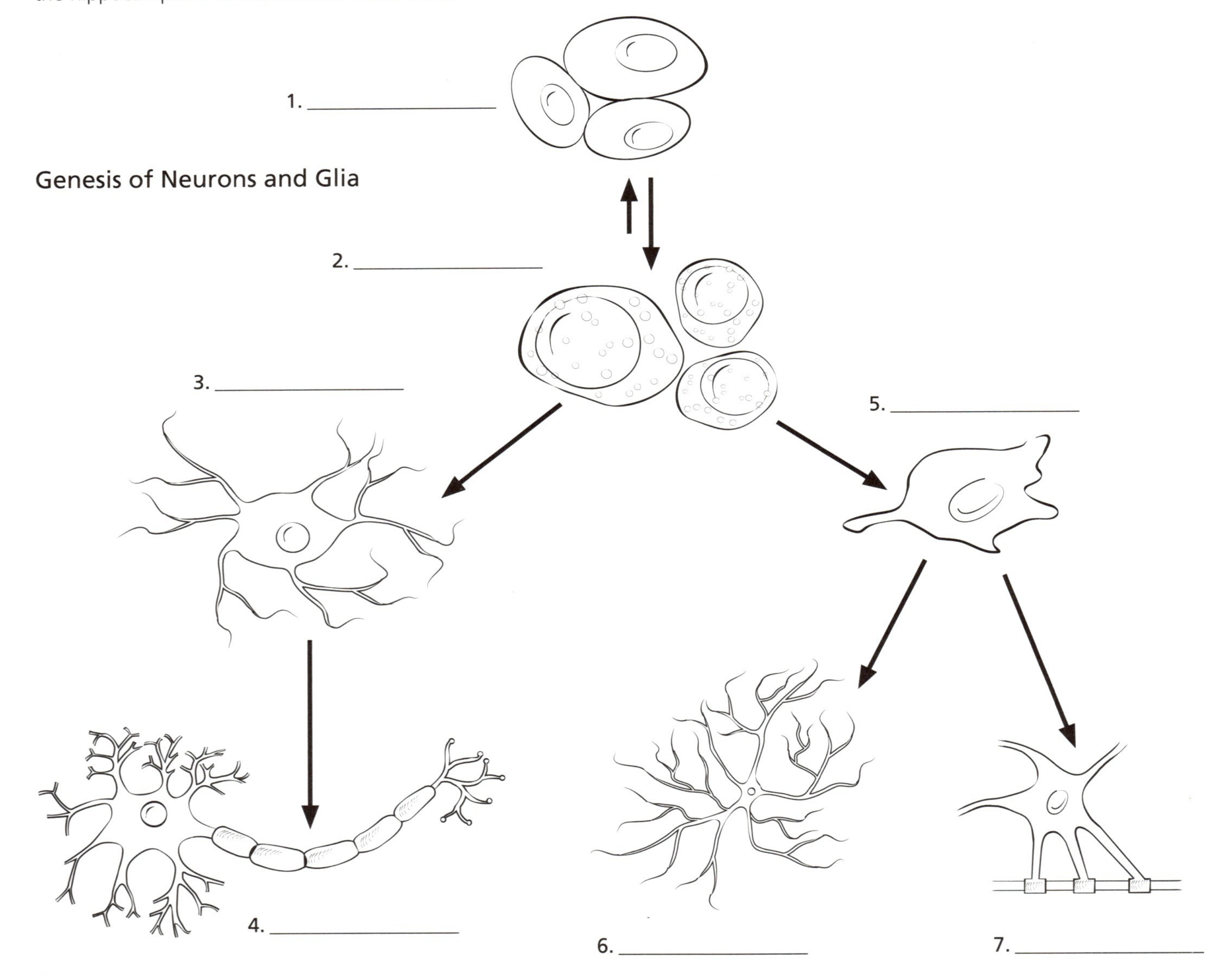

Answers

1. stem cell, 2. progenitor cell, 3. neuroblast, 4. neuron, 5. glioblast, 6. astrocyte, 7. oligodendrocyte

Spinal Cord Development

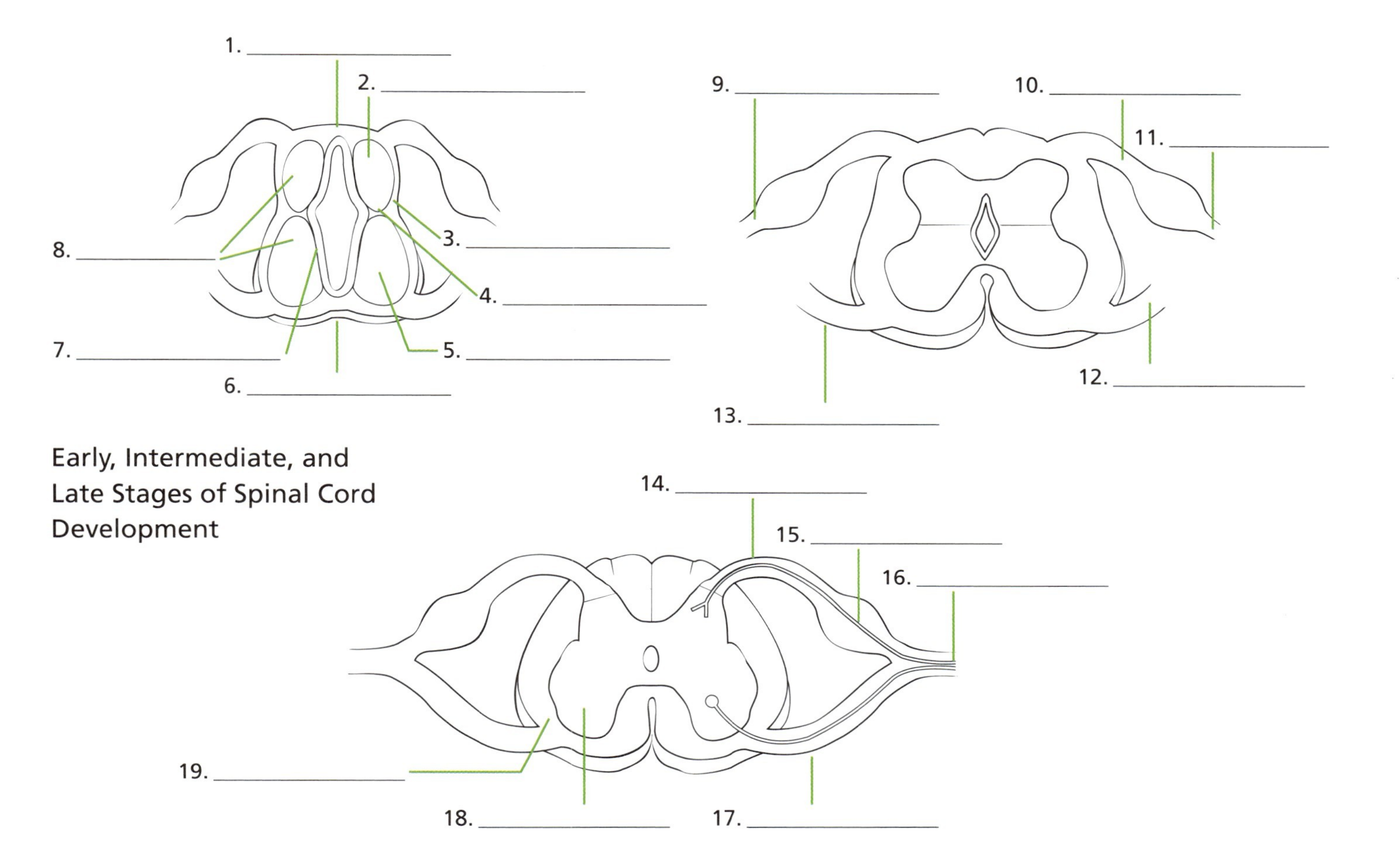

Early, Intermediate, and Late Stages of Spinal Cord Development

The developing spinal cord is made up of three layers: the ventricular, mantle, and marginal layers. The ventricular layer contains undifferentiated, proliferating cells. The mantle layer is formed from the neuroepithelial cells that make up the wall of the neural tube. It contains the primary cell bodies of neurons and ultimately becomes the gray matter. The associated axons will grow into the marginal layer, which becomes white matter. By four weeks post conception, immature neurons known as neurocytes accumulate in specific areas relative to the sulcus limitans, the structure that originates from the neural tube and divides the marginal layer into basal and alar plates. Dorsal neurocytes form the alar plate, which is associated with sensory pathways. Ventral neurocytes form the basal plate, which is associated with motor pathways. The developing spinal cord has functional components: general somatic afferent, general somatic efferent, general visceral afferent, and general visceral efferent. Somatic input and output originate from the skin and deep tissues, whereas visceral input and output originate from the internal organs. Afferent (incoming) neurons transmit information from the periphery and visceral structures to the brain, while efferent (outgoing) neurons transmit information from the CNS to skeletal muscle, smooth muscle, and glands.

Answers

1. roof plate, 2. alar plate, 3. marginal layer, 4. sulcus limitans, 5. basal plate, 6. floor plate, 7. ventricular layer, 8. mantle layer, 9. dorsal root, 10. dorsal horn, 11. sensory pathway, 12. motor pathway, 13. ventral horn, 14. dorsal root, 15. dorsal root ganglion, 16. spinal nerve, 17. ventral root, 18. gray matter, 19. white matter

Neural Crest Development and Function

During the fourth embryonic week of development, cells in the lateral margin of the neural groove derived from the ectoderm migrate to the margins of the neural tube to form the neural crest. Neural crest cells migrate to many different locations and differentiate into many different cell types in the embryo based on local biochemical signaling. Examples of different cell types include smooth muscle, osteoblasts, adipocytes, chondrocytes, melanocytes, and neurons. The migration pathways of neural crest cells are also controlled by signaling factors in the cellular milieu. The neural crest is made up of five components: cranial, trunk, vagal, sacral, and cardiac. The cranial neural crest regions are made up of the three primary subdivisions of the embryonic brain: the forebrain, midbrain, and hindbrain. Neural crest cells in the trunk region become ganglia and aortic nerve clusters. Vagal and sacral neural crest cells eventually form the parasympathetic ganglia of the gut. Another type, cardiac neural crest cells, produces musculoskeletal tissues of the large arteries, as well as cartilage, connective tissue, and neurons.

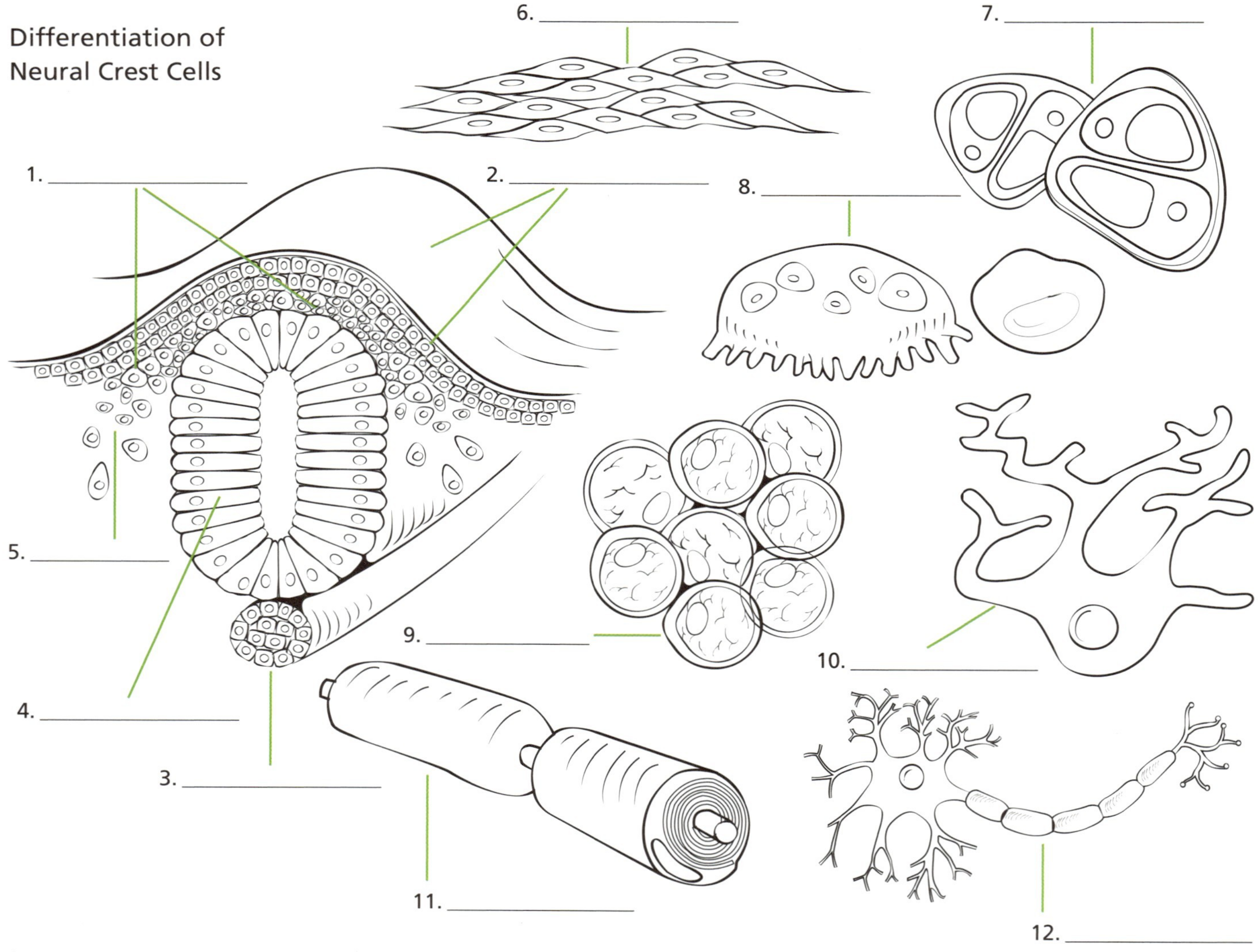

Answers

1. neural crest, 2. ectoderm, 3. notochord, 4. neural tube, 5. neural crest cells, 6. smooth muscle cells, 7. osteoblasts/osteoclasts, 8. adipocytes, 9. chondrocytes, 10. melanocytes, 11. Schwann cells, 12. neurons

Formation of the Brain Vesicles

The neural tube is the embryonic precursor to the CNS. It forms the three primary vesicles of the brain when it closes, including the forebrain (prosencephalon), midbrain (mesencephalon), and hindbrain (rhombencephalon) during the fourth embryonic week. Vesicles are small subdivisions within the developing brain that correspond to different structures and functions. From the three primary vesicles come the five secondary vesicles and their corresponding brain regions, a development that occurs during week six after conception. The telencephalon includes regions of the brain such as the cortex, basal ganglia, hippocampus, and olfactory system. The diencephalon includes the thalamus, pineal gland, retina, and optic nerve. The mesencephalon is the midbrain, the metencephalon is the pons and cerebellum, and the myelencephalon is the medulla. Regions of the forebrain, or prosencephalon, control body temperature, reproductive function, eating, and sleeping. The midbrain, which includes the diencephalon and mesencephalon, controls vision, hearing, motor control, and alertness. The hindbrain includes the metencephalon and myelencephalon, and it controls respiration, equilibrium, and autonomic functions such as breathing and heart rate.

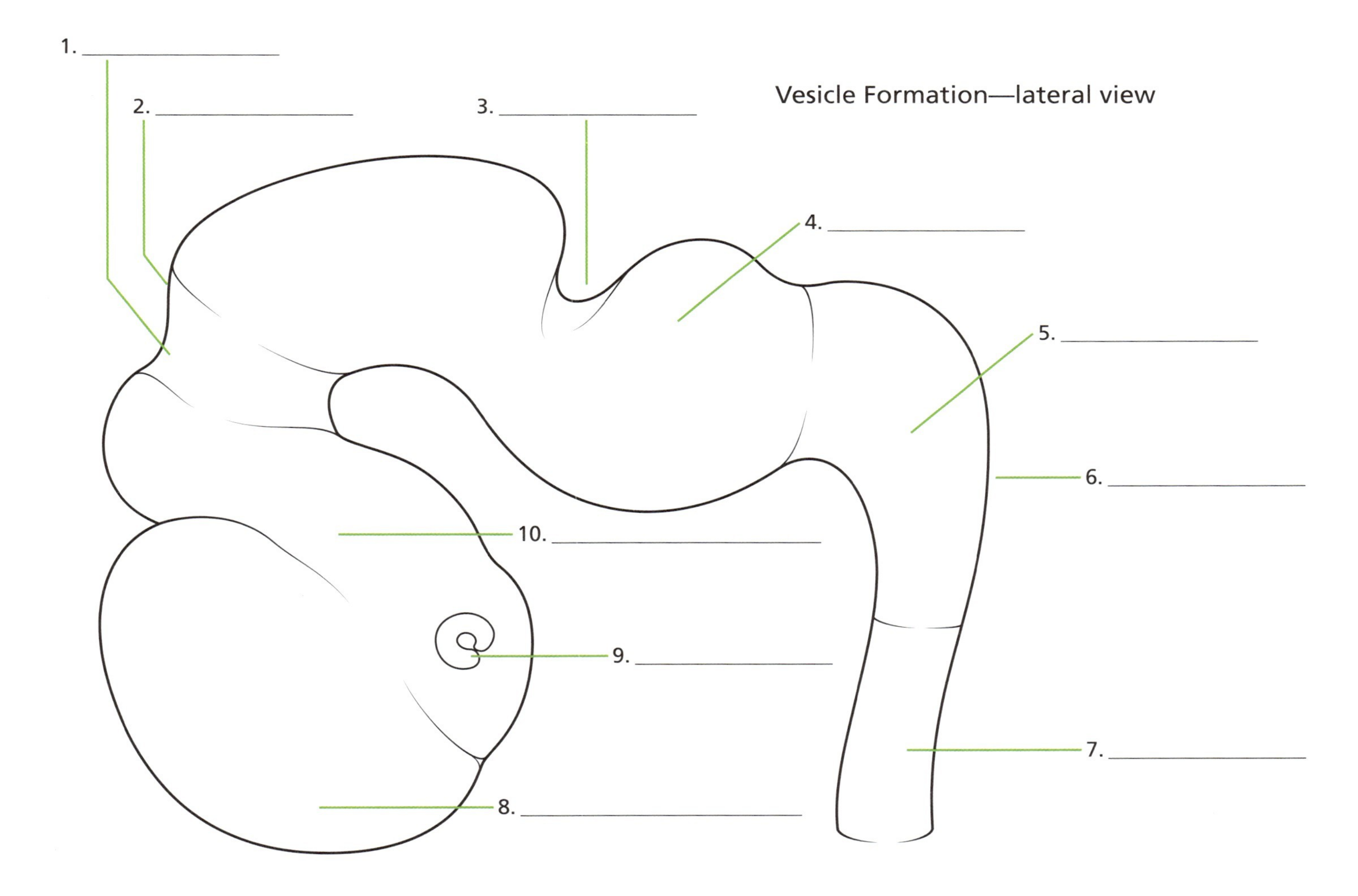

Answers

1. mesencephalon (midbrain), 2. cephalic flexure, 3. pontine flexure, 4. metencephalon (pons), 5. myelencephalon (medulla), 6. cervical flexure, 7. spinal cord, 8. telencephalon (cerebral hemispheres), 9. optic vesicle, 10. diencephalon

Brainstem Development

The brainstem is the posterior part of the brain that is joined to the spinal cord. It includes the medulla oblongata (myelencephalon), pons (metencephalon), and midbrain (mesencephalon). The brainstem contains the cranial nerves and provides the main sensory and motor innervation to the head and neck. The brainstem also plays a role in regulating respiratory and cardiac functions, as well as maintaining consciousness. It is formed by the third, fourth, and fifth vesicles in the developing brain during the first five weeks of life. By birth, the brainstem is one of the most fully developed regions of the brain. The lower brainstem develops with respect to the basal plate, alar plate, and sulcus limitans. The functional components of the cranial nerve nuclei, the collections of neurons associated with cranial nerves in the brainstem, include general somatic afferent, general somatic efferent, general visceral afferent, and general visceral efferent neurons. Afferent neurons transmit information from the periphery and visceral structures to the brain, while efferent neurons transmit information from the CNS to skeletal muscle, smooth muscle, and glands.

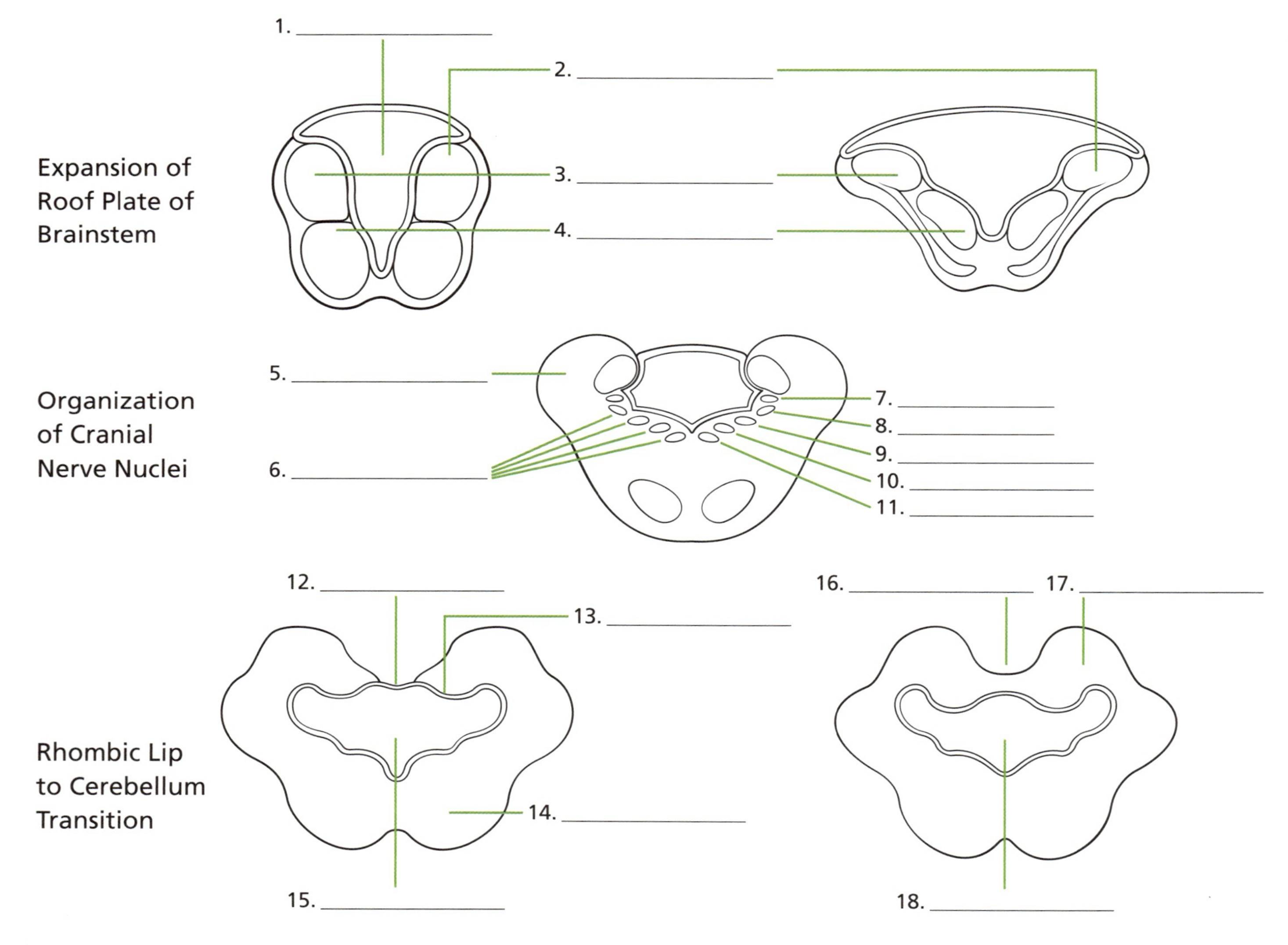

Answers

1. fourth ventricle, 2. alar plate, 3. sulcus limitans, 4. basal plate, 5. cerebellum, 6. cranial nerve motor nuclei, 7. special and general sensory afferent neurons, 8. special and general visceral afferent neurons, 9. general visceral efferent neuron, 10. special visceral efferent neuron, 11. general sensory efferent neuron, 12. roof plate, 13. rhombic lip, 14. pons, 15. fourth ventricle, 16. vermis, 17. cerebellar hemisphere, 18. fourth ventricle

Cerebellum Development

Early in the embryonic stage (between conception and gestational week eight), the cerebellar plate begins as a bulge on the rhombencephalon, which is the primary vesicle that represents the hindbrain. The rhombic lip is established, producing granule cells during development. Neuron progenitors migrate from the ventricular zone, an embryonic region of the cerebral cortex that is the origin of the majority of neurons in the brain. The neuroepithelium of the fourth ventricle (ventricular zone), the portion of the embryonic ectoderm that gives rise to the nervous system, is the source of all inhibitory, GABAergic neurons, including Purkinje cells, neurons that send inhibitory signals to deep cerebellar nuclei and constitute the sole motor coordination output in the cerebellum. The rhombic lip is the source of all excitatory, glutamatergic neurons, including cerebellar nuclei and external germinal zone neurons, from which granule cells originate. The mammalian cerebellum is divided into three lobes and ten lobules.

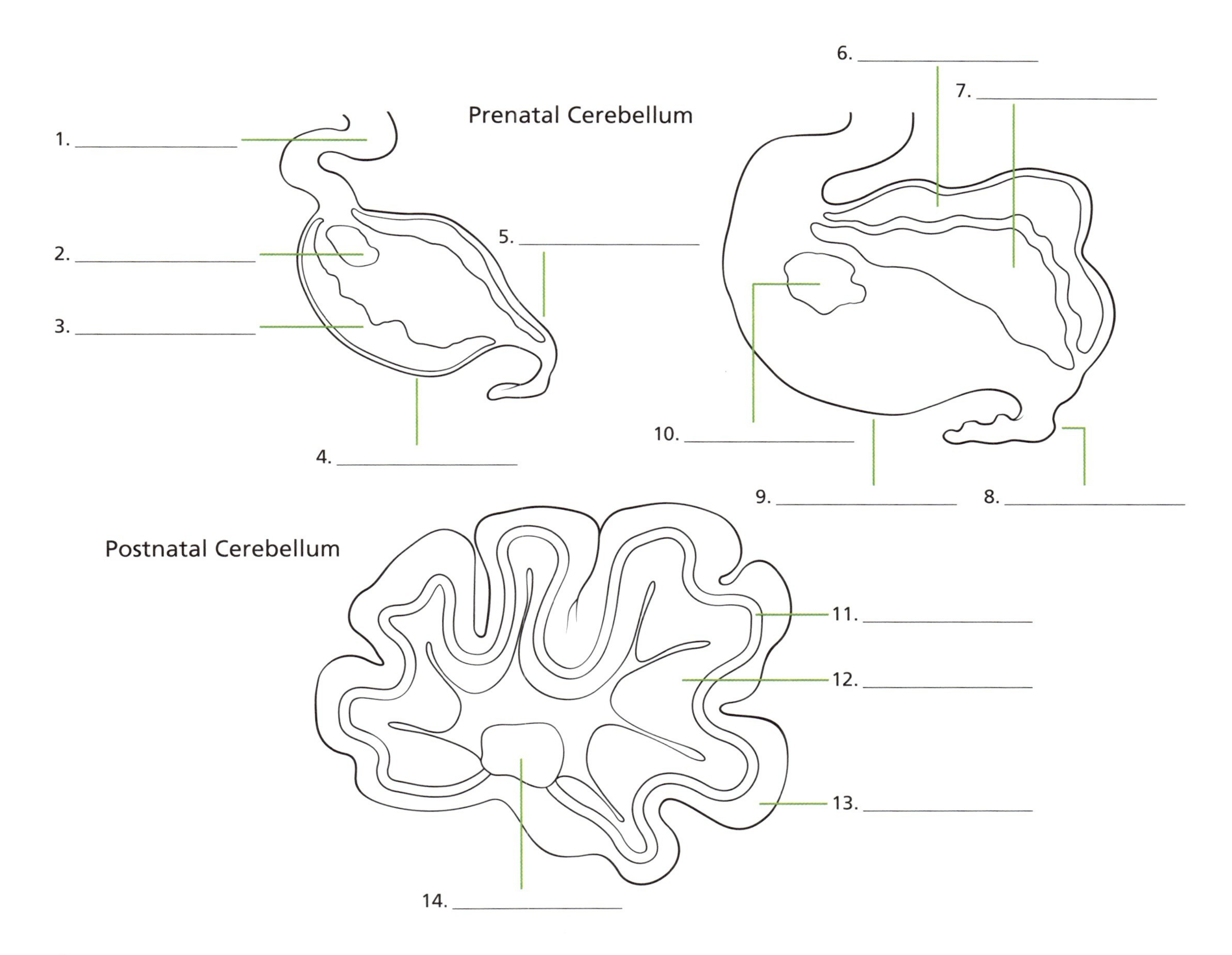

Answers

1. mesencephalon, 2. nuclear transitory zone, 3. Purkinje cell precursors, 4. ventricular zone, 5. rhombic lip, 6. external geminal zone, 7. Purkinje cell clusters, 8. rhombic lip, 9. ventricular zone, 10. cerebellar nuclei neurons, 11. Purkinje cell layer, 12. granular layer, 13. molecular layer, 14. cerebellar nuclei neurons

Cerebral Hemisphere Development

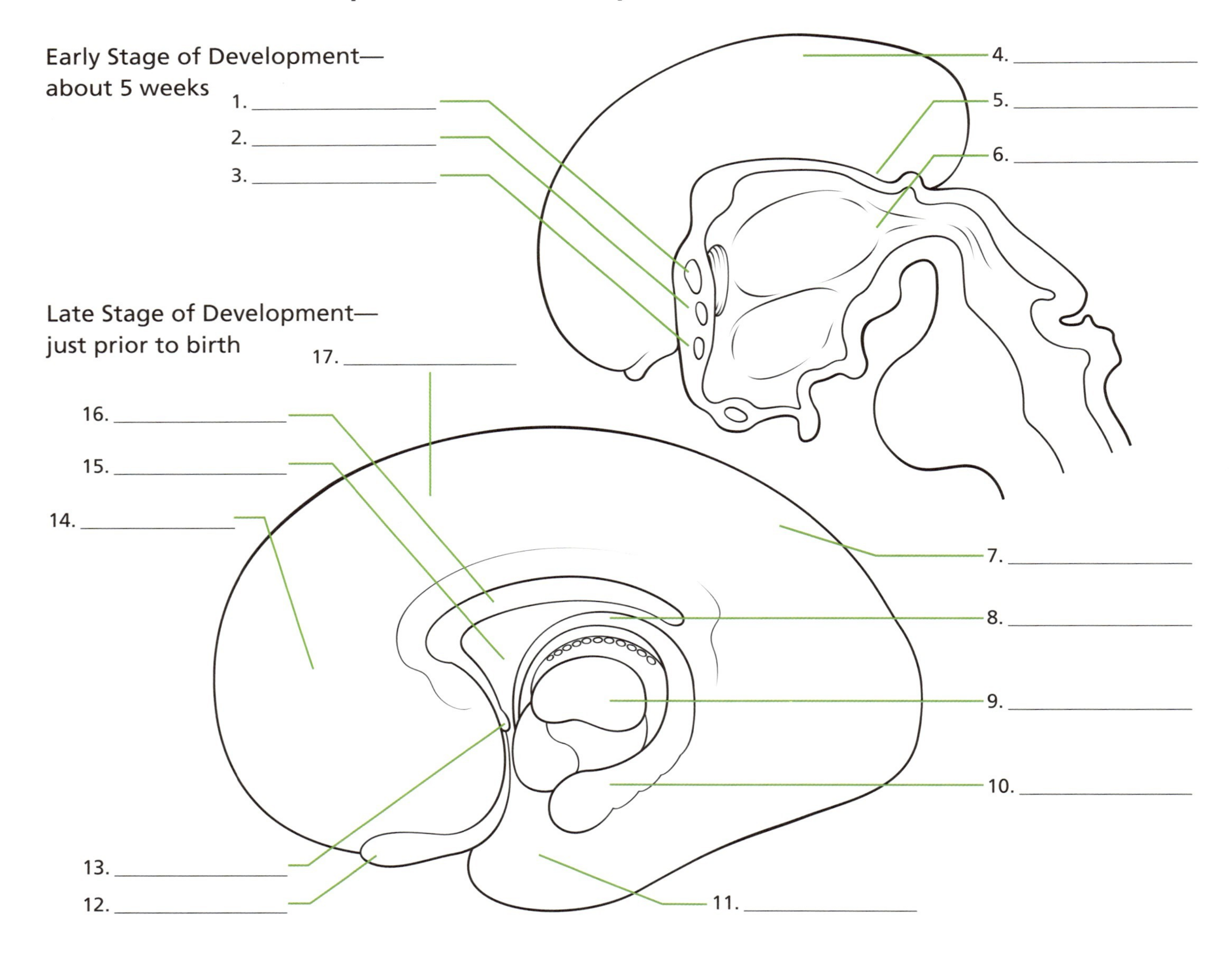

The cerebral hemispheres begin to develop during the fifth week following conception. Development of the cerebral hemispheres results in fusion with the diencephalon and occurs from the most anterior part of the neural plate from a specialized part of the embryonic ectoderm known as the telencephalon. The cavity inside the neural tube becomes the ventricular system, and the prosencephalon, or forebrain, becomes the cerebral hemispheres and cortex. The medial longitudinal fissure divides the brain into the left and right hemispheres. Although it is commonly believed that different functions are associated with one hemisphere or the other ("right-brained" vs. "left-brained"), most brain functions are distributed across both hemispheres. The brain is also divided into four lobes, each with corresponding structures based on the function. The frontal lobe develops first as the cerebral hemispheres grow anteriorly, then laterally and superiorly to form the parietal lobe, and posteriorly and inferiorly to form the occipital and temporal lobes. The temporal, frontal, occipital, and parietal lobes are named for the corresponding bones of the skull and are separated by landmarks known as fissures and sulci, which are grooves in the surface of the cerebral hemispheres. The grooves on the surface of the cerebral hemispheres are thought to arise from early development of white matter tracts and tension that results from the expansion of brain material.

Answers

1. corpus callosum, 2. commissure of the fornix, 3. anterior commissure, 4. cerebral hemisphere, 5. tela choroidea, 6. diencephalon, 7. cerebral hemisphere, 8. fornix, 9. diencephalon, 10. hippocampal formation, 11. anterior temporal lobe, 12. olfactory bulb, 13. anterior commissure, 14. frontal lobe, 15. septum pellucidum, 16. corpus callosum, 17. parietal lobe

Cerebral Cortex Development

The cerebral cortex is the outer layer of the brain that develops during the second trimester of pregnancy. The cortex forms during embryonic development as the hemispheres mature. The parietal cortex forms from the dorsal and lateral areas of the neural tube, whereas the temporal and occipital cortices form from the posterior and ventral areas of the neural tube. The cortex is divided into two sections: the neocortex and the allocortex. The neocortex makes up most of the cerebral cortex and is divided into six layers, whereas the allocortex is much smaller with fewer cell layers. Each layer of the neocortex is composed of different cell types, including pyramidal, stellate, and granular neurons. These different cell types perform specific functions based on their structure and location in the brain. For example, pyramidal cells grow out into other regions and form the internal capsule; they receive input from the thalamus. The cerebral cortex is divided into two different types of cortices, known as primary and association, which are based on the level of functioning required. Primary cortices are responsible for simpler functions such as sensory input, while association cortices are involved in more complex functions such as creativity, abstraction, and language.

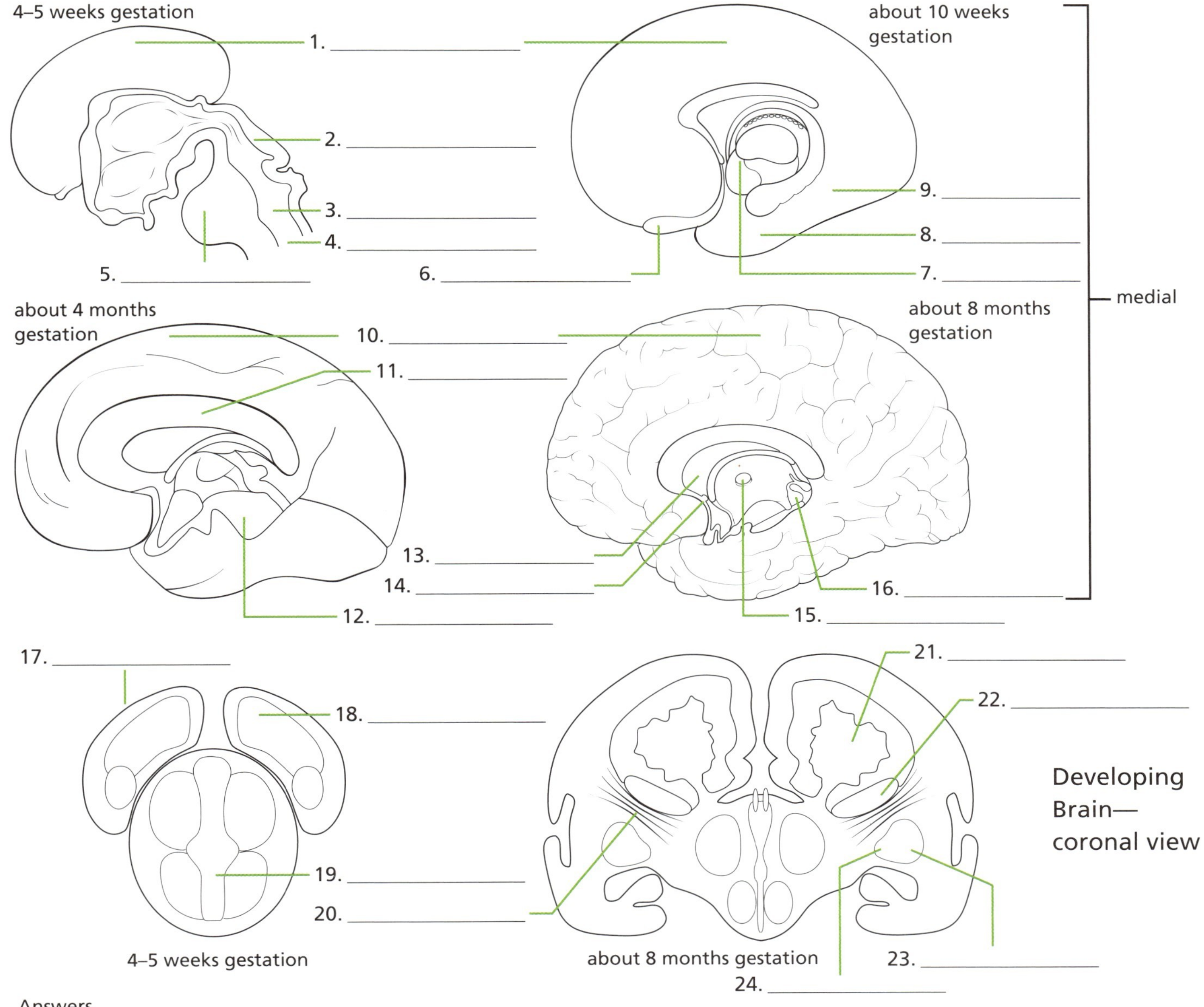

Answers

1. cerebral cortex, 2. midbrain, 3. medulla, 4. spinal cord, 5. pons, 6. olfactory bulb and tubercle, 7. interventricular foramen, 8. pyriform cortex, 9. entorhinal cortex, 10. cerebral cortex, 11. corpus callosum, 12. midbrain, 13. septum pellucidum, 14. anterior commissure, 15. massa intermedia, 16. posterior commissure, 17. cerebral cortex, 18. lateral ventricle, 19. third ventricle, 20. internal capsule, 21. choroid plexus of lateral ventricle, 22. caudate nucleus, 23. putamen, 24. globus pallidus

Macroscopic Structure of the Thoracic Spinal Cord

The spinal cord is a vital component of the CNS, facilitating information transfer from the body to the brain, and vice versa. It is a columnar structure of nervous tissue that emerges through the foramen magnum of the skull and descends through the vertebral foramina of the spine's vertebrae. In adults, the spinal cord is 15–20 inches (40–50 cm) long and 0.4–0.6 inches (1–1.5 cm) thick. It is divided into cervical, thoracic, lumbar, and sacral regions, corresponding to the vertebrae groups of the same names. Thirty-one paired spinal nerves emerge from the cord and course through the intervertebral foramina toward their distal targets. These nerves carry sensory, motor, and autonomic information. In the lumbar region, the cord begins to taper, forming the conus medullaris. Inferior to this, spinal nerves yet to exit the vertebral column descend in parallel, forming the cauda equina. The spinal cord is housed in the same three meninges that protect the brain (the pia, the arachnoid, and the dura), and is further surrounded and cushioned by adipose tissue in the epidural space. A central canal (which is continuous with the ventricles of the brain) travels the length of the spinal cord and is filled with cerebrospinal fluid, as is the subarachnoid space surrounding the cord.

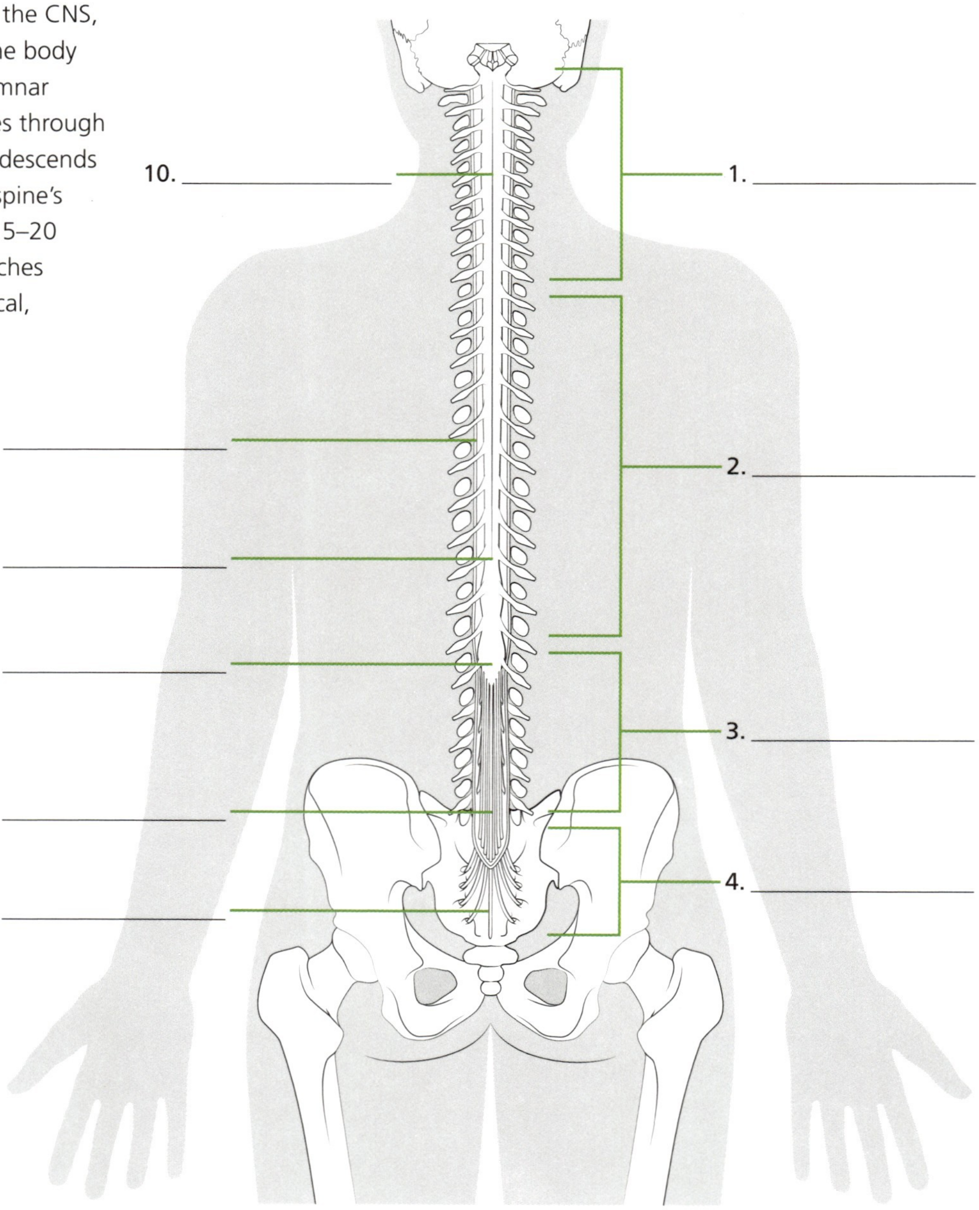

1. ______
2. ______
3. ______
4. ______
5. ______
6. ______
7. ______
8. ______
9. ______
10. ______

Spinal Column—posterior view

Answers

1. cervical spinal nerves, 2. thoracic spinal nerves, 3. lumbar spinal nerves, 4. sacral spinal nerves, 5. coccygeal ligament, 6. cauda equina, 7. conus medullaris, 8. lumbosacral enlargement, 9. meninges, 10. cervical enlargement

Cross-sectional Anatomy of the Thoracic Spinal Cord

Viewed in transverse section, the spinal cord contains white matter in its periphery, gray matter in the shape of an "H" on the inside, and a central canal that is filled with cerebrospinal fluid. The two sets of projections of the gray matter are called the dorsal (posterior) horns and ventral (anterior) horns. The horns form columns of gray matter throughout the entire length of the spinal cord. The two dorsal horns are separated by the posterior median sulcus, while the two ventral horns are separated by the anterior median fissure. The gray commissure is a thin strip of white matter that surrounds the central canal and connects the two halves of the spinal cord. The dorsal columns mainly contain the ascending pathways for the dorsal column–medial lemniscal pathways, while the ventral horns contain descending vestibular, ventral corticospinal, reticulospinal, and tectospinal tract axons. The lateral columns contain ascending pathways of the lateral spinothalamic and spinoreticular tract axons. The first-order neuronal cell bodies are contained in the dorsal root ganglion neurons, which are subserved by a dorsal and a ventral spinal nerve root.

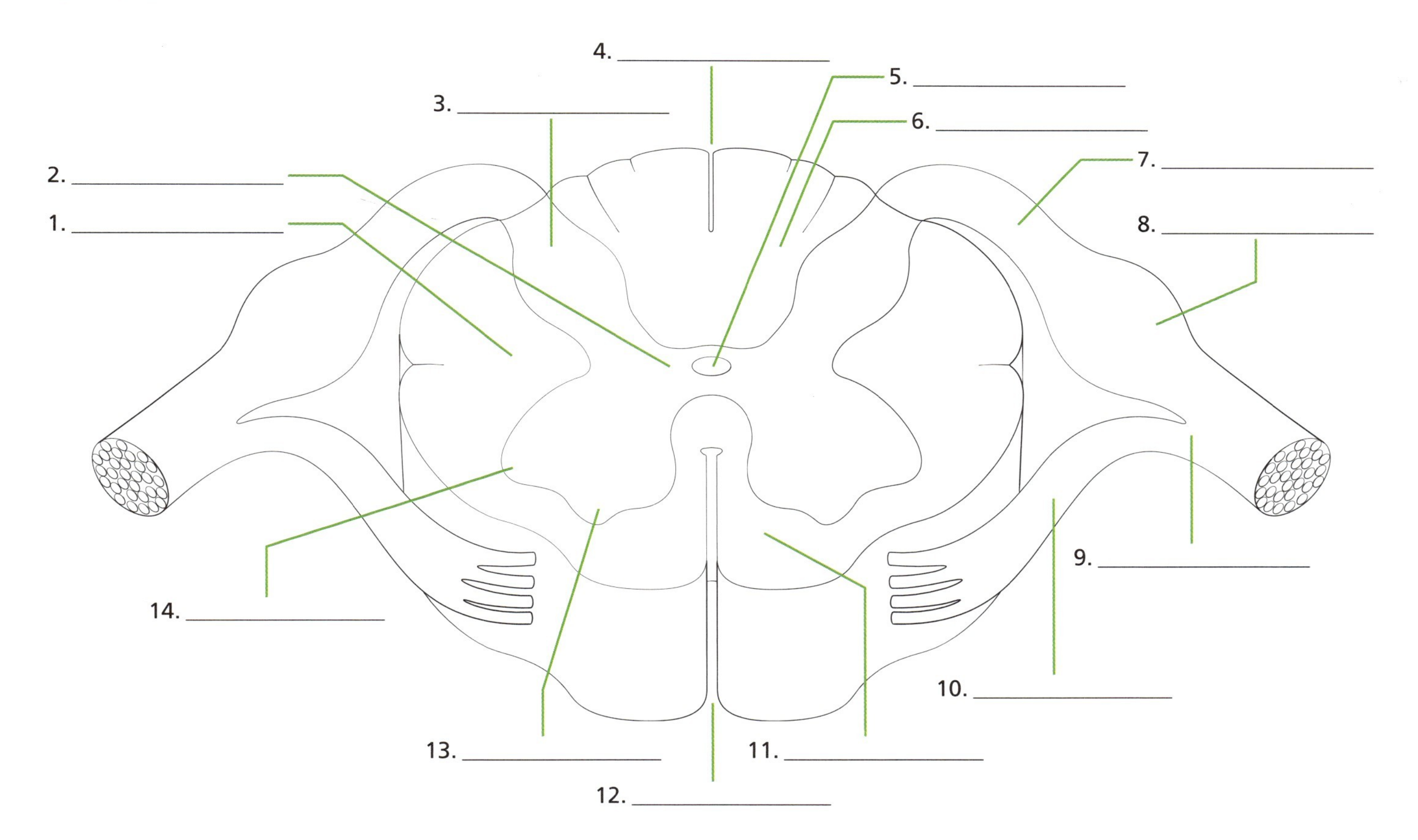

Thoracic Spinal Cord—
cross section, ventral view

Answers

1. lateral column, 2. gray commissure, 3. dorsal horn, 4. posterior median sulcus, 5. central canal, 6. dorsal column, 7. dorsal root of spinal nerve, 8. dorsal root ganglion, 9. spinal nerve, 10. ventral root of spinal nerve, 11. ventral column, 12. anterior median fissure, 13. ventral horn, 14. lateral horn

Spinal Cord Structure at Different Levels

The amount of gray matter is greatest in segments of the spinal cord that deal with sensory and motor control of limbs. These areas are expanded at the cervical enlargement, which supplies nerves to the shoulder girdles and upper limbs, and at the lumbosacral enlargement, which provides nerves to the pelvis and lower limbs. Cross sections of the spinal cord show the differing characteristic appearances of the gray and white matter in the cervical, thoracic, lumbar, and sacral regions. Additional lateral horns are visible in the thoracic area of the spinal cord, which is primarily involved with functions of the sympathetic division of the autonomic motor system (changes in cardiac, pulmonary, hepatic, and gastrointestinal activities).

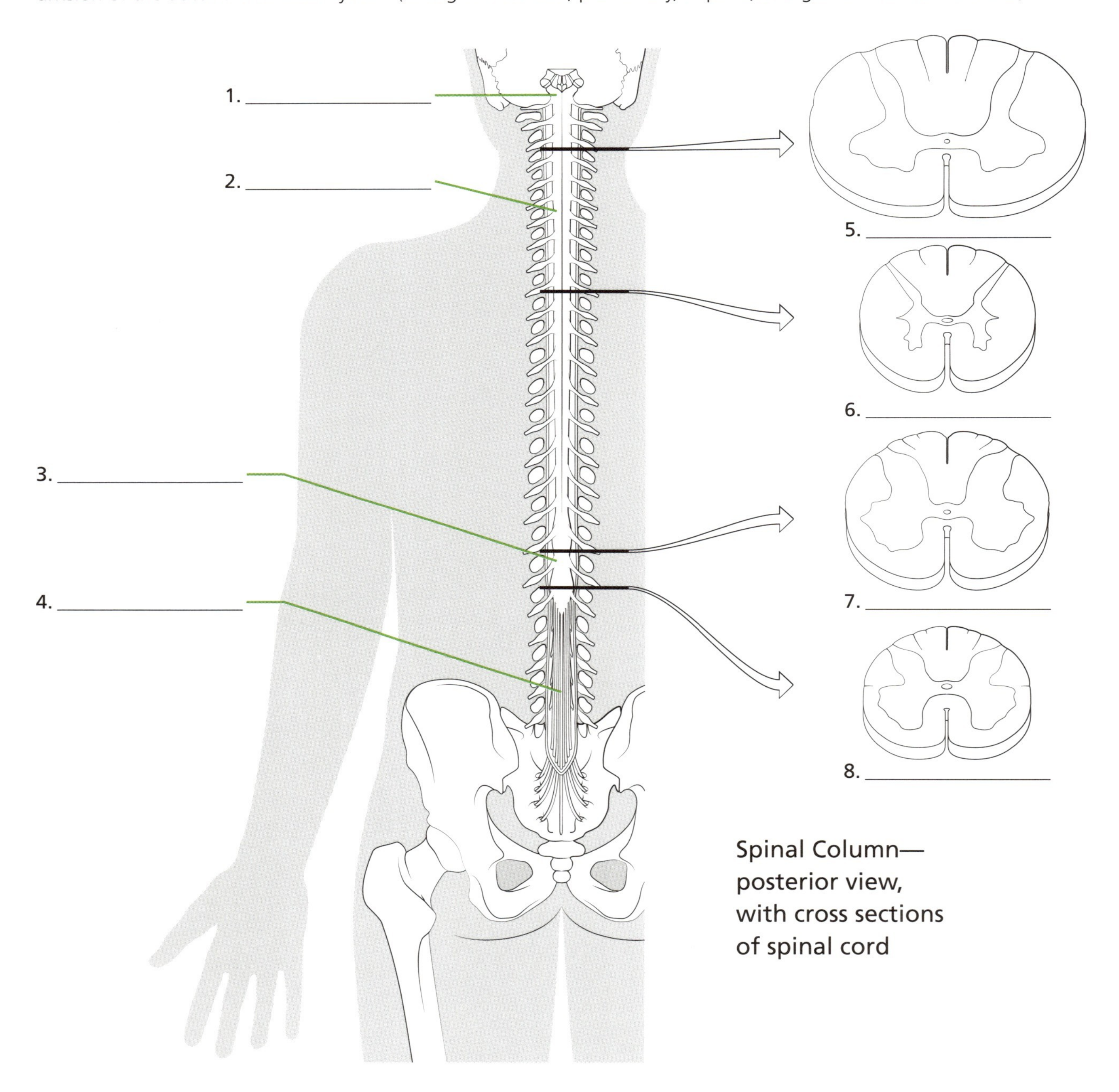

Spinal Column—posterior view, with cross sections of spinal cord

Answers

1. brainstem, 2. cervical enlargement, 3. lumbosacral enlargement, 4. cauda equina, 5. cervical, 6. thoracic, 7. lumbar, 8. sacral

Ascending Nerve Tracts: Dorsal Column Pathway

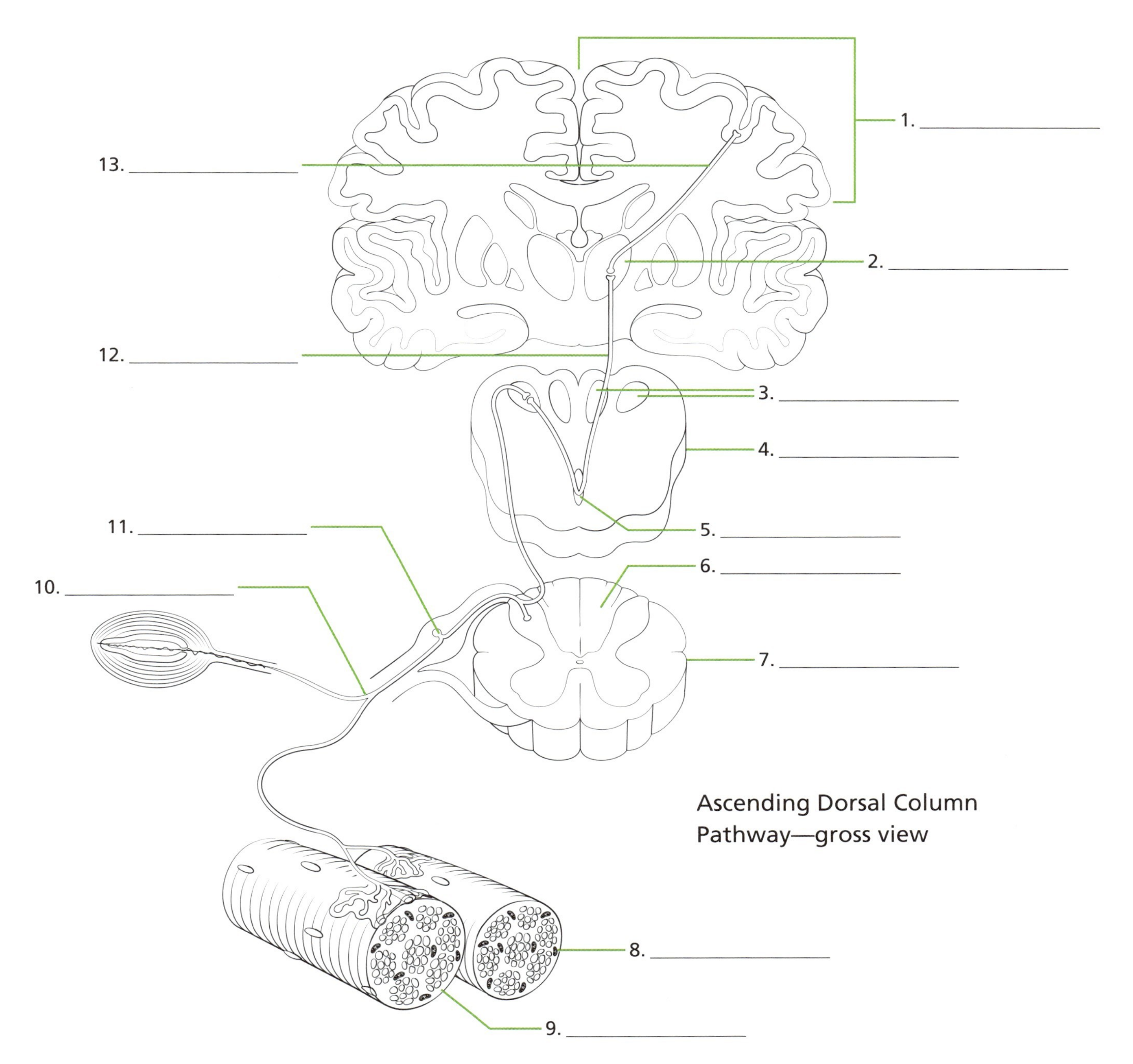

Ascending Dorsal Column Pathway—gross view

Also referred to as the dorsal lemniscal pathway, the dorsal column pathway mediates conscious proprioception, sense of tactile discrimination, and sense of vibration and forms recognition signals. The first-order neuronal soma is localized in the dorsal root ganglion at all spinal levels, and the signal then enters the dorsal columns of the spinal cord. The signal travels up the spinal cord to reach the gracile and cuneate nuclei of the caudal medulla, where it synapses with neurons in the ventral posterolateral nucleus (VPL). The signal then reaches the postcentral gyrus/somatosensory cortex (Brodmann areas 3, 1, and 2). The fact that the nerve signals cross over at the medulla means that those from the right side of the body are decoded by the left side of the brain, and those from the left side of the body are decoded by the right side of the brain.

Answers

1. primary somatosensory cortex, 2. thalamus, 3. dorsal column nuclei, 4. medulla oblongata, 5. decussation of medial lemniscus, 6. dorsal columns, 7. spinal cord, 8. proprioceptors, 9. mechanoreceptors, 10. first-order neuron, 11. dorsal root ganglion, 12. second-order neuron, 13. third-order neuron

Ascending Nerve Tracts: Spinothalamic Tract

The spinothalamic tract carries information from the skin related to pain, temperature, and very light (poorly localized) touch. Thinly myelinated and unmyelinated nerve fibers convey these signals, and the neuronal cell bodies of these axons are located in the dorsal root ganglion. As these axons enter the dorsal spinal cord, they give rise to collateral branches that ascend and descend several levels via the Lissauer's tract and then synapse with second-order neurons in the substantia gelatinosa of Rolando or the nucleus proprius. The axons then decussate one or two spinal nerve segments above the point of entry in the spinal cord via the anterior white commissure to the anterolateral corner of the spinal cord. The axons then ascend to the rostral ventromedial medulla in the brainstem, travel up rostrally, and form synapses with several third-order neuronal nuclei in the thalamus, including the medial dorsal, ventral posterior, and ventral medial posterior nuclei. From here, signals ascend to the cingulate cortex, the primary somatosensory cortex, and the insular cortex.

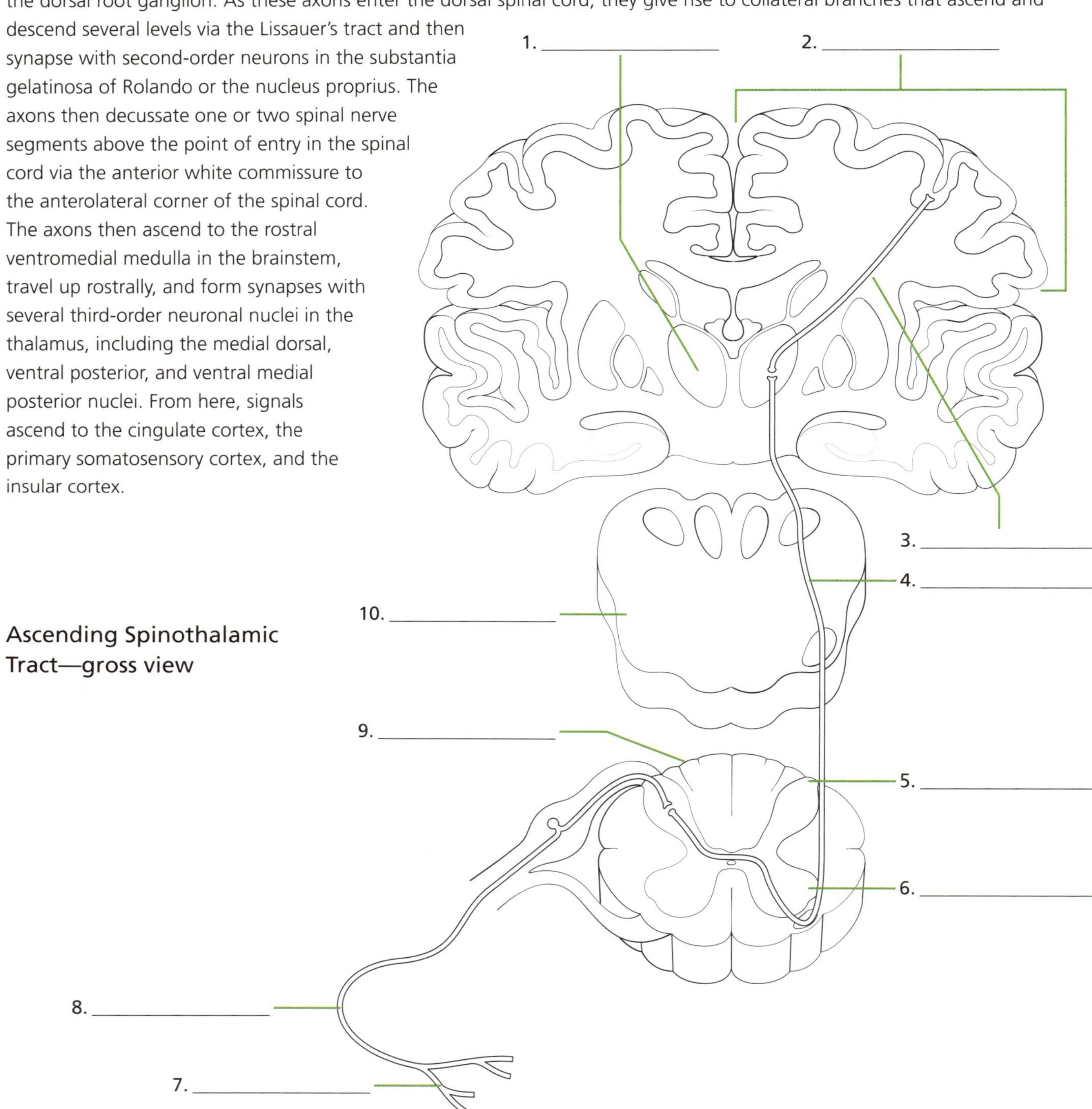

Ascending Spinothalamic Tract—gross view

Answers

1. thalamus, 2. primary somatosensory cortex, 3. third-order neuron, 4. second-order neuron, 5. Lissauer's tract, 6. anterolateral quadrant, 7. nociceptors or thermoceptors, 8. first-order neuron, 9. spinal cord, 10. medulla oblongata

Ascending Nerve Tracts: Dorsal Spinocerebellar Tract

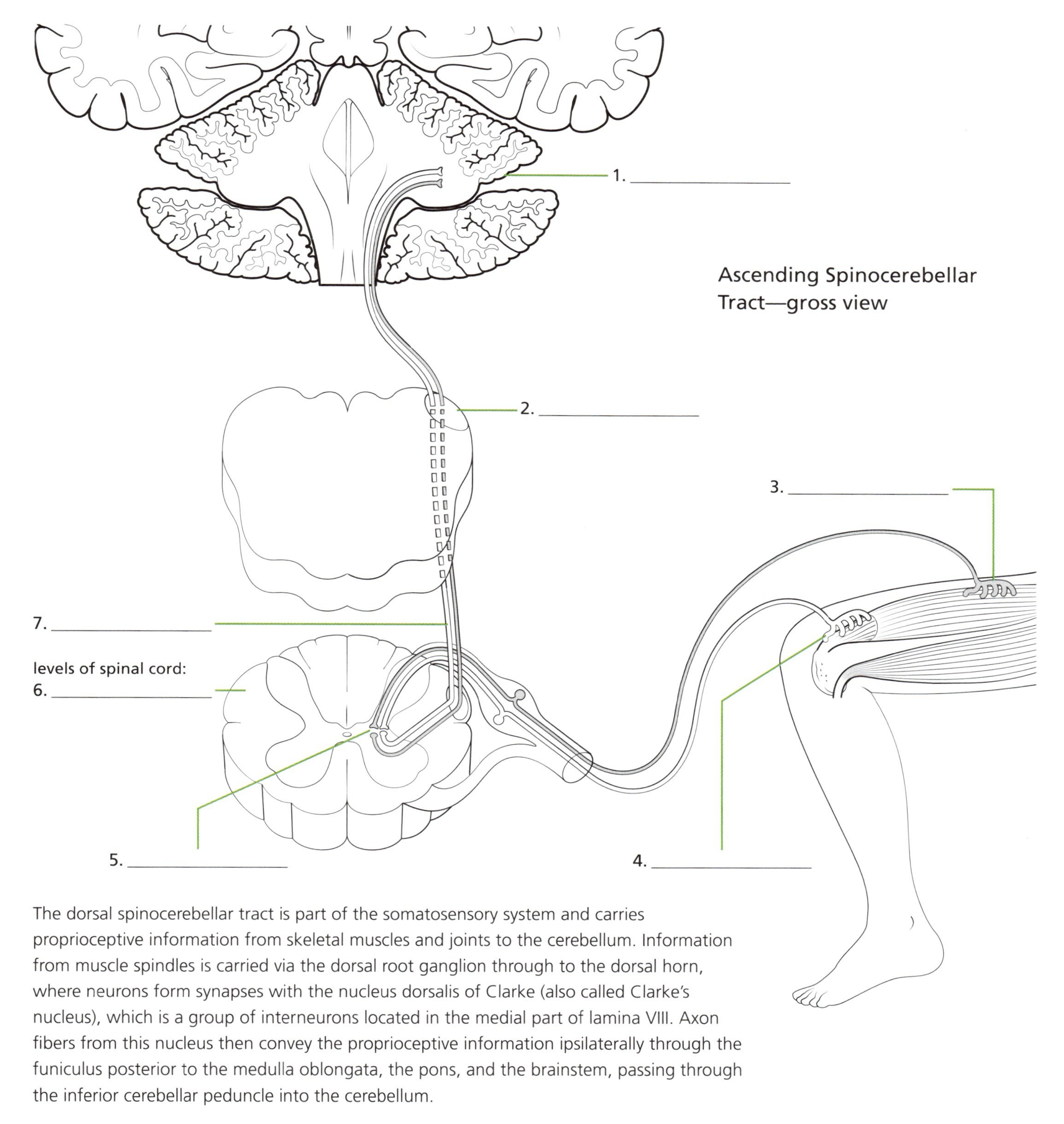

Ascending Spinocerebellar Tract—gross view

The dorsal spinocerebellar tract is part of the somatosensory system and carries proprioceptive information from skeletal muscles and joints to the cerebellum. Information from muscle spindles is carried via the dorsal root ganglion through to the dorsal horn, where neurons form synapses with the nucleus dorsalis of Clarke (also called Clarke's nucleus), which is a group of interneurons located in the medial part of lamina VIII. Axon fibers from this nucleus then convey the proprioceptive information ipsilaterally through the funiculus posterior to the medulla oblongata, the pons, and the brainstem, passing through the inferior cerebellar peduncle into the cerebellum.

Answers

1. cerebellum, 2. inferior cerebellar peduncle, 3. muscle spindle, 4. Golgi tendon organ, 5. nucleus dorsalis of Clarke, 6. C8–L2, 7. posterior spinocerebellar tract

Ascending Nerve Tracts: Ventral Spinocerebellar Tract

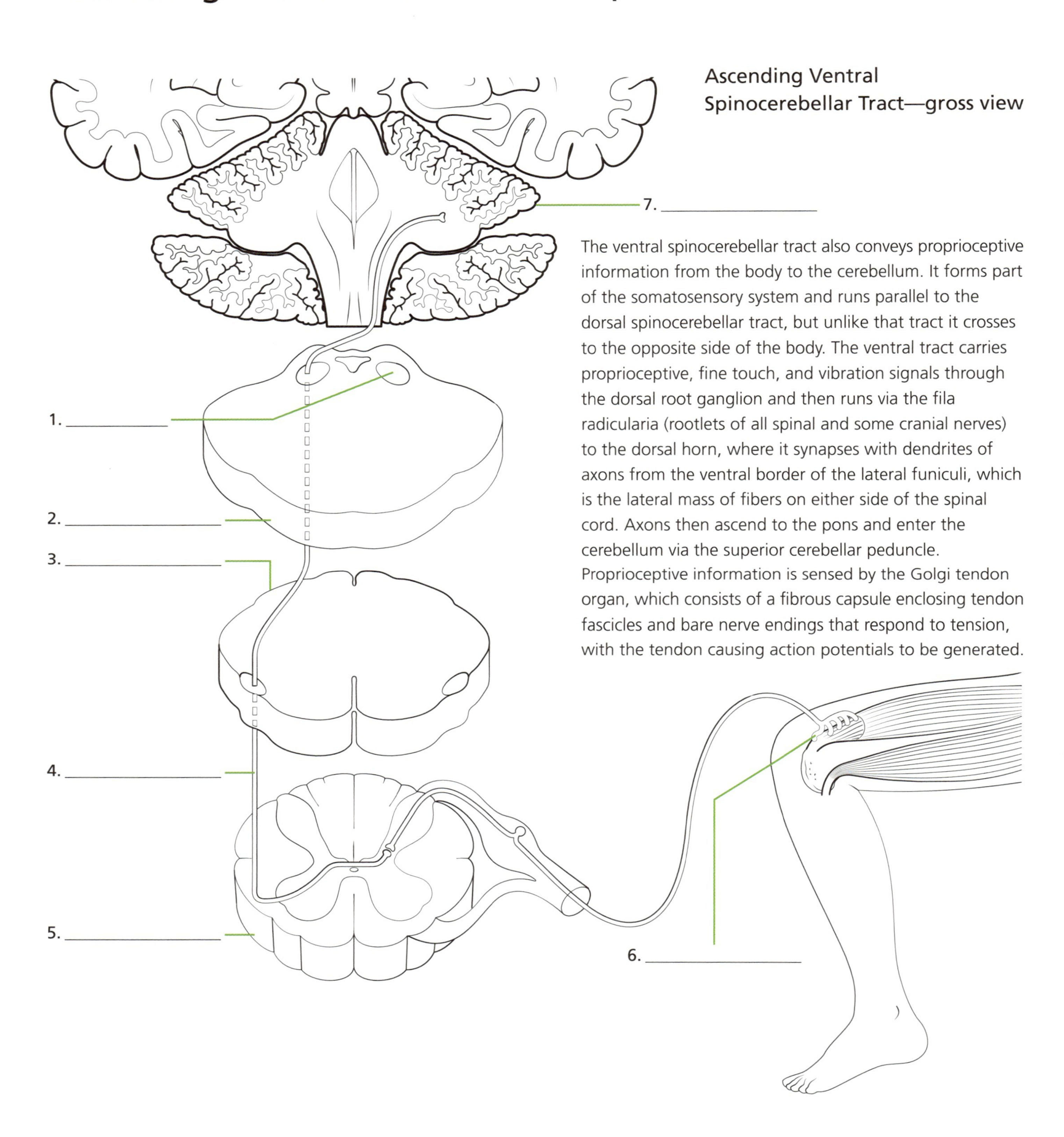

Ascending Ventral Spinocerebellar Tract—gross view

The ventral spinocerebellar tract also conveys proprioceptive information from the body to the cerebellum. It forms part of the somatosensory system and runs parallel to the dorsal spinocerebellar tract, but unlike that tract it crosses to the opposite side of the body. The ventral tract carries proprioceptive, fine touch, and vibration signals through the dorsal root ganglion and then runs via the fila radicularia (rootlets of all spinal and some cranial nerves) to the dorsal horn, where it synapses with dendrites of axons from the ventral border of the lateral funiculi, which is the lateral mass of fibers on either side of the spinal cord. Axons then ascend to the pons and enter the cerebellum via the superior cerebellar peduncle. Proprioceptive information is sensed by the Golgi tendon organ, which consists of a fibrous capsule enclosing tendon fascicles and bare nerve endings that respond to tension, with the tendon causing action potentials to be generated.

Answers

1. superior cerebellar peduncle, 2. pons, 3. medulla, 4. anterior (ventral) spinocerebellar tract, 5. spinal cord, 6. Golgi tendon organ, 7. cerebellum

Ascending Nerve Tracts: Cuneocerebellar Tract

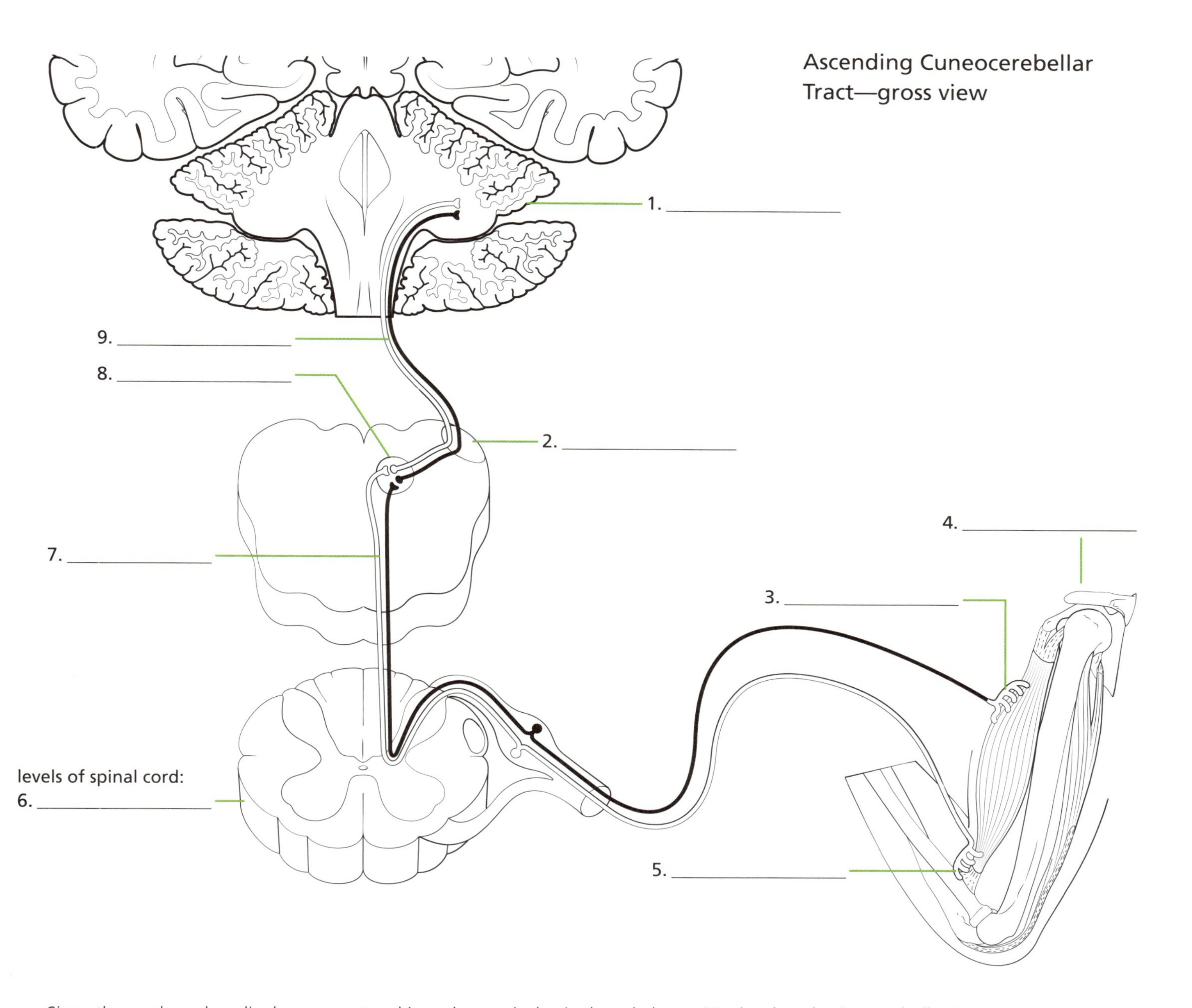

Since the nucleus dorsalis does not extend into the cervical spinal cord above C8, the dorsal spinocerebellar tract cannot convey proprioceptive signals from the upper limbs. Therefore, the ventral cuneocerebellar tract carries information about proprioception from the Golgi tendon organ at the junction of the muscle and tendon of the upper limbs and neck. The proprioceptive signals ascend in the ipsilateral fasciculus cuneatus (a bundle of nerve fibers on the same side of the body) of the spinal cord and innervate the accessory cuneate nucleus (equivalent to the nucleus dorsalis of Clarke). The cuneocerebellar tract enters the cerebellum via the inferior cerebellar peduncle and terminates topographically in the anterior and posterior lobes dedicated to the upper extremities. This pathway is the rostral equivalent of the posterior spinocerebellar tract and analogous to the dorsal spinocerebellar tract for the upper limbs.

Answers

1. cerebellum, 2. inferior cerebellar peduncle, 3. muscle spindle, 4. upper limb, 5. Golgi tendon organ, 6. C7–C1, 7. fasciculus cuneatus, 8. accessory cuneate nucleus, 9. cuneocerebellar tract

Descending Nerve Tracts: Corticospinal Tract

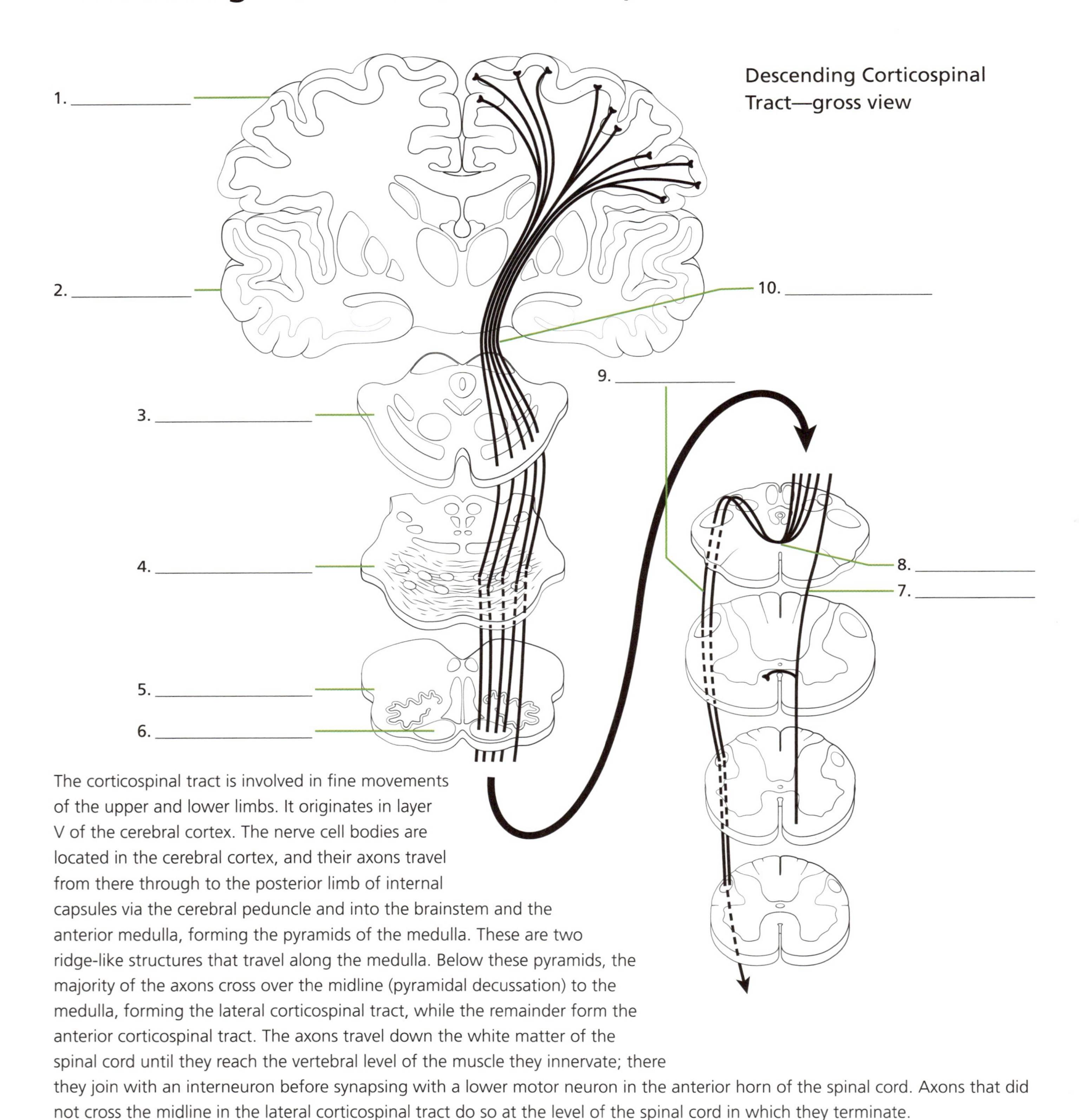

The corticospinal tract is involved in fine movements of the upper and lower limbs. It originates in layer V of the cerebral cortex. The nerve cell bodies are located in the cerebral cortex, and their axons travel from there through to the posterior limb of internal capsules via the cerebral peduncle and into the brainstem and the anterior medulla, forming the pyramids of the medulla. These are two ridge-like structures that travel along the medulla. Below these pyramids, the majority of the axons cross over the midline (pyramidal decussation) to the medulla, forming the lateral corticospinal tract, while the remainder form the anterior corticospinal tract. The axons travel down the white matter of the spinal cord until they reach the vertebral level of the muscle they innervate; there they join with an interneuron before synapsing with a lower motor neuron in the anterior horn of the spinal cord. Axons that did not cross the midline in the lateral corticospinal tract do so at the level of the spinal cord in which they terminate.

Answers

1. motor cortex, 2. forebrain, 3. midbrain, 4. pons, 5. medulla, 6. pyramid, 7. anterior corticospinal tract, 8. pyramidal decussation, 9. lateral corticospinal tract, 10. corticospinal tract

Descending Nerve Tracts: Vestibulospinal Tract

Descending Vestibulospinal Tract—gross view

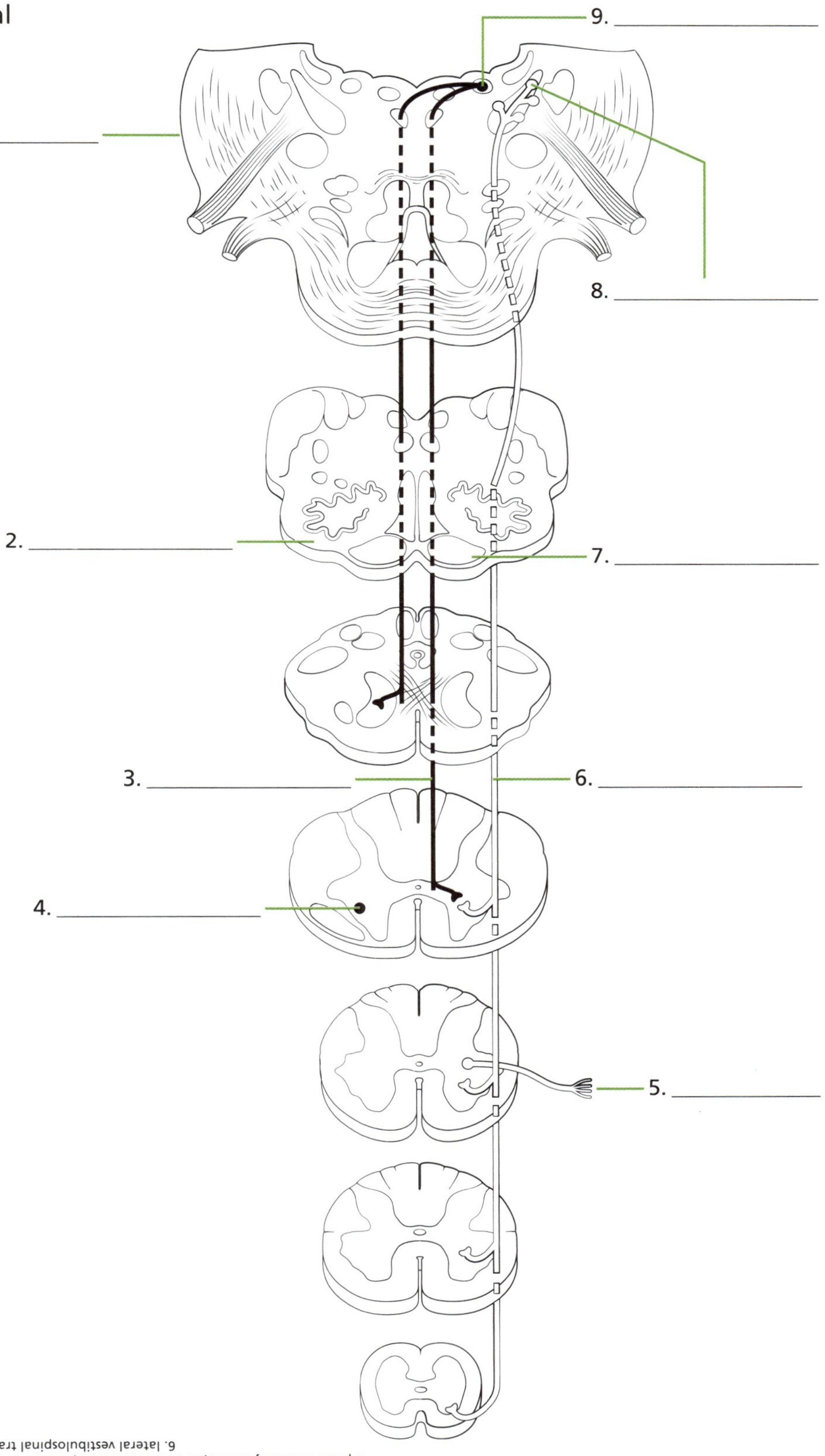

9. ____________
1. ____________
8. ____________
2. ____________
7. ____________
3. ____________
6. ____________
4. ____________
5. ____________

The lateral vestibulospinal tract originates in the lateral vestibular nucleus in the pons and descends uncrossed (ipsilateral) in the anterior portion of the lateral funiculus of the spinal cord. It descends the entire length of the spinal cord and terminates at the interneurons of laminae VII and VIII (regions in the dorsal horn), while some neurons terminate on the dendrites of alpha motor neurons in the same laminae, activating motor neurons that innervate extensor muscles in the trunk and limbs. Like other descending motor pathways, the vestibulospinal tract relays information from a nucleus to motor neurons. The vestibular nuclei receive information regarding changes in orientation of the head through the vestibulocochlear nerve, a nerve that connects the inner ear to the brain. The main function of this tract is to control the muscles that maintain upright posture and balance.

The medial vestibulospinal tract originates in the medial vestibular nucleus (the cranial nuclei of the vestibular nerve), which extends from the inferior olivary nucleus of the medulla oblongata to the caudal portions of the pons. The tract then joins with the contralateral medial longitudinal fasciculus and descends in the anterior funiculus of the spinal cord to the cervical cord segments, terminating on neurons of laminae VII and VIII. The main function of this tract is to adjust the position of the head in response to changes in posture.

Answers

1. pons–medulla junction, 2. medial lemniscus, 3. medial vestibulospinal tract, 4. anterior horn cell neuron III, 5. motor endplates, 6. lateral vestibulospinal tract, 7. pyramid, 8. lateral vestibular nucleus, 9. medial vestibular nucleus

Descending Nerve Tracts: Reticulospinal Tract

The two reticulospinal tracts originate in the brainstem reticular formation in the pons and medulla. They carry motor and autonomic information, as well as modulating pain signals. The lateral tract arises from the medullary reticular formation, and the medial tract arises from the pontine reticular formation. The medullary reticulospinal tract originates at the nucleus reticularis gigantocellularis, a segment of the reticular formation containing giant neuronal cells, and it courses ipsilaterally through the anterior funiculus of the spinal cord. It exits at all spinal levels, terminating in laminae VII and IX of the spinal cord. The pontine reticulospinal tract originates in the caudal pontine reticular nucleus and the oral pontine reticular nucleus, and it courses ipsilaterally through the longitudinal fasciculus. It passes through to the anterior funiculus of the spinal cord and exits ipsilaterally at all spinal levels, terminating at laminae VII and VIII of the spinal cord.

The medullary projections are responsible for inhibiting extensor spinal reflex activity, while the pontine projections are involved in facilitating extensor spinal reflexes.

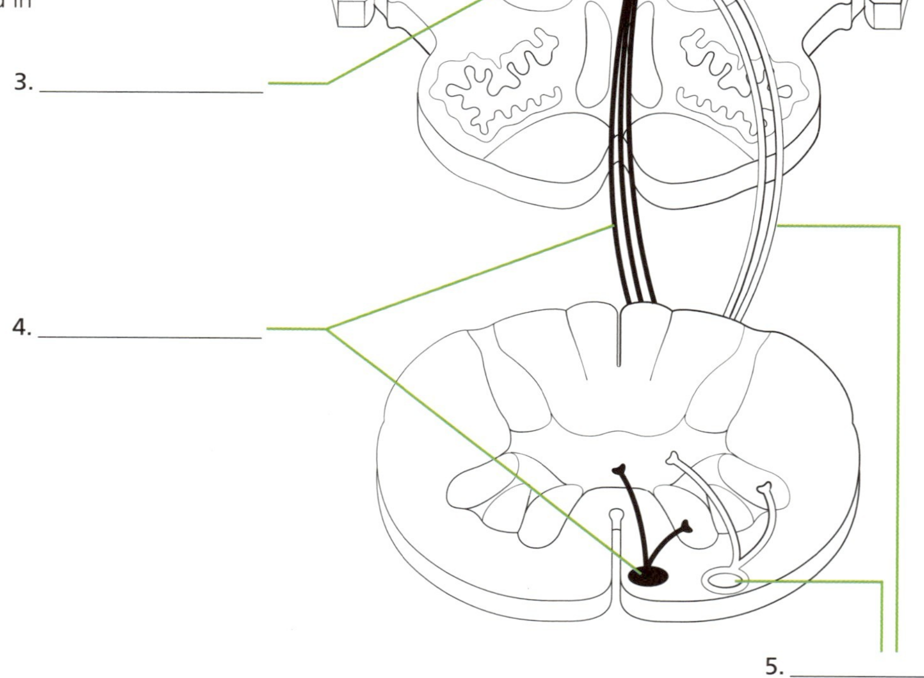

Descending Reticulospinal Tract—gross view

Answers

1. pons, 2. medulla, 3. nucleus reticularis gigantocellularis, 4. pontine reticulospinal tract, 5. medullary reticulospinal tract, 6. medullary reticular formation, 7. pontine reticular formation

Descending Nerve Tracts: Rubrospinal Tract

The rubrospinal tract arises from neurons located in the midbrain. It originates in the magnocellular red nucleus (neurons located in the magnocellular layer of the lateral geniculate nucleus of the thalamus) and immediately crosses to the contralateral side of the midbrain. It then descends in the lateral part of the brainstem tegmentum, known as the ventral tegmental decussation. The axons then descend past the inferior colliculus in the midbrain and continue through the medial lemniscus to the pons. Axons then descend through to the lateral reticular nucleus of the medulla and through the lateral funiculus of the spinal cord close to the lateral corticospinal tract, finally innervating spinal neurons at all levels of the cord. The function of this tract is to facilitate flexor motor neurons and inhibit extensor motor neurons, and it is an alternative tract by which voluntary motor commands can be sent to the spinal cord. However, it is a major pathway only in animals and plays a relatively minor role in humans. In addition, the red nucleus receives inputs from the cerebellum, and it is therefore assumed to take part in transmitting signals for learned motor commands from the cerebellum to the musculature.

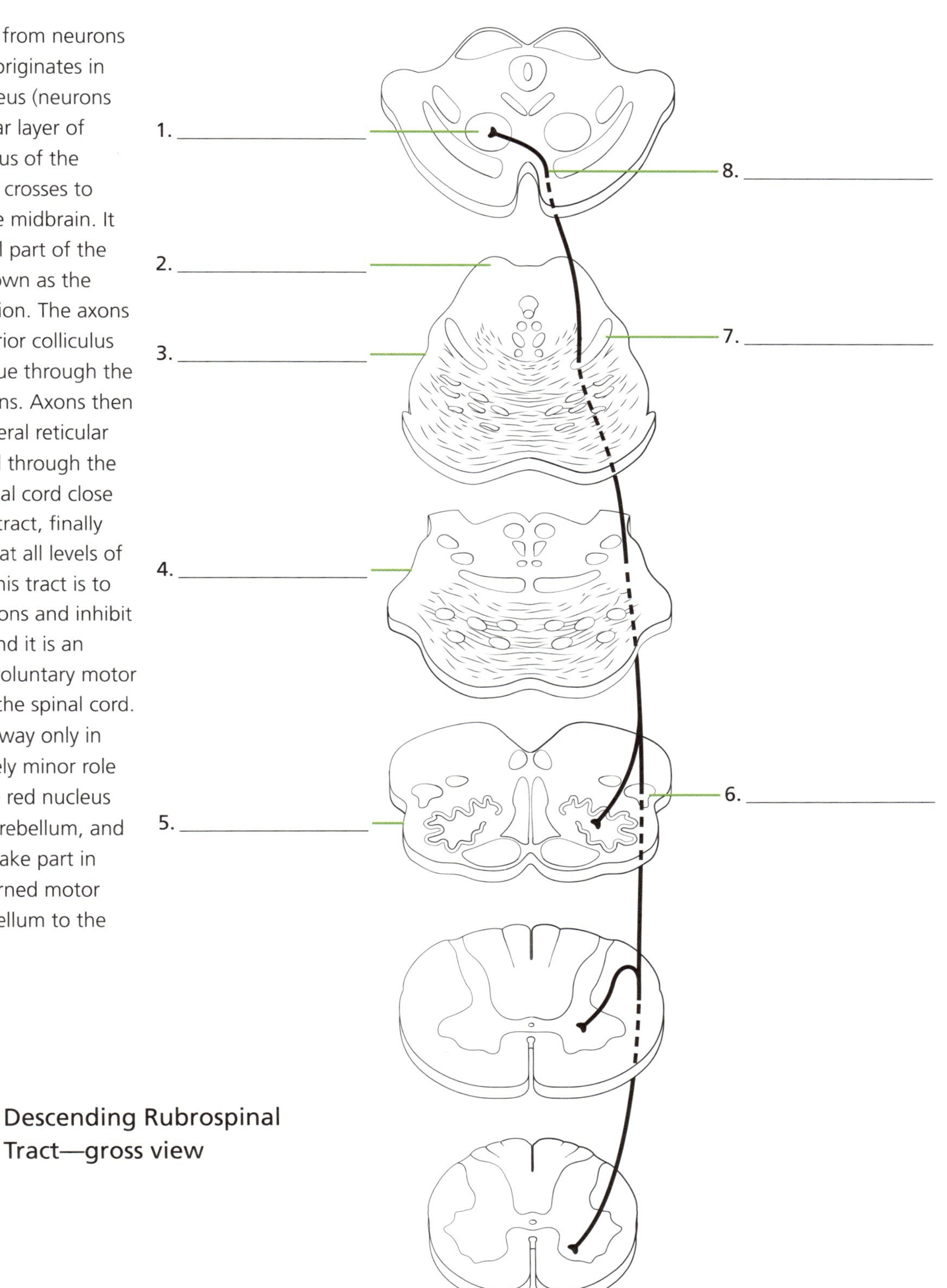

Descending Rubrospinal Tract—gross view

Answers

1. red nucleus, 2. inferior colliculus, 3. midbrain, 4. pons, 5. medulla, 6. lateral reticular nucleus, 7. medial lemniscus, 8. ventral tegmental decussation

Structure of the Spinal Nerves

The spinal cord originates immediately below the brainstem and extends from the first cervical vertebra (C1) to the first lumbar vertebra (L1). It has 31 pairs of spinal nerves, with eight pairs in the cervical segment, twelve pairs in the thoracic segment, five pairs in the lumbar segment, five pairs in the sacral segment, and one pair in the coccygeal segment.

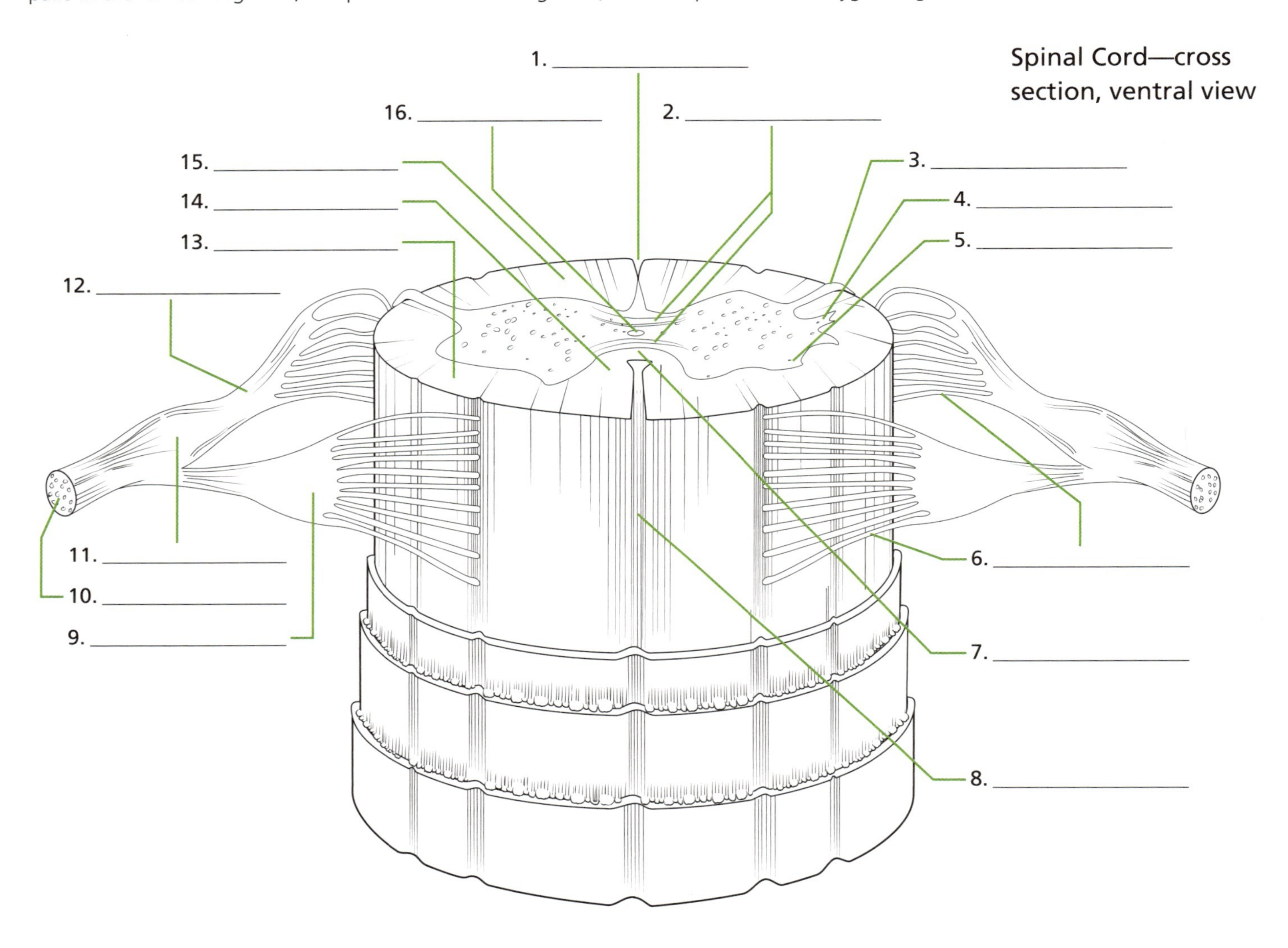

The spinal nerves extend from the spinal canal through the intervertebral foramen to the body, carrying information to and from the brain and body. The intervertebral foramen is the opening between each pair of vertebrae. Afferent sensory axons are in the dorsal roots of the spinal cord and transmit sensory information from the periphery to the spinal cord. Efferent motor axons travel through the ventral roots of the spinal cord and transmit motor information from the brain to the periphery.

The primary function of the spinal cord is to serve as a conduit for the motor and sensory tracts. The ascending sensory tracts and the descending motor tracts are located in the white matter of the spinal cord. However, the spinal cord contains its own neural centers for coordinating spinal reflexes. These reflexes are caused by sensory information being relayed from receptors in the muscles to interneurons in the gray matter.

Answers

1. posterior median sulcus, 2. gray commissure, 3. gray matter—posterior horn, 4. gray matter—lateral horn, 5. gray matter—anterior horn, 6. nerve rootlets, 7. anterior white commissure, 8. anterior median fissure, 9. ventral root, 10. spinal nerve, 11. dorsal root ganglion, 12. dorsal root, 13. white matter—lateral column, 14. white matter—ventral column, 15. white matter—dorsal column, 16. central canal

Spinal Nerves and Plexuses

A nerve plexus is a network of nerve fibers that innervate the same region of the body. There are six major nerve plexuses, with other minor plexuses found deeper in the body. The cervical plexus is located between C1 and C4, the first and fourth cervical segments in the neck, and serves motor and sensory functions for the head, neck, and shoulders. The brachial plexus is composed of nerves originating from cervical segments C5 to C8 and the first thoracic segment, T1. The brachial plexus proceeds through the neck and axilla (armpit) into the shoulder and arms. This plexus of nerves provides motor and sensory functions to the shoulders, arms, and hands. The lumbosacral plexus originates from T12, the twelfth thoracic segment, to S5, the fifth sacral segment. It is generally divided into three divisions, called the lumbar, sacral, and pudendal plexuses. The pudendal plexus is also described as the coccygeal plexus and is a small bundle of nerves located at the coccyx bone. This plexus includes sacral nerves S4 and S5 and the coccygeal spinal nerve, Co1. The network of interwoven nerve fibers that make up these plexuses can easily be damaged and cause pain and weakness at anatomically distal regions.

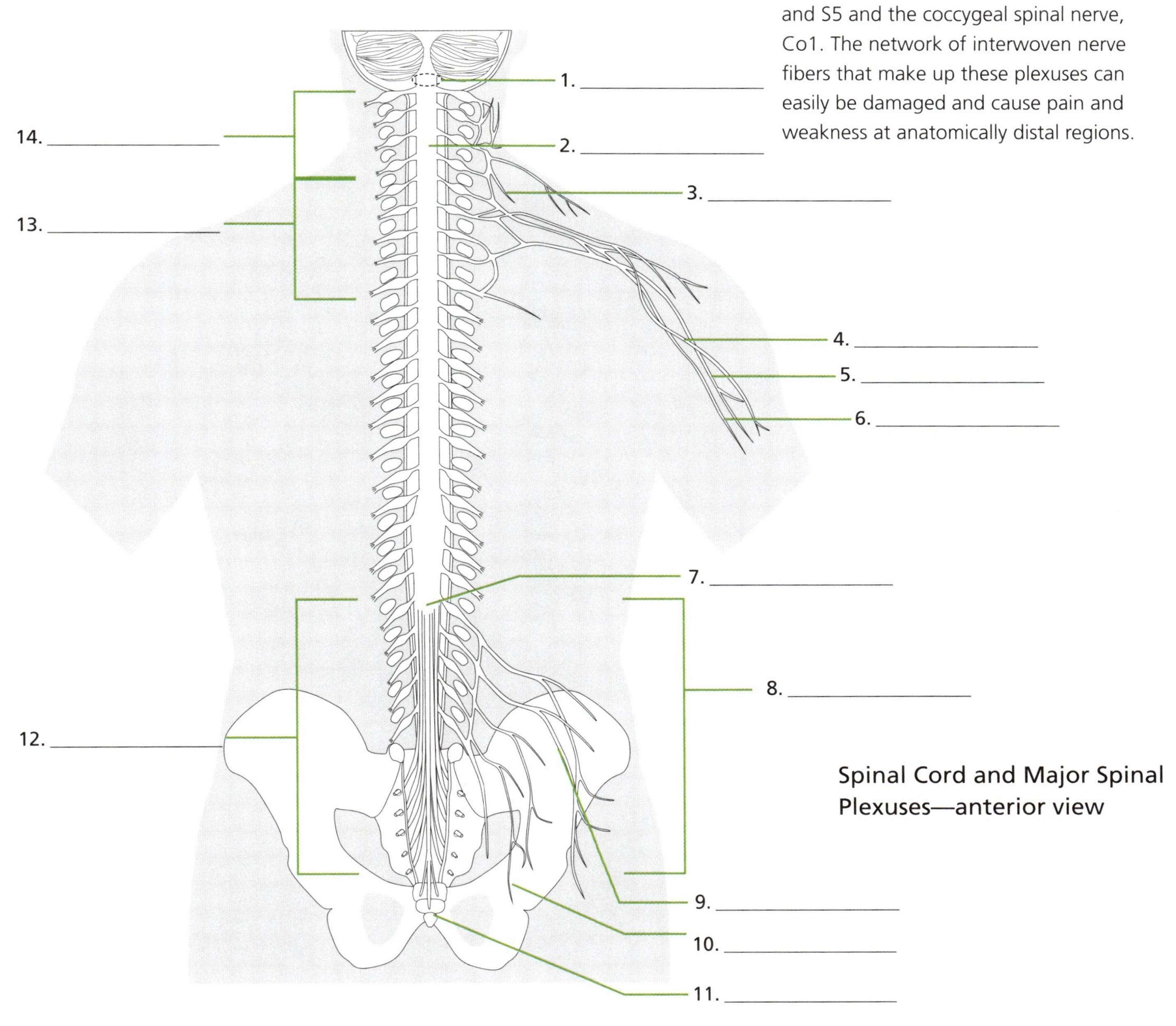

Spinal Cord and Major Spinal Plexuses—anterior view

Answers

1. foramen magnum, 2. spinal cord, 3. phrenic nerve, 4. radial nerve, 5. median nerve, 6. ulnar nerve, 7. conus medullaris, 8. cauda equina, 9. femoral nerve, 10. sciatic nerve, 11. Co1 (coccygeal nerve), 12. lumbosacral plexus, 13. brachial plexus, 14. cervical plexus

Dermatomes

A dermatome is a region of the skin that is innervated by a single spinal nerve that relays sensations, such as pain and touch, to the brain. In total, there are 30 spinal nerves, including eight cervical, twelve thoracic, five lumbar, and five sacral, and one corresponding dermatome for each.

The pattern of dermatome skin regions in the thorax and limbs differs. The thorax dermatomes are stacked like disks, with each disk of skin innervated by a different thoracic or lumbar nerve fiber. The dermatomes of the arms and legs, however, run longitudinally. The face is innervated by a cranial nerve, the trigeminal nerve, and not by a spinal nerve.

The viral disease shingles affects a single dermatome and presents as a painful rash. It is caused by the presence of the herpes zoster virus in a specific spinal nerve. For example, the presence of a painful rash along the dorsal aspect of the arms would be due to an infection of the C6 or C7 spinal nerves.

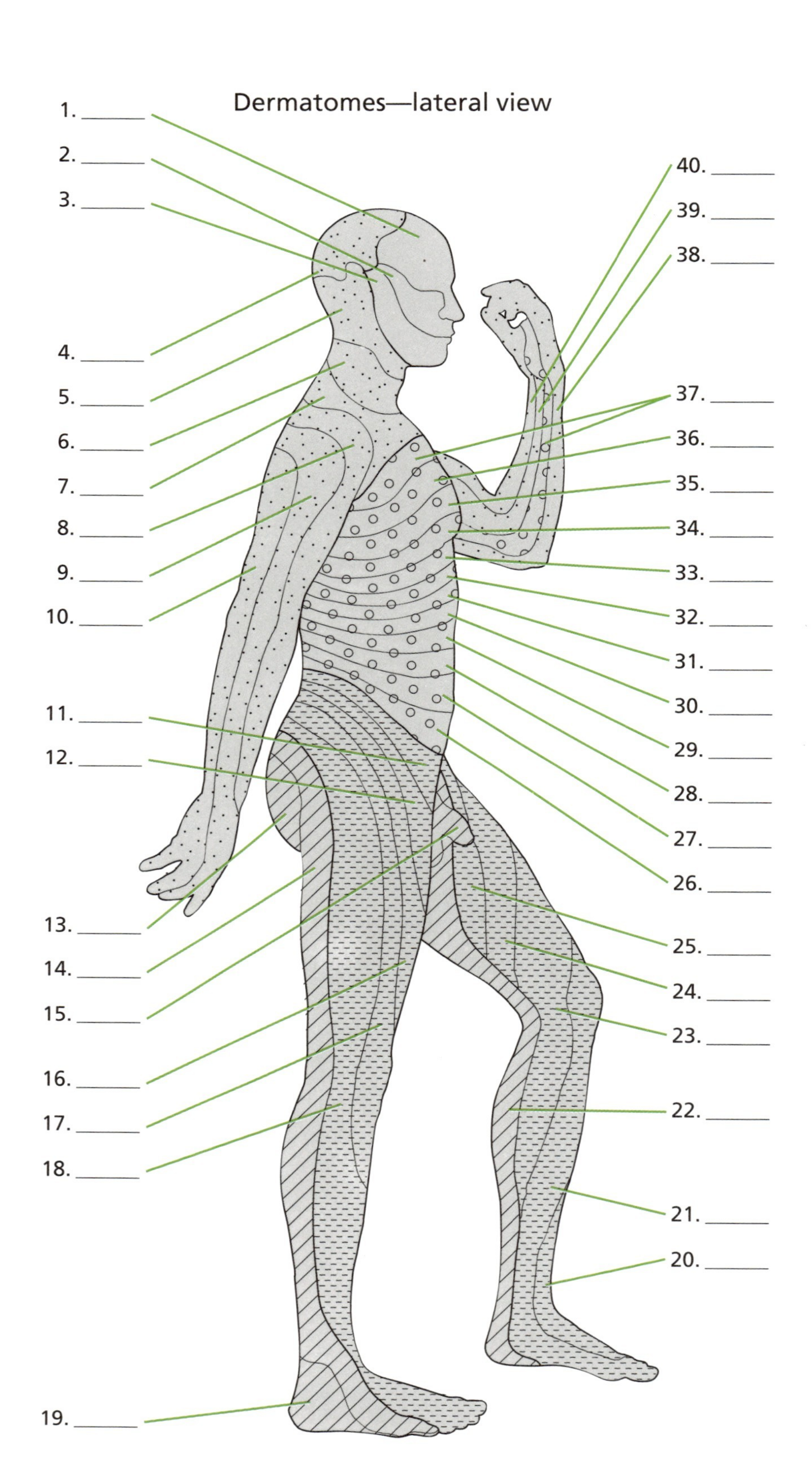

Answers

1. V1, 2. V2, 3. V3, 4. C2, 5. C3, 6. C4, 7. C5, 8. C6, 9. C7, 10. C8, 11. L1, 12. L2, 13. S2, 14. S1, 15. S3, 16. L3, 17. L4, 18. L5, 19. S2, 20. L5, 21. L4, 22. S2, 23. L3, 24. L2, 25. L1, 26. T12, 27. T11, 28. T10, 29. T9, 30. T8, 31. T7, 32. T6, 33. T5, 34. T4, 35. T3, 36. T2, 37. T1, 38. C8, 39. C5, 40. C6

Myotomes

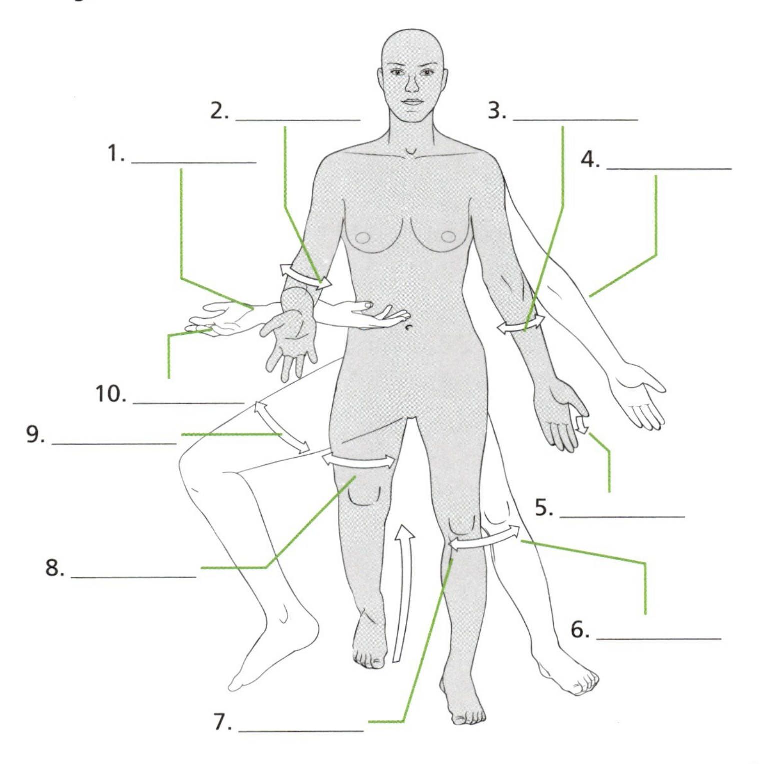

A myotome is a group of muscles that are innervated by the motor fibers of a specific spinal nerve. For example, to abduct the fingers of the hand laterally or medially requires a single myotome, the T1 myotome. However, most muscles are innervated by more than one spinal nerve root.

Spinal nerves contain thousands of nerve fibers (sensory and motor) that are connected to the spinal cord via spinal nerve roots. Within the brachial plexus (C5, C6, C7, C8, and T1), the nerve roots braid together to form three trunks (superior, inferior, and middle), which then divide into anterior and posterior divisions. From these six divisions arise the nerve cords that innervate specific muscles. Anatomically, the median nerve has root branches in C5, C6, C8, and T1, whereas the ulnar nerve has roots in C8 and T1, with a slight contribution from C7.

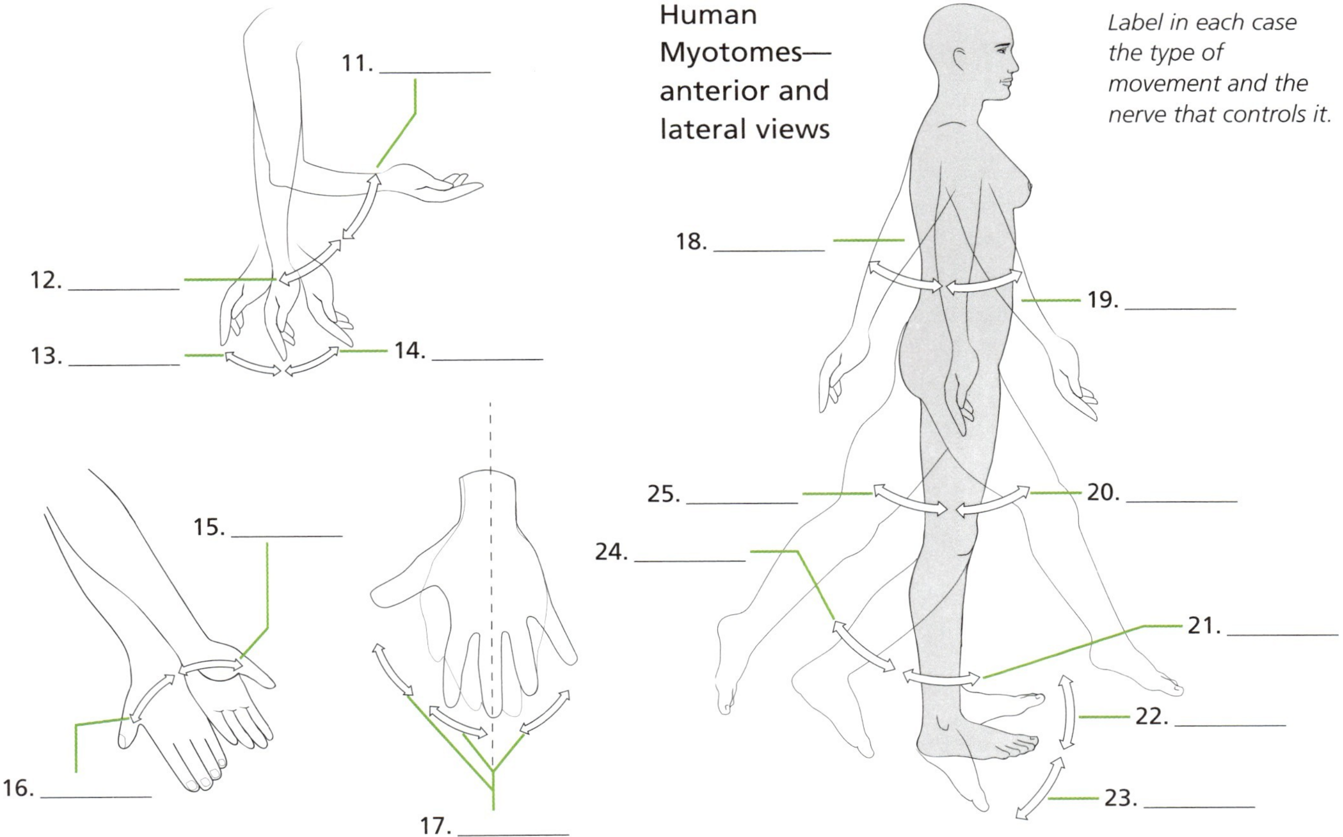

Human Myotomes—anterior and lateral views

Label in each case the type of movement and the nerve that controls it.

Answers

1. lateral rotation of the shoulder C5, 2. medial rotation of the shoulder C6, C7, C8, 3. abduction of the shoulder C5, 4. adduction of the shoulder C6, C7, C8, 5. finger extension C7, C8, 6. abduction of the hip L5, S1, 7. adduction of the hip L1, L2, L3, 8. medial rotation of the hip L1, L2, L3, 9. lateral rotation of the hip L5, S1, 10. finger flexion C7, C8, 11. flexion of the elbow C5, C6, 12. extension of the elbow C6, C7, 13. wrist extension C6, C7, 14. wrist flexion C6, C7, 15. forearm supination C6, 16. forearm pronation C7, C8, 17. finger abduction and adduction T1, 18. extension of the shoulder C6, C7, C8, 19. flexion of the shoulder C5, 20. flexion of the hip L2, L3, 21. extension of the knee L3, L4, 22. dorsiflexion of the ankle L4, L5, 23. plantarflexion of the ankle S1, S2, 24. flexion of the knee L5, S1, 25. extension of the hip L4, L5

Skin Sensory Organs—Meissner's, Pacinian, Ruffini's, and Merkel

Mechanoreceptors are sensory receptors that are classified based on the nature of the stimuli they transduce. They are located in the skin and are classed as either encapsulated or unencapsulated, which refers to whether or not the neurons are surrounded by a fluid-filled capsule. Each mechanoreceptor has mechanically gated ion channels that respond to physical pressures, such as those caused by touch, stretching, and sound.

There are four mechanoreceptors that respond to tactile stimuli, three of which are encapsulated (Meissner's corpuscles, Pacinian corpuscles, and Ruffini's endings) and one of which is unencapsulated (Merkel disks). Meissner's corpuscles are located in the upper dermis of glabrous (smooth) skin, such as that found on the fingertips, and respond to fine touch and pressure. Merkel disks are also located in the upper layers but are found in both glabrous and hairy skin. They are densely distributed in the fingertips and lips and respond to soft touches or light pressure. Ruffini's endings are deeper in the dermis of glabrous and hairy skin and respond to stretching and warmth. Finally, Pacinian corpuscles are also located deep in the dermis of glabrous and hairy skin and respond to vibrations and pressure.

Smooth Skin and Subcutaneous Layers—sectional view

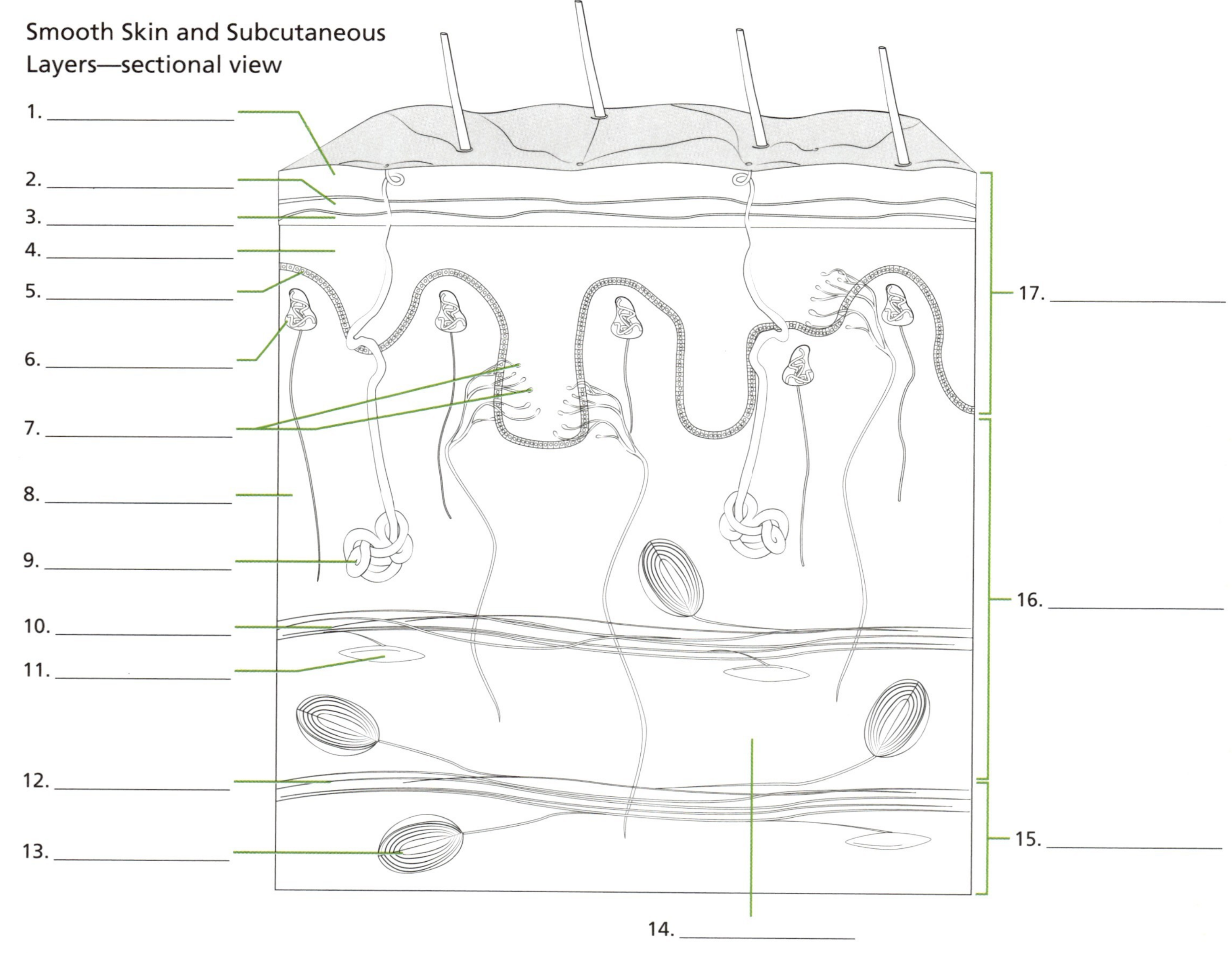

Answers

1. stratum corneum, 2. stratum lucidum, 3. stratum granulosum, 4. stratum spinosum, 5. stratum basale, 6. Meissner's corpuscle, 7. Merkel disks, 8. papillary, 9. sweat gland, 10. nerve fibers, 11. Ruffini's endings, 12. nerve fibers, 13. Pacinian corpuscle, 14. reticular, 15. subcutis (hypodermis), 16. dermis, 17. epidermis

Skin Sensory Organs—Pain Receptors and Hair Follicles

Nociceptors are sensory nerve endings that cause the perception of pain when a noxious stimulus triggers a response. Nociceptors are located everywhere, externally and internally, particularly in the skin and muscles, respectively. Nociceptors can be further distinguished based on the nature of the noxious stimulus: There are thermal (respond to extreme heat or cold), mechanical (excessive pressure or skin incisions), chemical (environmental irritants), and polymodal (two or more sensory functions) nociceptors.

Mechanoreceptors are sensory receptors that respond to mechanical pressures on glabrous (smooth) and hairy skin. In hairy skin, there are also follicular receptors in the hair follicles that respond to deflection of the hairs on the skin.

Hairy Skin and Subcutaneous Layers—sectional view

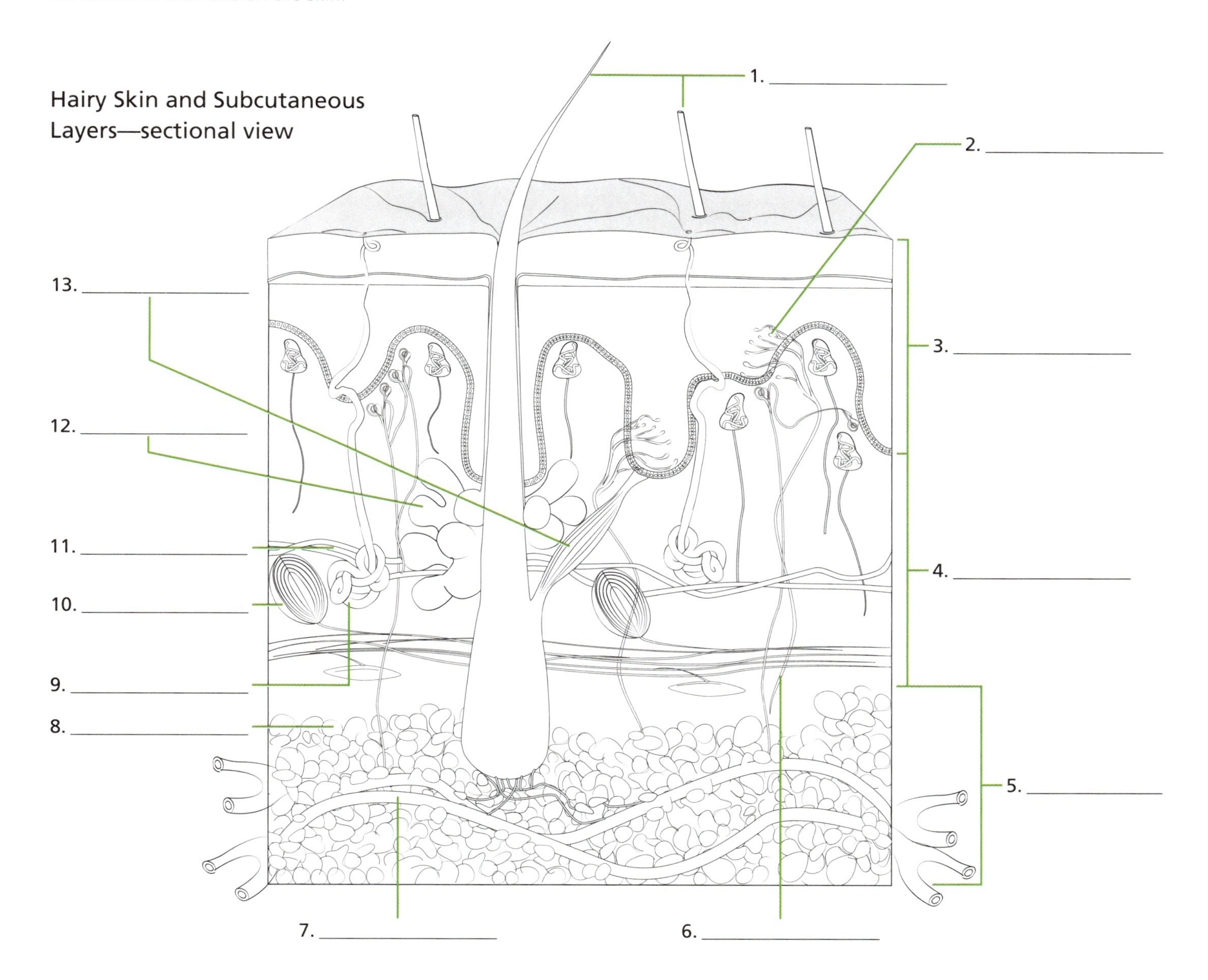

Answers

1. hair, 2. sensory nerve endings, 3. epidermis, 4. dermis, 5. subcutis (hypodermis), 6. nerve, 7. arteriole, 8. fat, collagen, and fibroblasts, 9. sweat gland, 10. Pacinian corpuscle, 11. capillaries, 12. sebaceous gland, 13. arrector pili muscle

Muscle Stretch Receptors

Muscle stretch receptors are proprioceptive mechanoreceptors that respond to the generated force of movement. Muscle stretch receptors that are located within the belly of the muscle are called muscle spindles, while receptors at the tendon insertion point with the muscle are called the Golgi tendon organs.

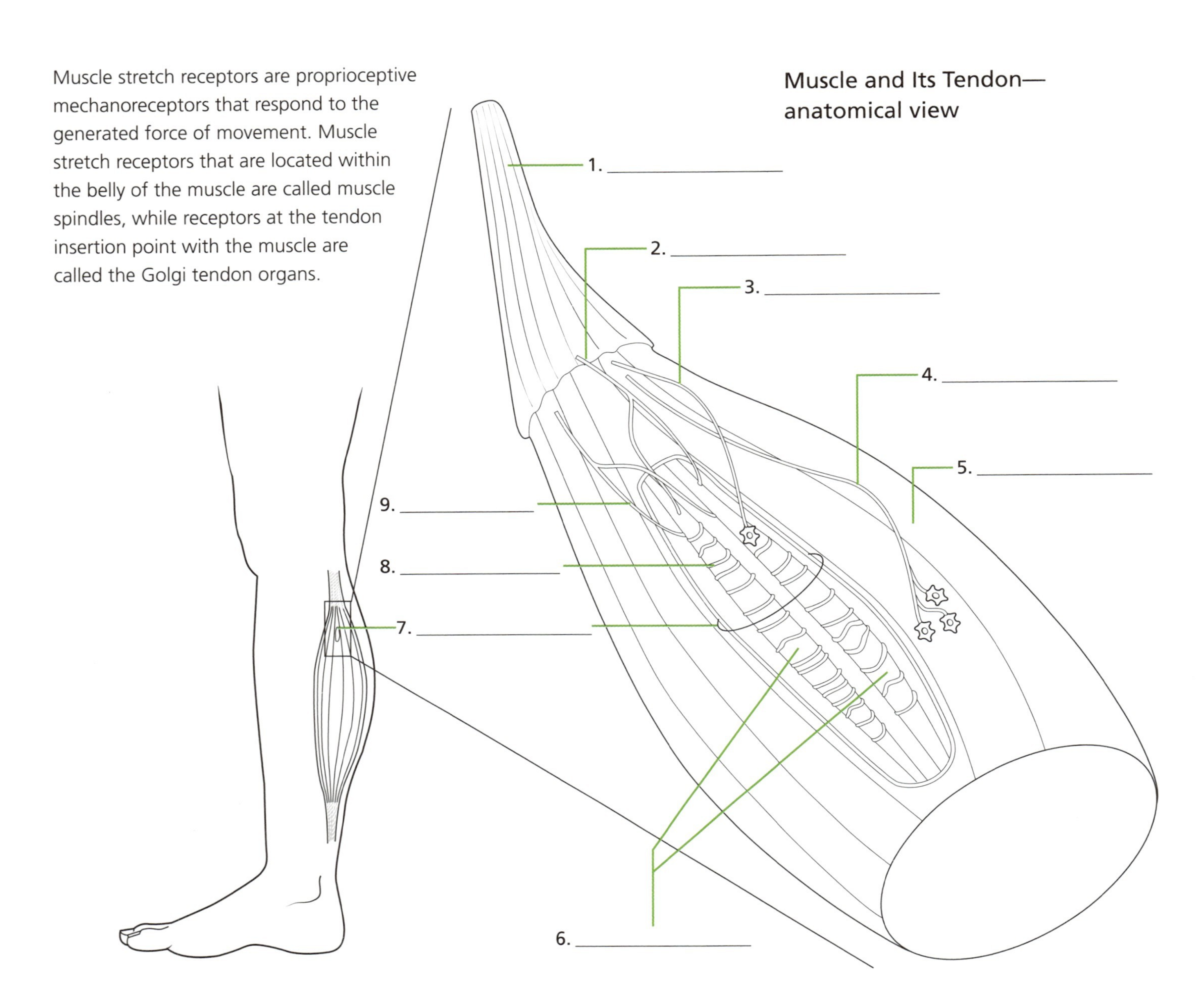

Muscle contraction is controlled by efferent alpha motor neurons that innervate extrafusal muscle fibers. Two types of sensory endings in the muscle spindle detect this muscle contraction—type Ia and type II. Type Ia sensory endings detect the rate of muscle stretch, and type II sensory endings detect position of a static muscle. Type Ia and type II afferent sensory fibers transit to the medulla oblongata, where information on muscle distension is relayed. The brain processes this information and determines the position of body parts involved. For the CNS to modify sensitivity and maintain muscle spindle tautness, an efferent gamma motor neuron innervates the muscle spindle.

The motor neurons work in conjunction with the muscle stretch receptors to provide the connectivity necessary for proprioception and coordinated muscle activity.

Answers

1. tendon, 2. gamma efferent motor fiber, 3. type II sensory fiber, 4. alpha efferent motor fiber, 5. extrafusal muscle fiber, 6. intrafusal muscle fiber, 7. connective tissue capsule, 8. muscle spindle, 9. type Ia sensory fiber

Tendon Organ Receptors

Proprioception is the sense of knowing the position of the body in a three-dimensional space. This sense relies on information from the muscle spindles in the belly of the muscles, as well as the Golgi tendon organs at the insertion point of the tendon into the muscle. The Golgi tendon organ is a neurotendinous spindle that is enclosed in a fibrous capsule; it provides sensory information primarily on muscle tension and joint position.

The Golgi tendon organ is made up of collagen fibrils interwoven with branches of the type 1b sensory afferent neuron in the tendon. When the muscle generates force, the stretched collagen fibrils activate the terminals of the type 1b sensory afferent neuron. Each Golgi tendon organ is innervated by a single type 1b sensory afferent neuron, which relays muscle tension information to the spinal cord and then on to the cerebellum and cerebral cortex.
The sensory feedback from the Golgi tendon organ assists in the regulation of movement during locomotion.

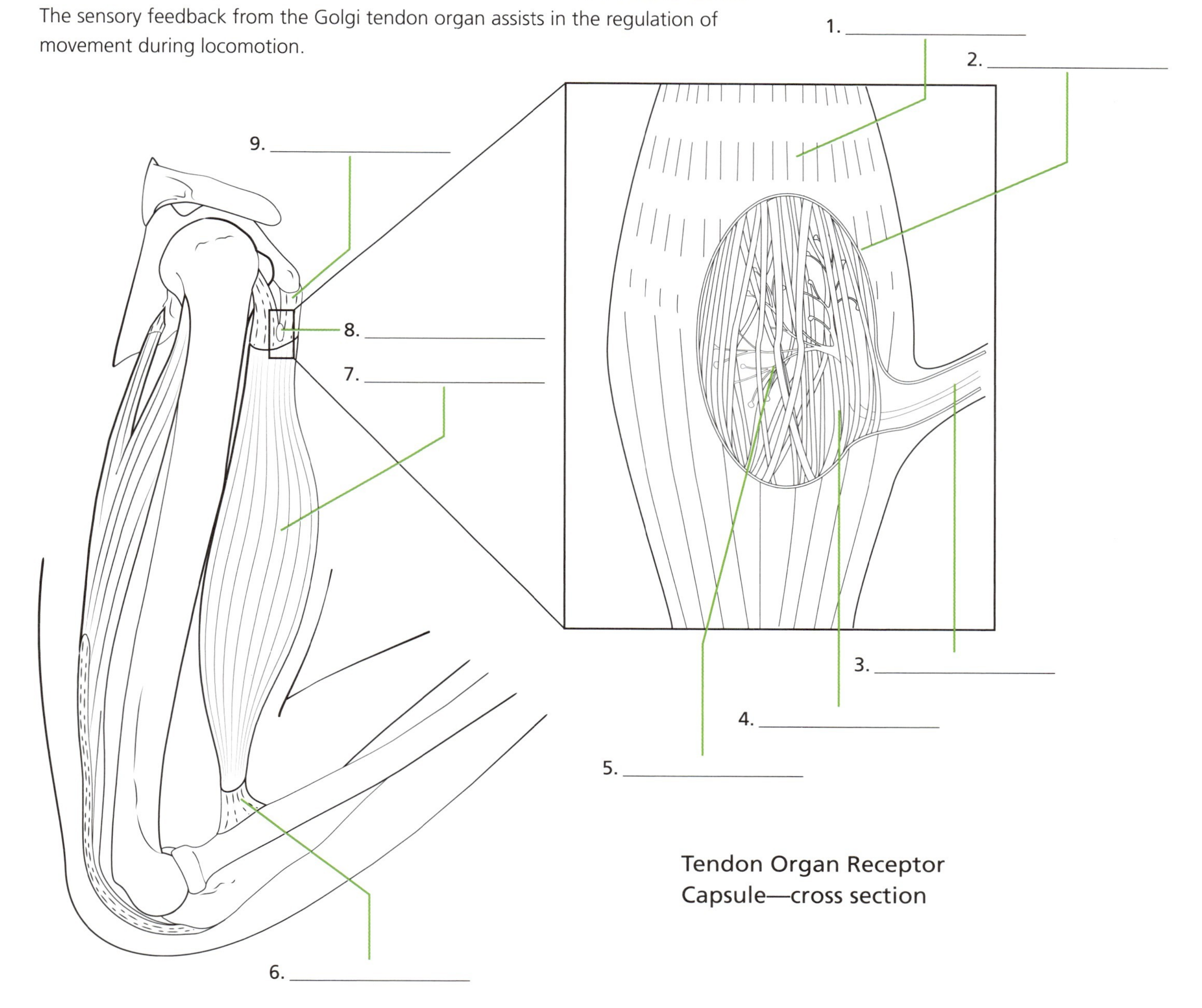

Tendon Organ Receptor
Capsule—cross section

Answers

1. extrafusal muscle fibers, 2. capsule, 3. type lb afferent neuron, 4. collagen fiber, 5. type lb afferent sensory neuron, 6. tendon, 7. extrafusal fiber, 8. Golgi tendon organ, 9. tendon

Joint Receptors

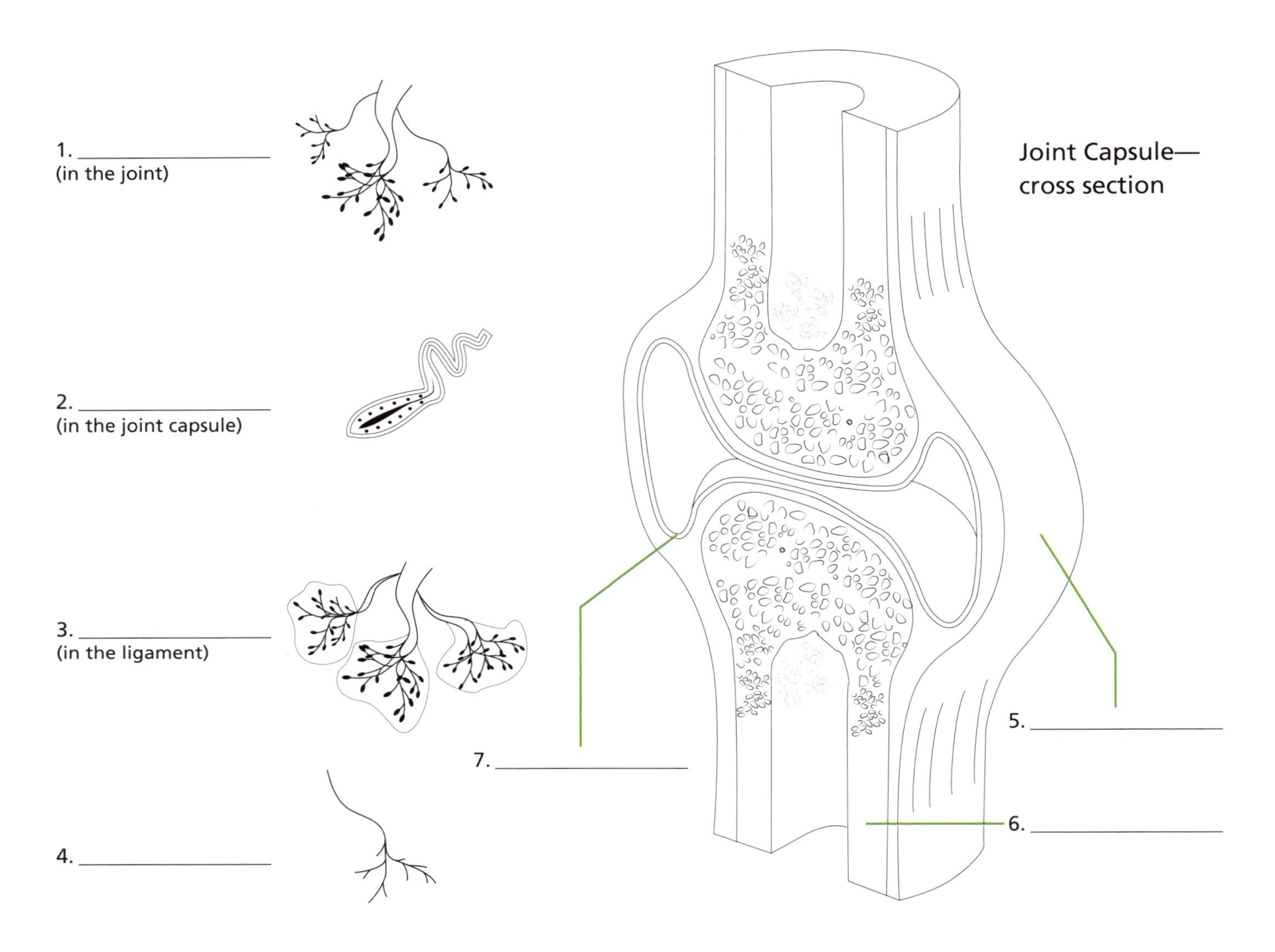

The somatosensory system allows the body to perceive physical sensations through proprioceptive receptors. Joint receptors monitor the stretch in the joint capsule and provide information on the position and movement of the limbs. This awareness is known as kinesthesia and is the means by which the brain senses dynamic movement.

Joint receptors are found in the synovial junctions and produce perceived signals of movement in the joint capsule and the ligaments. The joint receptors also protect the joint from overextension and associated injury. There are four types of joint receptors (types I, II, III, and IV), which are Ruffini-like receptors, Paciniform receptors, encapsulated Golgi-type endings, and free nerve endings, respectively.

The Ruffini-like type I receptors are found in the joint capsule and are active at rest as well as during movement. The Paciniform type II receptors are in the periosteum near the joint capsule and provide information on the velocity of joint movement. The Golgi-type endings (type III) are interwoven with the collagen fibers of the ligament and are activated at the extreme end of ligament stretching. Free nerve endings (type IV) are also located in the joint capsule; many of these endings act like nociceptors, whereby they are stimulated only by extreme deformation of the joint.

Answers

1. type I, 2. type II, 3. type III, 4. type IV, 5. joint capsule, 6. bone, 7. cartilage

Neuromuscular Junctions

A neuromuscular junction is a synapse between the presynaptic terminal of a motor neuron and the postsynaptic membrane of a striated muscle fiber. This excitation-contraction coupling is the connection between an electrical action potential and the mechanical contraction response of the muscle fiber.

The synaptic terminal of the motor neuron forms a synaptic cleft with the postsynaptic motor endplate. An action potential travels down the motor neuron axon, causing an influx of calcium (Ca^{2+}) ions via voltage-gated ion channels at the axon terminal. Acetylcholine is then released from presynaptic vesicles into the synaptic cleft and diffuses across the junction, binding to acetylcholine receptors on the muscle fiber. This causes the ion channels to open and sodium (Na^{+}) ions to enter and depolarize the plasma membrane of the muscle fiber. The resulting depolarization is known as an endplate potential and causes the release of Ca^{2+} ions from the sarcoplasmic reticulum of the muscle fiber. The sarcoplasmic reticulum is a specialized type of endoplasmic reticulum found only in muscle cells. Its function is to store and release Ca^{2+} ions, thus stimulating muscle contraction. To prevent continuous muscle contraction, acetylcholinesterase removes acetylcholine from the synaptic cleft by breaking it down to acetate and choline.

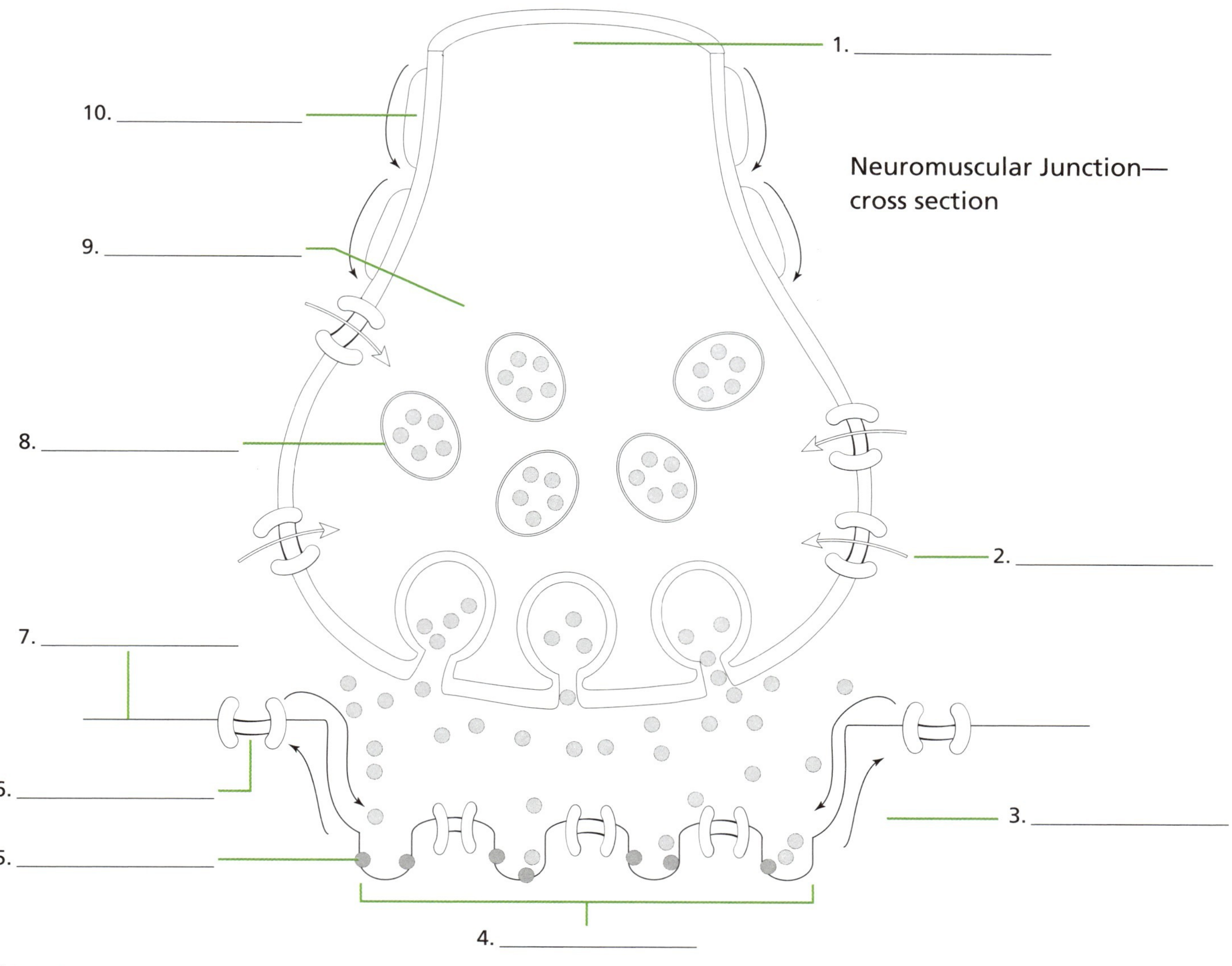

Answers

1. axon of motor neuron, 2. voltage-gated calcium channel, 3. muscle fiber, 4. motor endplate, 5. acetylcholinesterase, 6. acetylcholine receptor site, 7. plasma membrane of muscle fiber, 8. vesicle of acetylcholine, 9. axon terminal, 10. myelin sheath

Spinal Cord Reflexes—Stretch Reflex

Spinal reflexes are automatic sensory-driven reflex actions that are controlled by descending motor pathways from the brainstem and cortex. There are four main types of spinal reflexes: the stretch reflex, the withdrawal reflex, the crossed-extensor reflex, and the Golgi-tendon reflex.

The stretch reflex is an automatic monosynaptic reflex that causes the muscle spindle to regulate muscle length by stimulating contraction of the muscle. This is a key reflex in maintaining an upright posture. If an individual begins to sway to the right, the stretch reflex is activated on the left, and as a result the sway is counteracted.

The knee-jerk reflex test clinically tests for the stretch reflex. When a hammer is tapped against the patellar tendon, this causes the extensor muscle of the knee (quadriceps) to stretch abruptly. To counteract this abrupt stretching, the stretch reflex is activated, causing the quadriceps to contract and the lower leg to extend. The opposing muscle group, which in this case is the hamstrings, is prevented from contracting and working against the stretch reflex. This reciprocal inhibition is performed by an inhibitory interneuron in the spinal cord that innervates the motor neuron controlling the opposing muscle.

Knee and the Patellar Reflex—left lateral view

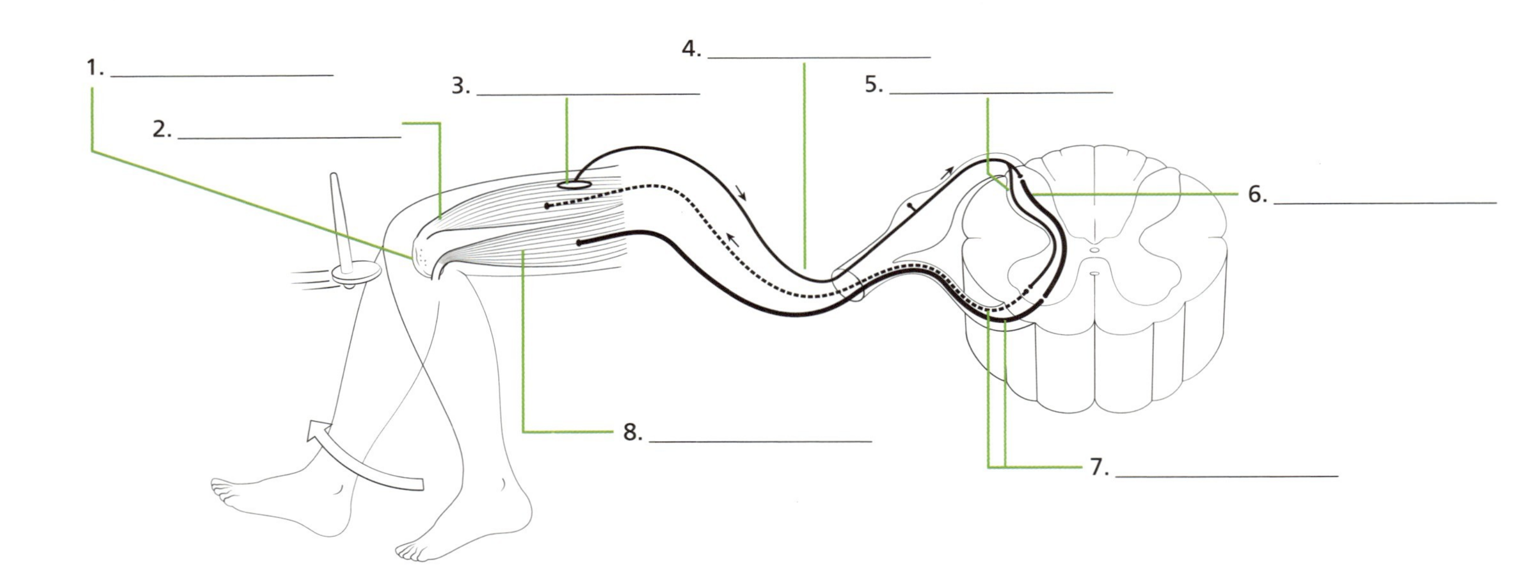

Answers

1. stretching of the muscle via patella tendon, 2. quadriceps muscle stretched, 3. muscle spindle fiber, 4. afferent neuron, 5. excitatory neuron, 6. inhibitory interneuron, 7. efferent neurons, 8. antagonistic flexor muscle

Spinal Cord Reflexes—Autogenic Inhibition Reflex

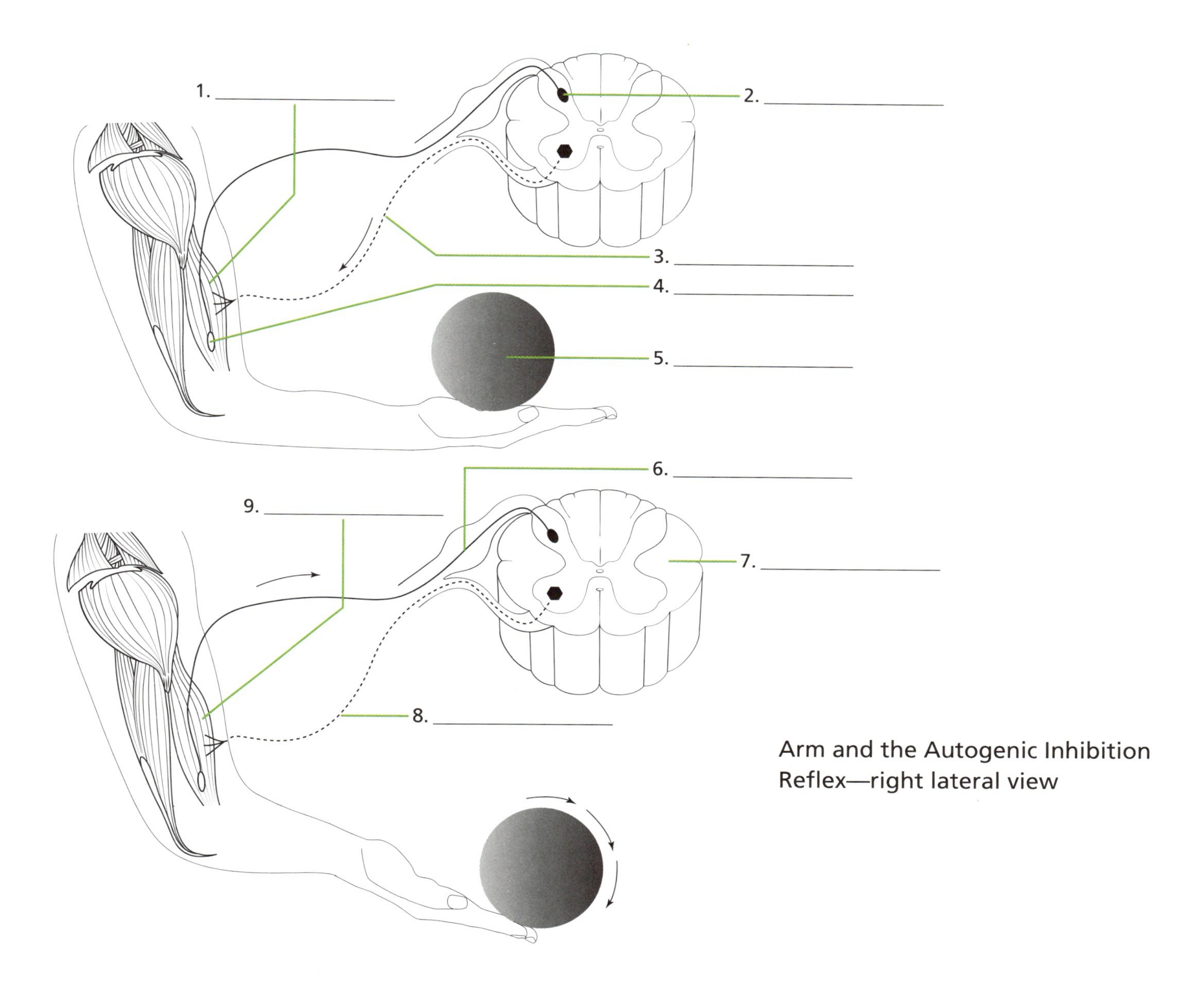

Arm and the Autogenic Inhibition Reflex—right lateral view

The muscle spindles and Golgi tendon organ are key receptors that allow us to be aware of movement and position of the body in a physical space. These receptors are also important components in certain motor control processes known as spinal reflexes.

The autogenic inhibition reflex is a spinal reflex that involves the Golgi tendon organ. When a heavy load applies excessive tension to a particular muscle, there is a higher chance that this may damage the particular muscle or tendon. When activated, the autogenic inhibition reflex stimulates the inhibitory interneuron in the spinal cord that innervates the motor neuron controlling that muscle.

Similar to the stretch reflex, the autogenic inhibition reflex also triggers action that coordinates the contraction of the opposing muscle. This ensures that the opposing muscle works in harmony with the muscle that has been inhibited from contracting.

Answers

1. muscle contracts, 2. inhibitory interneuron, 3. motor neuron, 4. Golgi tendon organ, 5. heavy load, 6. neuron from Golgi tendon fires, 7. spinal cord, 8. inhibited motor neuron, 9. muscle relaxes

Axonal Degeneration and Regeneration

Degeneration of an axon is a pathological event caused by neurodegenerative diseases and acute traumatic injury. It is generally impossible to repair damaged axons in the CNS after trauma. This is due to the presence of glial cells, which immediately move to the site of trauma and form a glial scar that prevents any possible future axonal regeneration. However, in the PNS the mechanism of programmed self-destruction (apoptosis) can sometimes be bypassed in an attempt to regenerate the damaged axon.

The severed nerve, at the proximal end, undergoes chromatolysis and begins to show signs of growth, known as growth cone sprouting. Chromatolysis is the disintegration of chromaphil material in response to external injury. This is a critical time point in the life of a neuron and is either a precursor of apoptosis or regeneration. At the distal end of the severed nerve, the axons and myelin sheath degenerate in a process called Wallerian degeneration. Meanwhile, the growth cones at the proximal end grow and extend under the influence of chemotactic factors released by the remnant Schwann cells. Macrophages are also known to release neurotrophic growth factors to encourage axonal regeneration until the axon has migrated to the correct location in the effector.

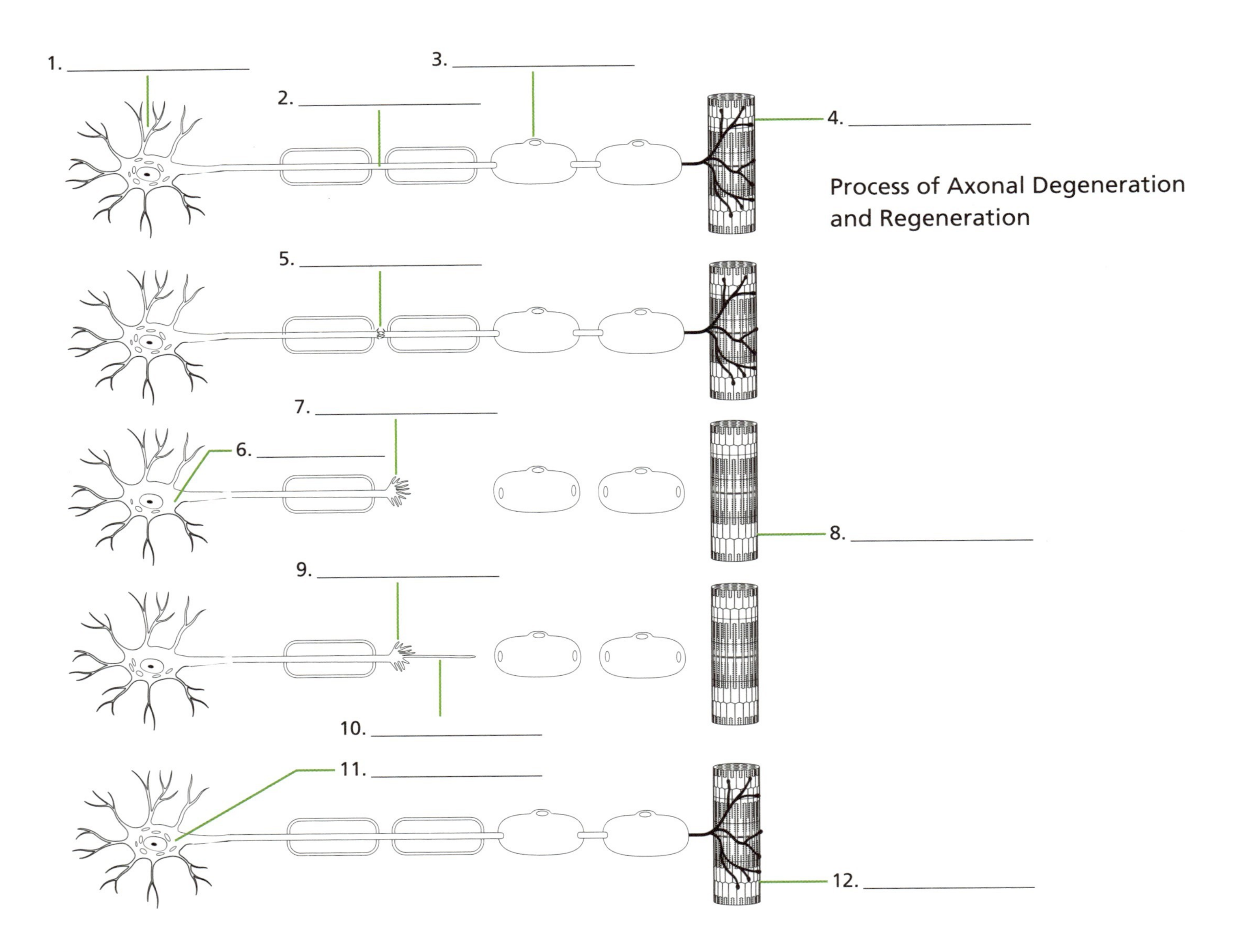

Process of Axonal Degeneration and Regeneration

Answers

1. cell body, 2. axon, 3. Schwann cell, 4. muscle fiber, 5. axon severed, 6. chromatolysis, 7. severed axon sprouting, 8. effector denervated, 9. sprouts degenerating, 10. axon regenerating, 11. chromatolysis no longer present, 12. target reinnervated

Autonomic Nervous System Structure

The autonomic nervous system consists of two sets of efferent neurons: preganglionic nerve fibers and postganglionic nerve fibers. The preganglionic fibers originate in the CNS and synapse onto clusters of neurons called ganglia. Postganglionic nerve fibers emerge from the ganglia and course to smooth muscles and glands throughout the body. The autonomic nervous system consists of two divisions: sympathetic and parasympathetic. The sympathetic nervous system has short preganglionic nerve fibers with ganglia close to the spinal cord, from which longer postganglionic nerve fibers emerge. The sympathetic division is generally involved in fight-or-flight scenarios, and its functions include increasing heart rate, dilating pupils, decreasing urine output, inhibiting peristalsis, and dilating blood vessels to skeletal muscles. The parasympathetic division has longer preganglionic nerve fibers with ganglia that are close to, or within, the effectors. The parasympathetic division is used to calm the body, "rest-and-digest," and balance the sympathetic division. Its functions include decreasing heart rate, constricting pupils, decreasing sweat output, increasing urine output, and constricting blood vessels to skeletal muscles. The sympathetic and parasympathetic nervous systems work in unison to maintain homeostatic balance.

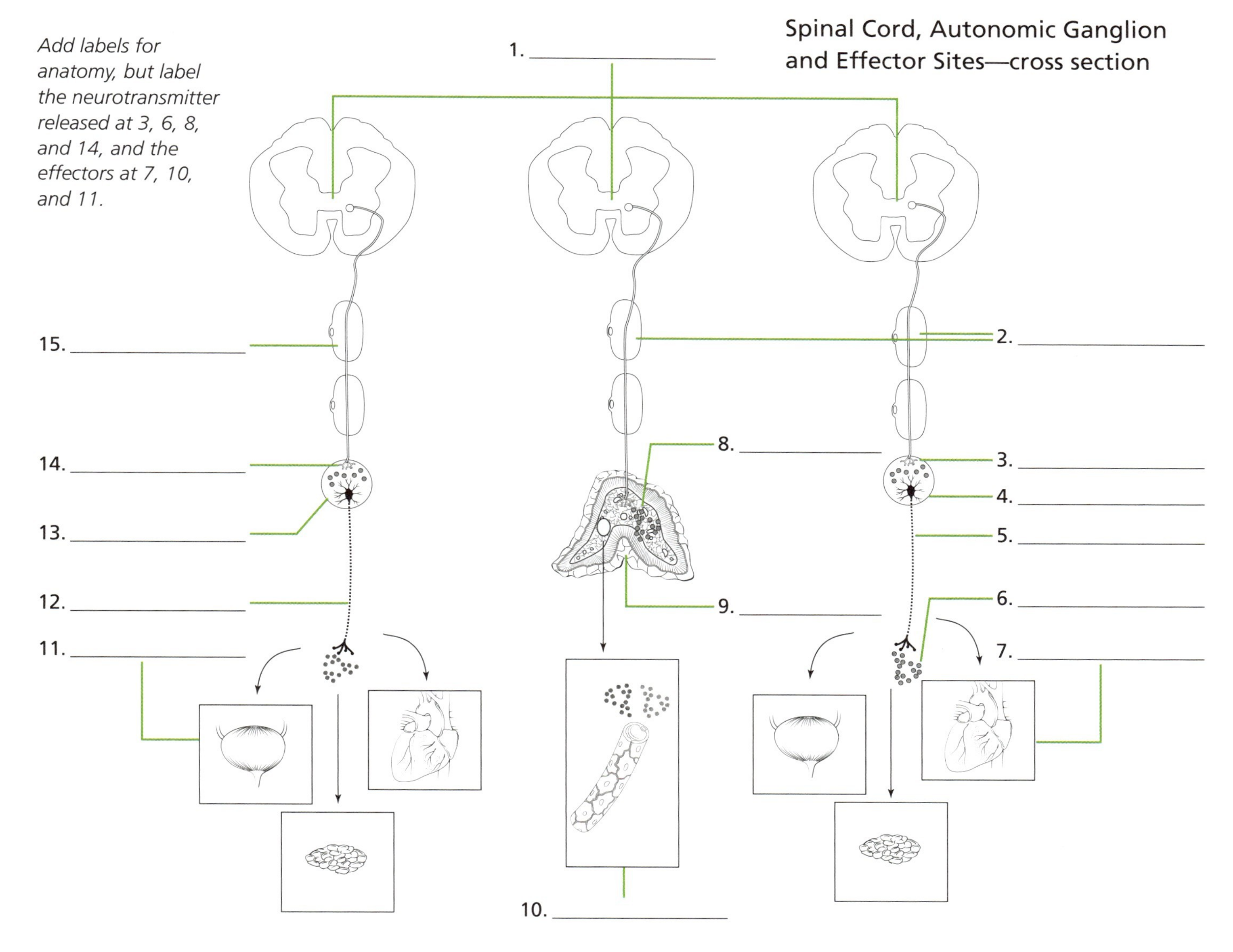

Answers

1. spinal cord, 2. sympathetic preganglionic neuron (myelinated), 3. acetylcholine, 4. autonomic ganglion, 5. sympathetic postganglionic neuron (unmyelinated), 6. norepinephrine, 7. norepinephrine effectors (glands/cardiac muscle/smooth muscle), 8. acetylcholine, 9. adrenal medulla, 10. effectors (release of epinephrine and norepinephrine), 11. acetylcholine effectors (glands/cardiac muscle/smooth muscle), 12. parasympathetic postganglionic neuron (unmyelinated), 13. autonomic ganglion, 14. acetylcholine, 15. parasympathetic preganglionic neuron (myelinated)

Autonomic Nervous System Neurotransmitters

Autonomic Nervous System Neurotransmitters

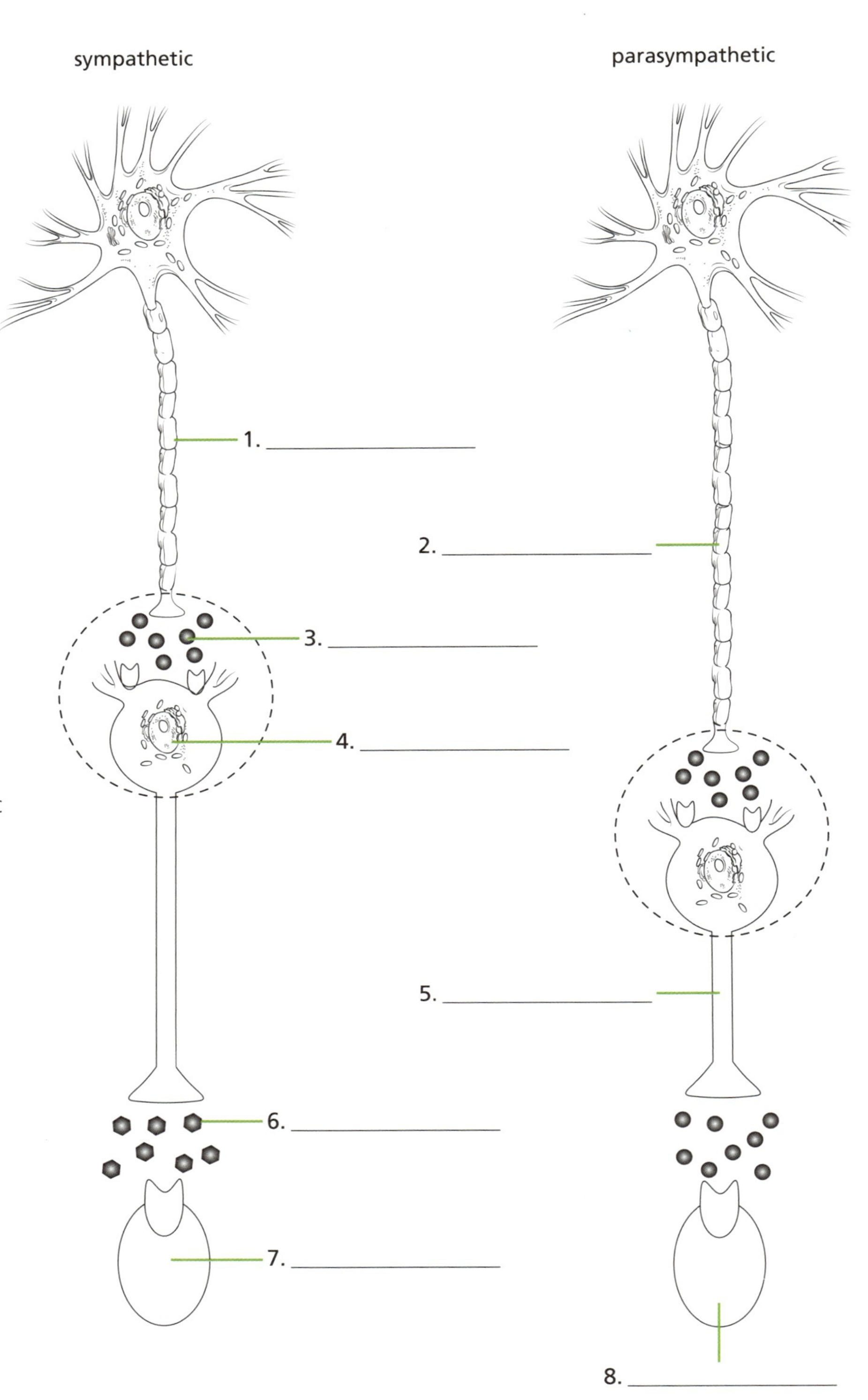

Both the sympathetic and parasympathetic divisions of the autonomic nervous system are classed as cholinergic nerves. This is because the preganglionic nerve fibers release acetylcholine at synapses in the ganglia, which then binds to nicotinic-type acetylcholine receptors expressed by the neurons in the ganglia that give rise to the postganglionic nerves. The postganglionic nerve fibers of the sympathetic division are either adrenergic (release norepinephrine) or cholinergic (release acetylcholine). Adrenergic postganglionic nerve fibers make up the majority of the sympathetic division and can act on either α- or β-type norepinephrine receptors. Some effector organs will have one or the other receptor, and some will have a ratio of both. The response by the effector will be based on the ratio of the α and β receptors and thus provides a broad range of control. The postganglionic fibers of the parasympathetic division are only cholinergic and act on muscarinic-type acetylcholine receptors.

Answers

1. preganglionic fibers (short), 2. preganglionic fibers (long), 3. acetylcholine, 4. nicotinic receptor, 5. postganglionic fibers, 6. norepinephrine, 7. α- or β-adrenergic receptor, 8. muscarinic receptor

Sympathetic Nervous System

The autonomic nervous system is divided into the sympathetic and parasympathetic divisions. The sympathetic division serves to regulate and modulate unconscious actions primarily related to the fight-or-flight scenario. The sympathetic nervous system comprises short preganglionic neurons that release acetylcholine at their synapses with longer postganglionic neurons, which extend throughout the body. Postganglionic sympathetic fibers release norepinephrine, which binds to adrenergic receptors at the target tissue.

The sympathetic division is responsible for physiological responses, such as increasing heart rate and blood pressure, dilating pupils, and increasing blood flow to skeletal muscle. During the fight-or-flight response, the preganglionic sympathetic fibers that end in the adrenal medulla activate epinephrine secretion. It is epinephrine, and norepinephrine, that assist in priming the body for action and survival. In addition, the sympathetic division also makes localized adjustments to the human body, helping to maintain homeostasis. These include increasing sweat output in response to increased body temperature, as well as stimulating the desire to urinate.

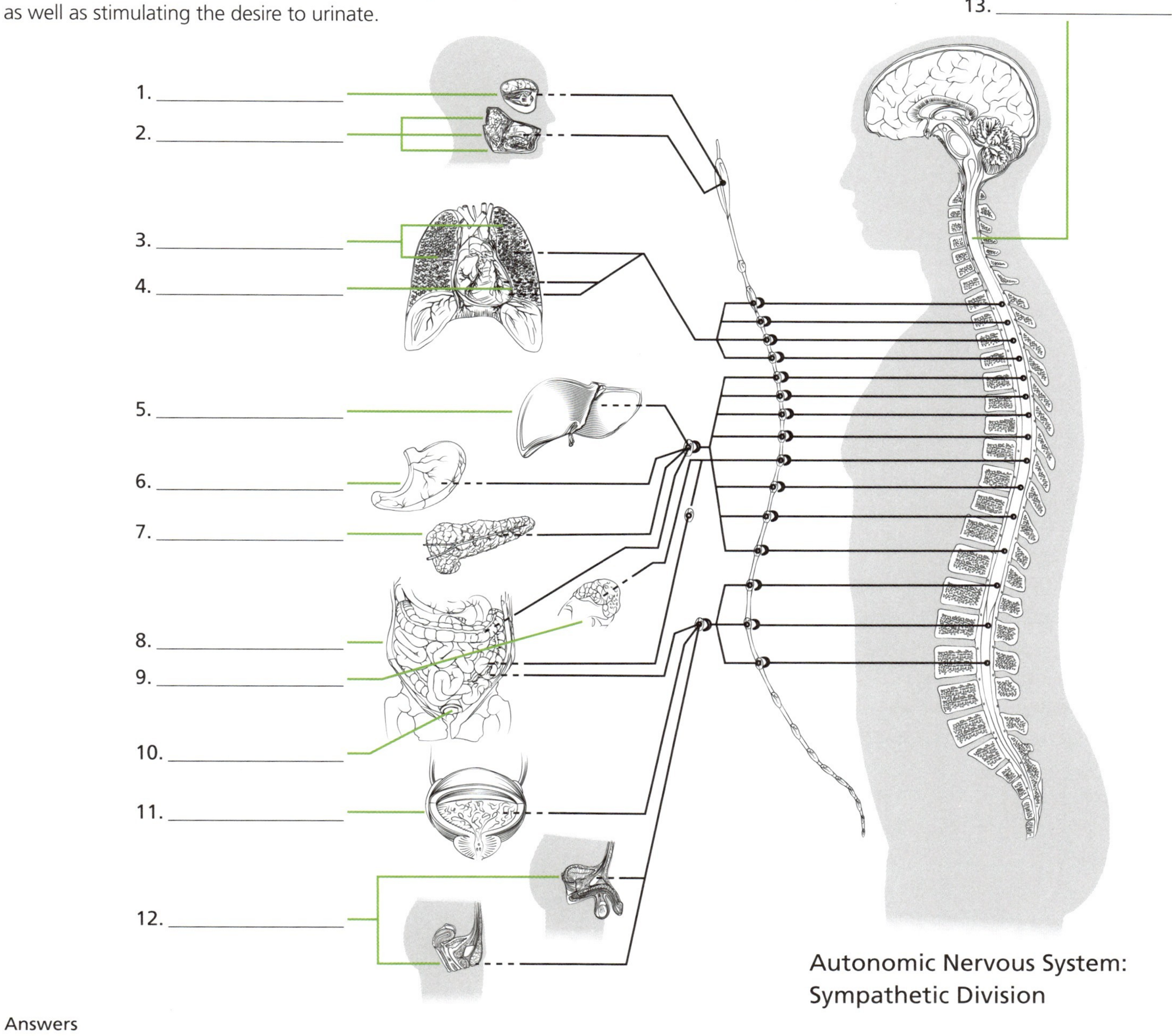

Autonomic Nervous System: Sympathetic Division

Answers

1. eye and lacrimal glands, 2. salivary glands, 3. lungs, 4. heart, 5. liver, 6. stomach, 7. pancreas, 8. small intestine, 9. adrenal medulla, 10. large intestine and rectum, 11. bladder, 12. reproductive organs, 13. spinal cord

Parasympathetic Nervous System

The parasympathetic division is one of two divisions of the autonomic nervous system (along with the enteric nervous system; see p. 69) that controls involuntary actions. It regulates organ and gland function during periods of rest, in what are typically referred to as rest-and-digest or feed-and-breed activities.

The preganglionic parasympathetic nerves arise in the CNS and include several key cranial nerves (CN III, CN VII, CN IX, CN X) as well as the three pelvic splanchnic nerves. These preganglionic nerves release acetylcholine at their synapses with postganglionic parasympathetic nerves elsewhere in the body. Postganglionic parasympathetic ganglia are typically very close to their target, where they also release acetylcholine to stimulate muscarinic receptors of the target organ.

The parasympathetic division aids in balancing out the sympathetic division by lowering blood pressure and heart rate. It also stimulates other key bodily functions, such as promoting saliva production, increasing muscular activity of the stomach and intestines to aid in digestion, eliminating waste by urination and defecation, and promoting lacrimation (the production of tears).

Autonomic Nervous System: Parasympathetic Division

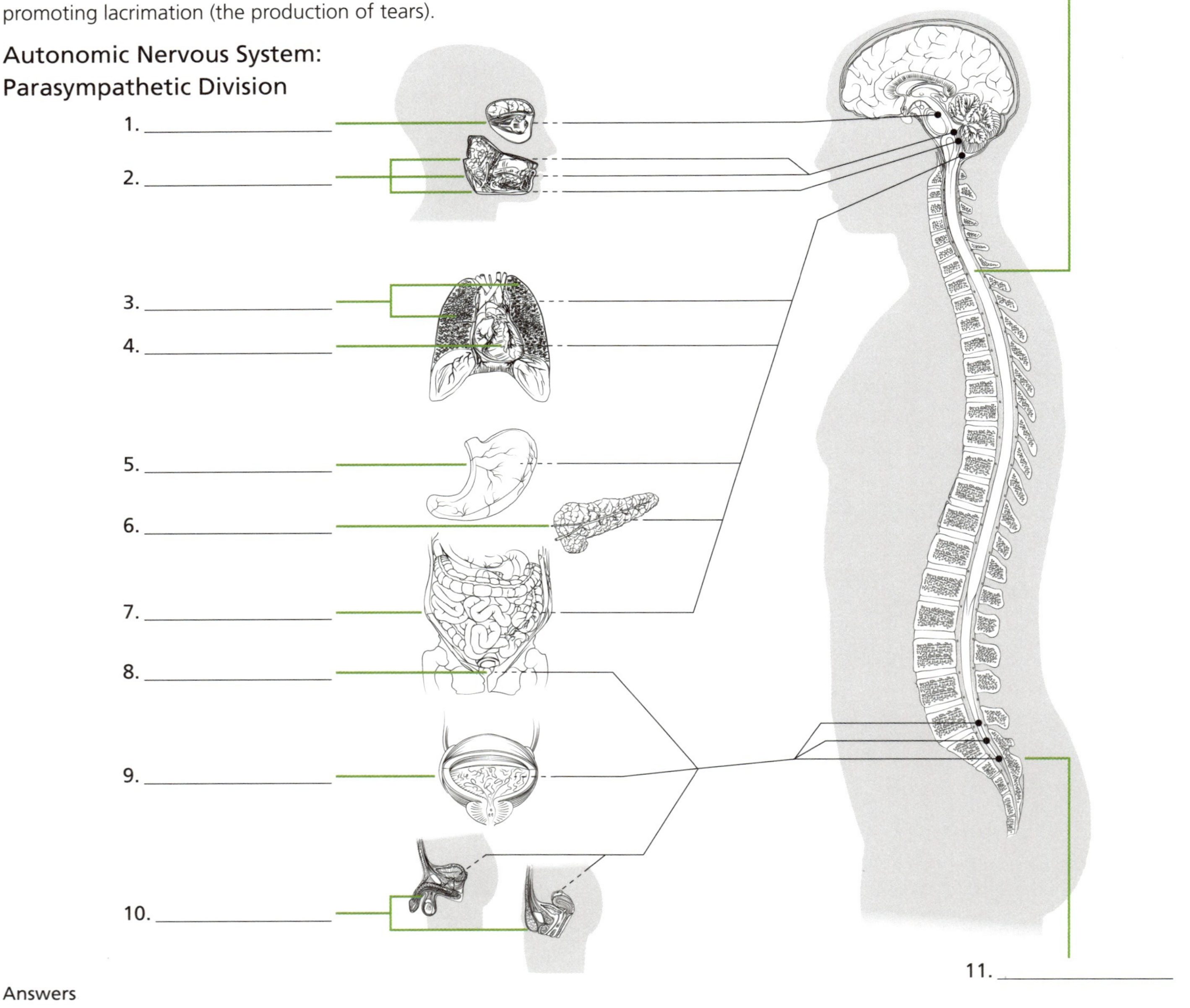

Answers

1. eye and lacrimal glands, 2. salivary glands, 3. lungs, 4. heart, 5. stomach, 6. pancreas, 7. small intestine, 8. rectum, 9. bladder, 10. reproductive organs, 11. sacrum, 12. spinal cord

Enteric Nervous System

Spinal Cord and Innervation of Gastrointestinal System—anatomical view

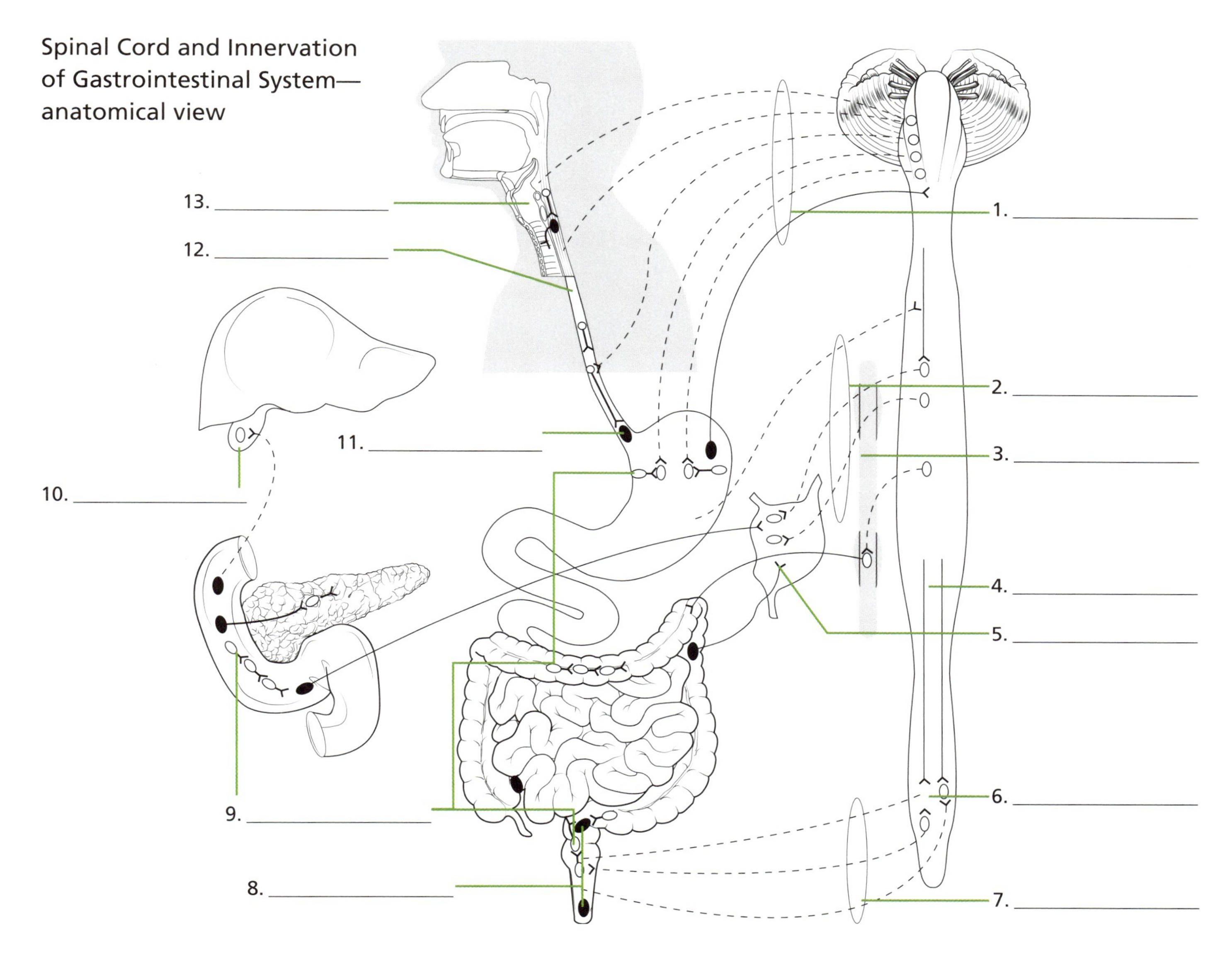

The enteric nervous system (ENS) is the nervous system of the gastrointestinal (GI) tract. It is a division of the autonomic nervous system and has a comparable number of neurons to that of the spinal cord. These are organized into two main neural networks, referred to as the myenteric plexus and the submucosal plexus. The two networks are made up of afferent and efferent nerves and interneurons that are able to coordinate and modulate reflex gut activity without any input from the CNS. This autonomic ability allows for normal digestive function and a robust system for responding to bulk and nutrient composition. The ENS relays information up and down the entire GI tract, from mouth to anus, in response to extrinsic and intrinsic factors. Receptors in the GI mucosa and muscle provide this information for reflexes in smooth muscle and secretory and endocrine cells.

However, the gut also has extensive two-way connections to the CNS through sympathetic and parasympathetic innervation. The sympathetic neurons inhibit GI secretion and any motor activity, such as peristalsis, the wave-like muscle contractions that propel food in the digestive tract. The parasympathetic connections, conversely, stimulate GI secretion and cause peristaltic movements when appealing food is seen, tasted, or smelled.

Answers

1. vagal pathways, 2. sympathetic pathways, 3. sympathetic nervous trunk, 4. thoracolumbar spinal cord, 5. celiac ganglia, 6. sacral spinal cord, 7. pelvic pathways, 8. sphincters, 9. enteric nervous system, 10. gall bladder, 11. sphincter, 12. esophagus, 13. trachea

Ventral (Anterior) Aspect of the Brainstem

The brainstem connects the spinal cord to higher areas of the brain and is anatomically divided into three parts: the midbrain, the pons, and the medulla.

The midbrain region is distinguished by two large cerebellar peduncles. At the midbrain level, the oculomotor nerve (CN III) emerges between the cerebral peduncles, whereas the trochlear nerve (CN IV) exits posteriorly.

The pons is identifiable by an anterior bulge that is composed of the pontine nuclei. These nuclei are involved in sleep regulation, mood, respiration, swallowing reflex and bladder control, eye movement, and many other functions. At the pons, the trigeminal nerve (CN V) is attached to the middle cerebellar peduncle, whereas the abducens nerve (CN VI) exits anteriorly at the junction of the pons and medulla.

The medulla has two distinctive pyramids on either side of the midline. These pyramids contain the corticospinal and corticobulbar tracts, which transmit motor impulses from the brain to the spinal cord and cranial nerves, respectively. The brainstem contains the nuclei and tracts of the cranial nerves.

Brainstem—ventral (anterior) view

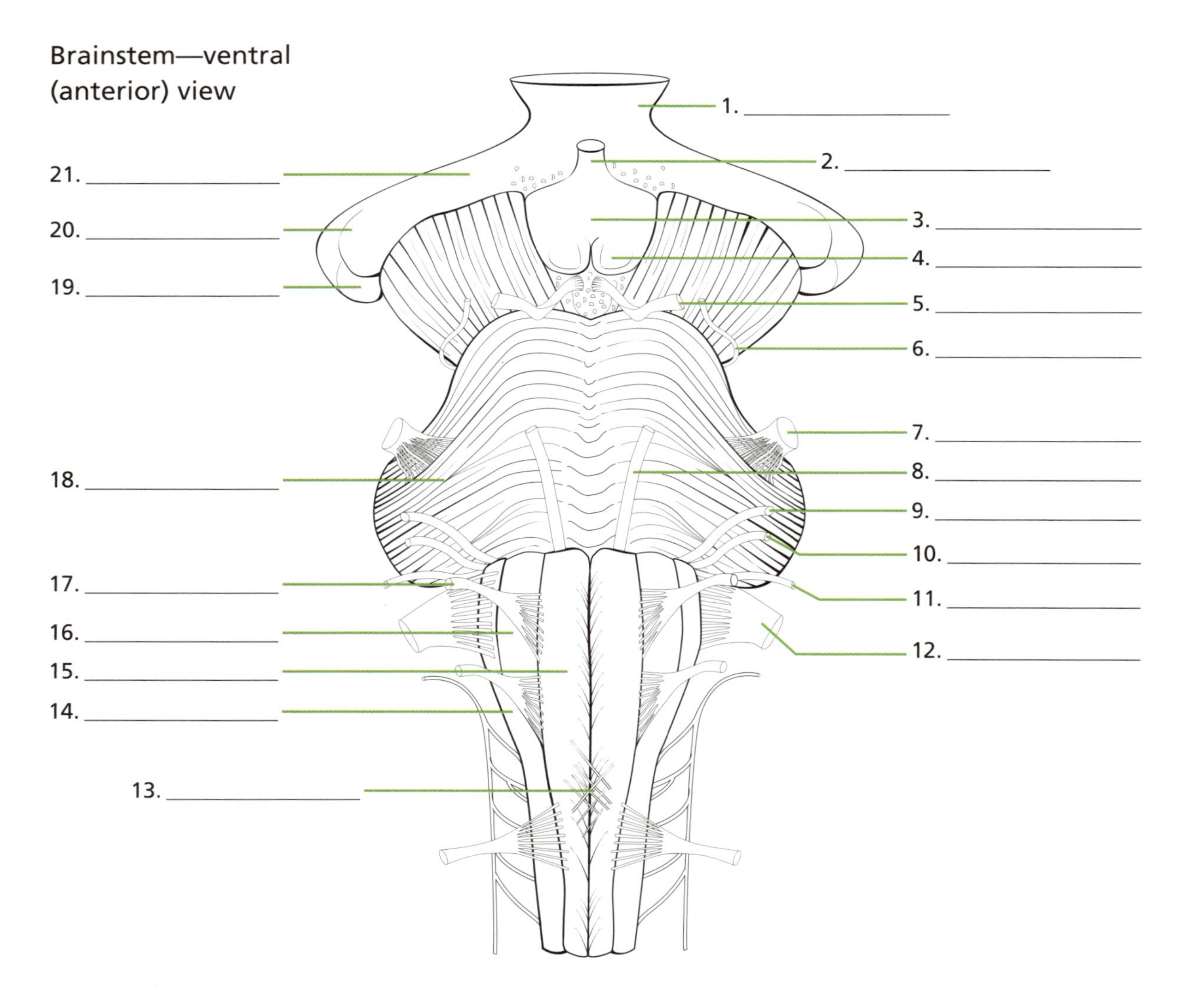

Answers

1. optic chiasm, 2. infundibulum, 3. tuber cinereum, 4. mammillary body, 5. oculomotor nerve (CN III), 6. trochlear nerve (CN IV), 7. trigeminal nerve (CN V), 8. abducens nerve (CN VI), 9. facial nerve (CN VII), 10. vestibulocochlear nerve (CN VIII), 11. glossopharyngeal nerve (CN IX), 12. vagus nerve (CN X), 13. decussation of pyramids, 14. circumolivary bundle, 15. pyramid, 16. olive, 17. hypoglossal nerve (CN XII), 18. junction of pons and middle cerebellar peduncle, 19. lateral geniculate nucleus, 20. medial geniculate nucleus, 21. optic tract

Dorsal (Posterior) Aspect of the Brainstem

In this illustration the cerebellum has been removed to show the location of the fourth ventricle. The brainstem contains the motor and sensory centers for ten of the twelve cranial nerve pairs that innervate the face and neck. The brainstem also provides a key pathway for motor and sensory nerve connections between the brain and the rest of the body. These pathways include the medial lemniscus pathway and the spinothalamic and corticospinal tracts.

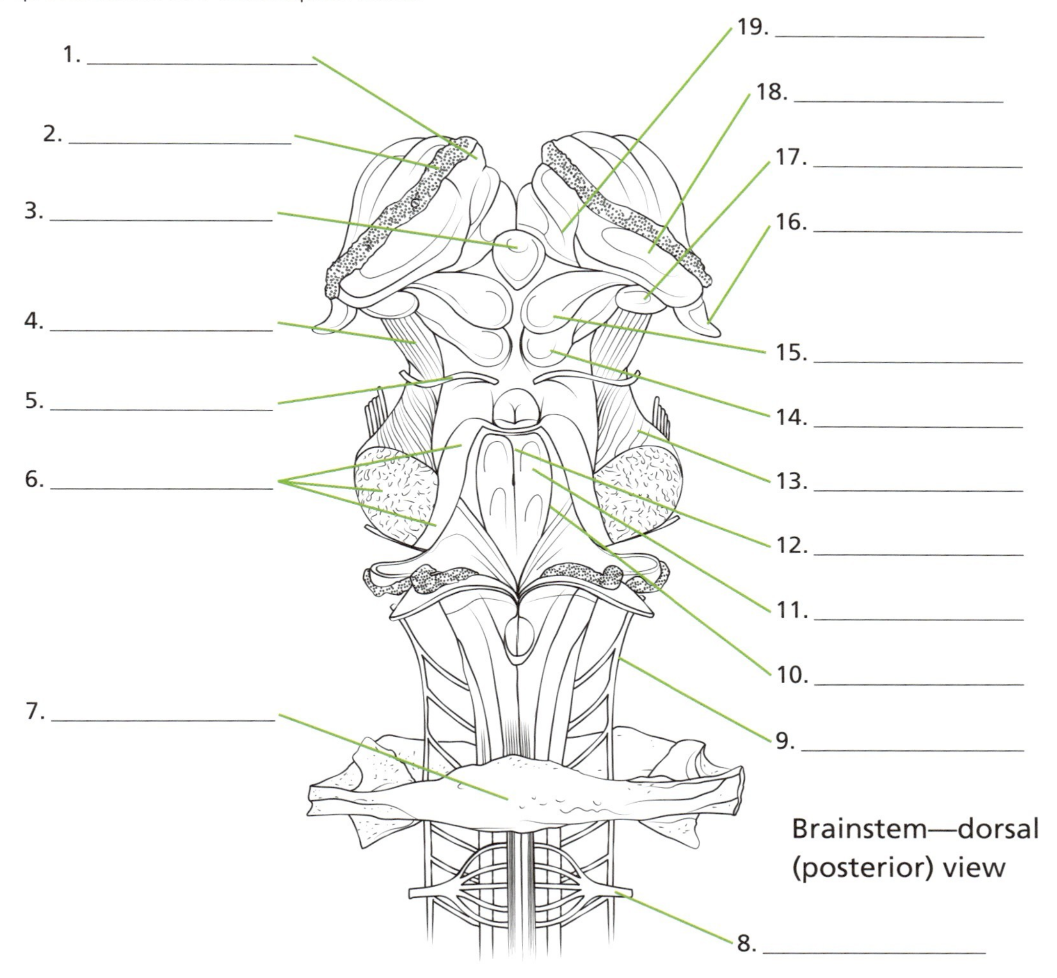

Brainstem—dorsal (posterior) view

The medial lemniscus is a sensory pathway that transmits impulses of fine touch and vibrations from the skin to the thalamus. The spinothalamic tract is also a sensory pathway from the skin to the thalamus, but instead transmits information about crude touch. The corticospinal tract is a descending motor pathway from the cerebral cortex that transmits fine-control movements.

The dorsal view of the brainstem shows a medial groove known as the dorsal median sulcus. The lateral columns to the median sulcus are the fasciculus gracilis, which provides conscious proprioception of the lower limbs and lower body. Lateral to that is the fasciculus cuneatus, which provides conscious proprioception as well as fine touch and fine vibrations from the upper limbs.

Answers

1. thalamus, 2. choroid plexus of lateral ventricle, 3. pineal body, 4. cerebral peduncle, 5. trochlear nerve (IV), 6. cerebellar peduncles, 7. atlas (C1), 8. first cervical nerve, 9. spinal accessory nerve (XI), 10. sulcus limitans, 11. facial colliculus, 12. dorsal median sulcus, 13. pons, 14. inferior colliculus, 15. superior colliculus, 16. lateral geniculate nucleus, 17. medial geniculate nucleus, 18. pulvinar, 19. habenula

Lateral Aspect of the Brainstem

The lateral aspect of the brainstem shows the structure and positions of the midbrain, pons, and medulla oblongata, as well as the relevant positions of the thalamus and hypothalamus.

The midbrain is the processing center of auditory and visual information. Converging auditory pathways provide information to the inferior colliculus from various regions of the brain, including the auditory cortex. The inferior colliculus relays this information to motor control centers of the superior colliculus and cerebellum. This auditory information contributes to the generation of reflex somatic and motor responses to sounds and sights. Reflex somatic responses are automatic reflexes, such as removing your hand from very hot surfaces. Notable features of the midbrain include the nuclei for the oculomotor nerve (CN III) and trochlear nerve (CN IV). These nerves are responsible for pupil constriction and eye movement (downward and laterally), respectively. The medulla oblongata is the autonomic center for the coordination of visceral functions; it also relays sensory information to the thalamus. Nerves that arise in the medulla oblongata include the glossopharyngeal nerve (CN IX), vagus nerve (CN X), spinal accessory nerve (CN XI), and hypoglossal nerve (CN XII).

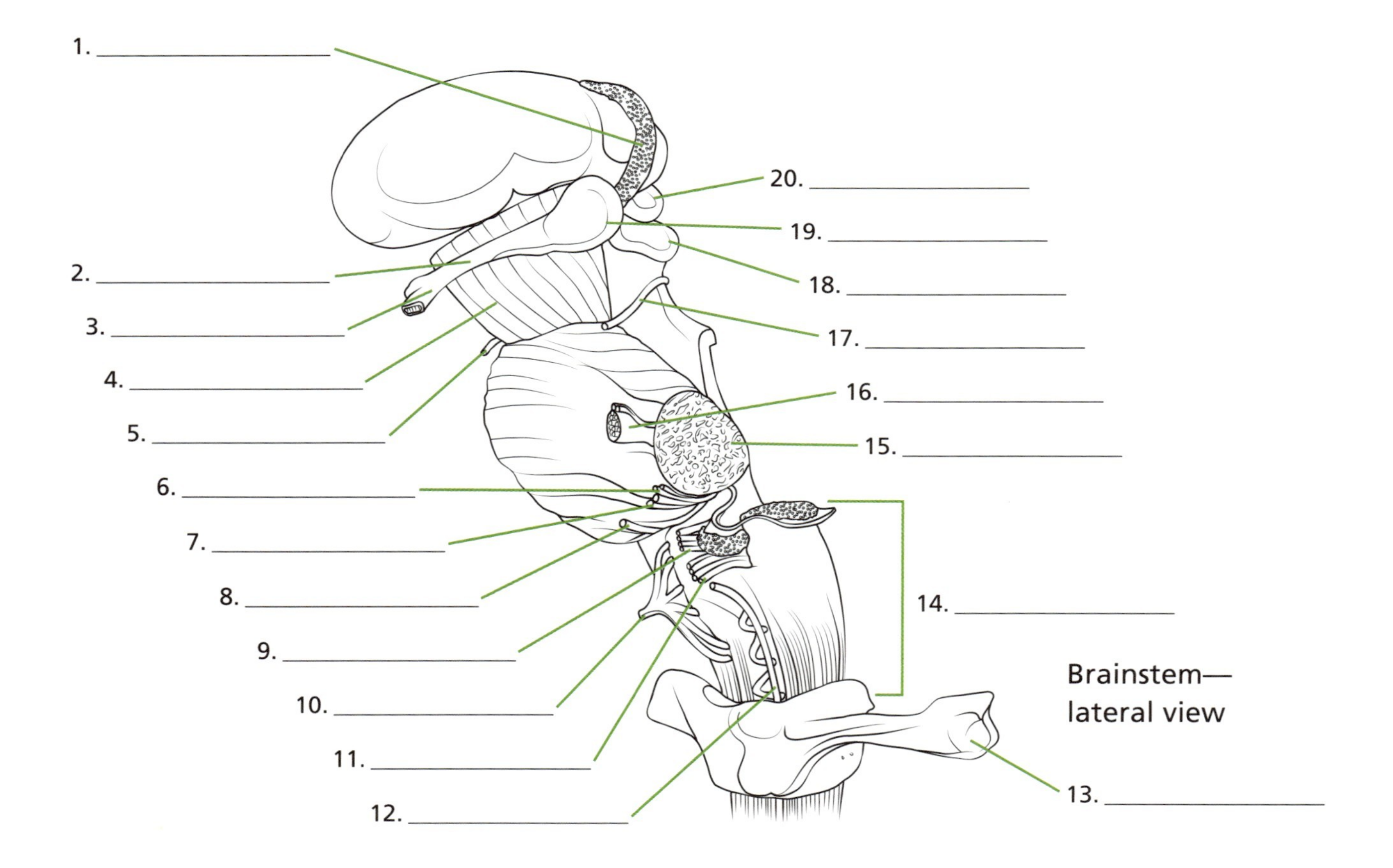

Answers

1. choroid plexus, 2. optic tract, 3. optic nerve (II), 4. cerebral peduncle, 5. oculomotor nerve (III), 6. vestibulocochlear nerve (VIII), 7. facial nerve (VII), 8. abducens nerve (VI), 9. glossopharyngeal nerve (IX), 10. hypoglossal nerve (XII), 11. vagus nerve (X), 12. spinal accessory nerve (XI), 13. atlas (C1), 14. medulla oblongata, 15. middle cerebellar peduncle, 16. trigeminal nerve (V), 17. trochlear nerve (IV), 18. inferior colliculus, 19. lateral geniculate nucleus, 20. superior colliculus

Internal Structure of the Brainstem

The brainstem plays a role in relaying information between the body and the brain. It has many basic functions, with centers that control involuntary functions, such as heart rate (cardiac center) and breathing (respiratory center). The brainstem contains ascending sensory pathways, including the cuneatus and gracilis nuclei for the sense of touch (arms and legs, respectively). It also includes the reticular formation, which filters incoming stimuli and assists in the circadian rhythm. The solitary nucleus is a series of nuclei that are embedded in the medulla and are involved in taste (facial nerve, CN VII) and respiratory and gastrointestinal reflexes (glossopharyngeal nerve, CN IX, and vagus nerve, CN X). The olivary nucleus is located on the anterior surface of the medulla. It consists of two parts: the inferior olivary nucleus, which is involved in motor learning, such as coordination of movement; and the superior olivary nucleus, which is involved in sound perception, such as perception of the time difference between sounds. The brainstem is incredibly important for traversing pathways that conduct information from the body to the brain and vice versa.

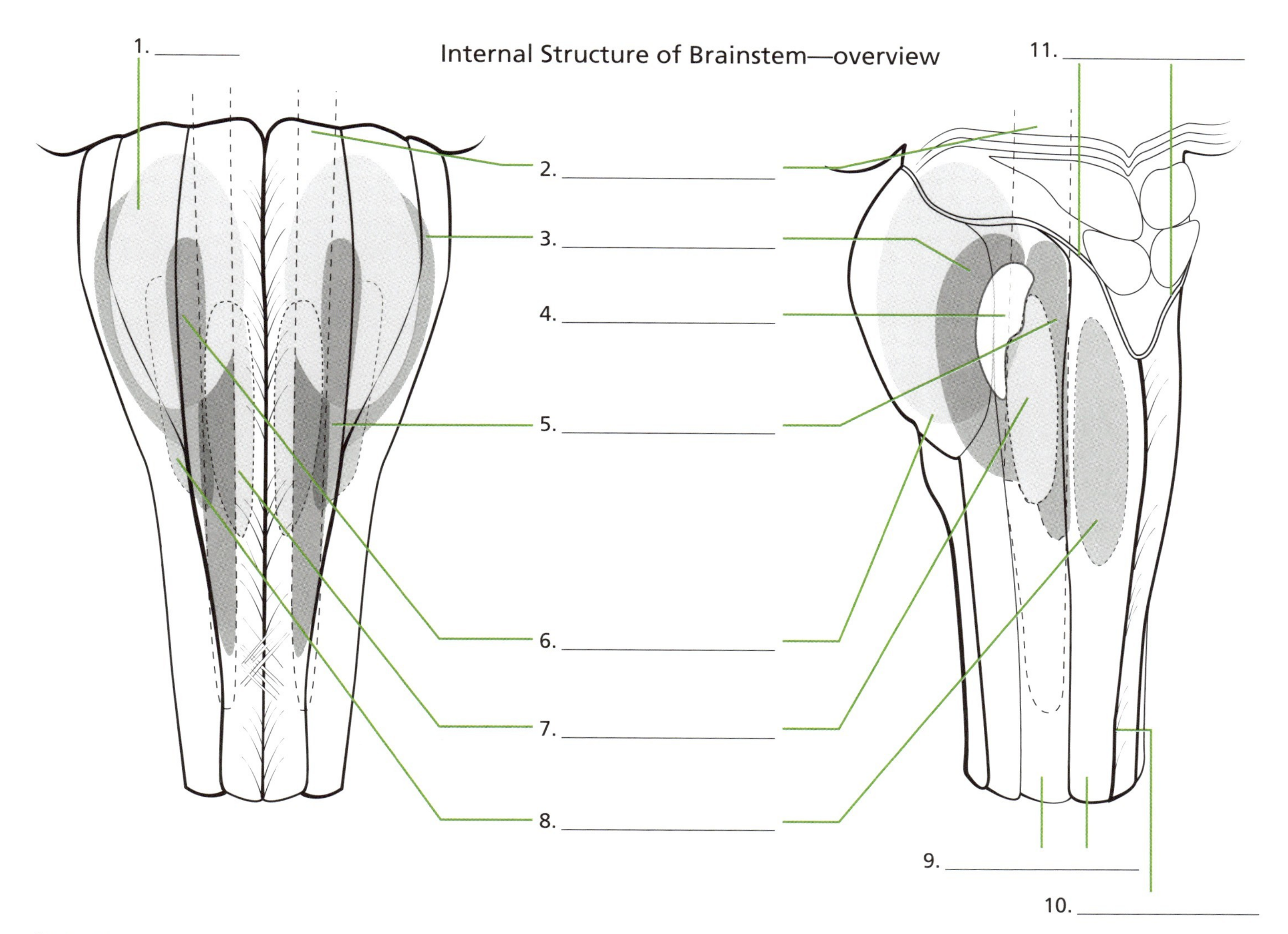

Answers

1. olive, 2. reticular formation, 3. cardiovascular centers, 4. respiratory rhythmicity center, 5. solitary nucleus, 6. olivary nucleus, 7. cuneate nucleus, 8. gracilis nucleus, 9. spinal cord, 10. posterior median sulcus, 11. attachment to membranous roof of fourth ventricle

Medulla Oblongata

The medulla is divided into two: the open, or superior, part; and the closed, or inferior, part. The pyramids are on each side of the midline and contain motor fibers of the corticobulbar and corticospinal tracts. The motor fibers of the corticobulbar and corticospinal tracts originate in the cerebral cortex and extend through the medulla oblongata to the cranial nerves and spinal cord, respectively. The transverse section of the medulla oblongata at the pyramidal decussation shows internal structures that have a similarity with the spinal cord. The ventral part shows the pyramids of the medulla, which are two white matter structures consisting of the corticospinal and corticobulbar tracts. These motor fiber tracts decussate—or cross—in the anterior median fissure of the medulla oblongata. This is known as the decussation of pyramids. Decussation is the crossing of nerve fibers from one lateral side to the other. This is believed to be an evolutionary trait that occurred when species evolved from invertebrates to vertebrates. The transverse section of the medulla at the sensory decussation shows a similar internal structure to the transverse section of the pyramidal decussation. A major difference is the gracilis nucleus and the cuneatus nucleus, which are very pronounced at this level. The fibers of the sensory decussation are responsible for fine touch and two-point discrimination (i.e., discerning two points of skin contact as two separate objects).

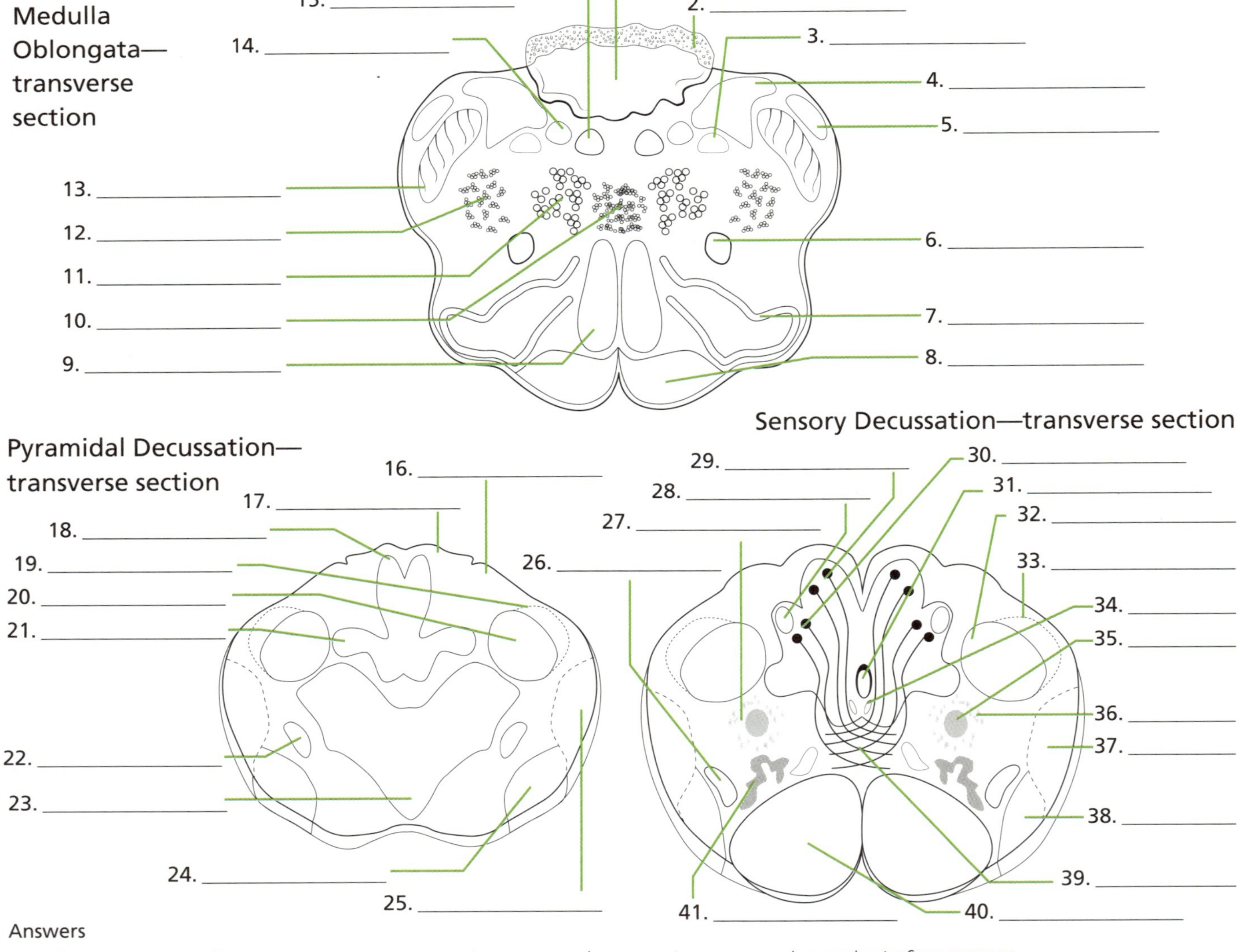

Answers

1. fourth ventricle, 2. choroid plexus, 3. solitary nucleus, 4. vestibular nuclear complex, 5. cochlear nuclei, 6. nucleus ambiguus, 7. inferior olivary nucleus, 8. pyramid, 9. medial lemniscus, 10. reticular formation—raphe nucleus, 11. reticular formation—medial nuclear group, 12. reticular formation—lateral nuclear group, 13. inferior cerebellar peduncle, 14. dorsal motor nucleus of vagus, 15. hypoglossal motor nucleus, 16. fasciculus gracilis, 17. fasciculus cuneatus, 18. gracilis nucleus, 19. spinal tract, 20. spinal nucleus, 21. cuneate nucleus, 22. spinothalamic tracts, 23. decussation of pyramids, 24. anterior spinocerebellar tract, 25. posterior spinocerebellar tract, 26. spinothalamic tracts, 27. reticular formation, 28. accessory cuneate nucleus, 29. gracilis nucleus, 30. cuneatus nucleus, 31. central canal, 32. spinal nucleus of trigeminal nerve, 33. spinal tract of trigeminal nerve, 34. medial longitudinal fasciculus, 35. fibers of vagus nerve, 36. nucleus ambiguus, 37. posterior spinocerebellar tract, 38. anterior spinocerebellar tract, 39. decussation of medial lemniscus, 40. pyramid, 41. inferior olivary nucleus

Pons

The pons is a distinctive anterior bulge rostral to the medulla. It is the region of the brainstem that contains key nuclei responsible for relaying information from the forebrain to the cerebellum and other regions of the brain. The cross section of the pons at the facial motor nucleus shows a number of key nuclei. There are two motor nuclei and one sensory nucleus. The abducens nucleus, located in a dorsomedial position close to the fourth ventricle, is a motor nucleus for the lateral movement of the eye. The facial nucleus of the facial nerve (CN VII), which can be found in the dorsomedial position close to the fourth ventricle, contains the motor nuclei that control facial musculature. The upper half of the pons—close to the midbrain—contains the main sensory and motor nuclei of the trigeminal nerve (CN V), which transmits somatosensory and motor information from the head. A characteristic feature of the pons is the presence of transverse pontine fibers, which join together as the middle cerebellar peduncle. Among these transverse fibers are the deep pontine nuclei, which are involved in motor activity and project fibers to the cortex. Overall, the pons is a continuation of the medulla and consists of continuous vertical fibers that descend from the cerebral cortex, terminating in either the pons or the medulla.

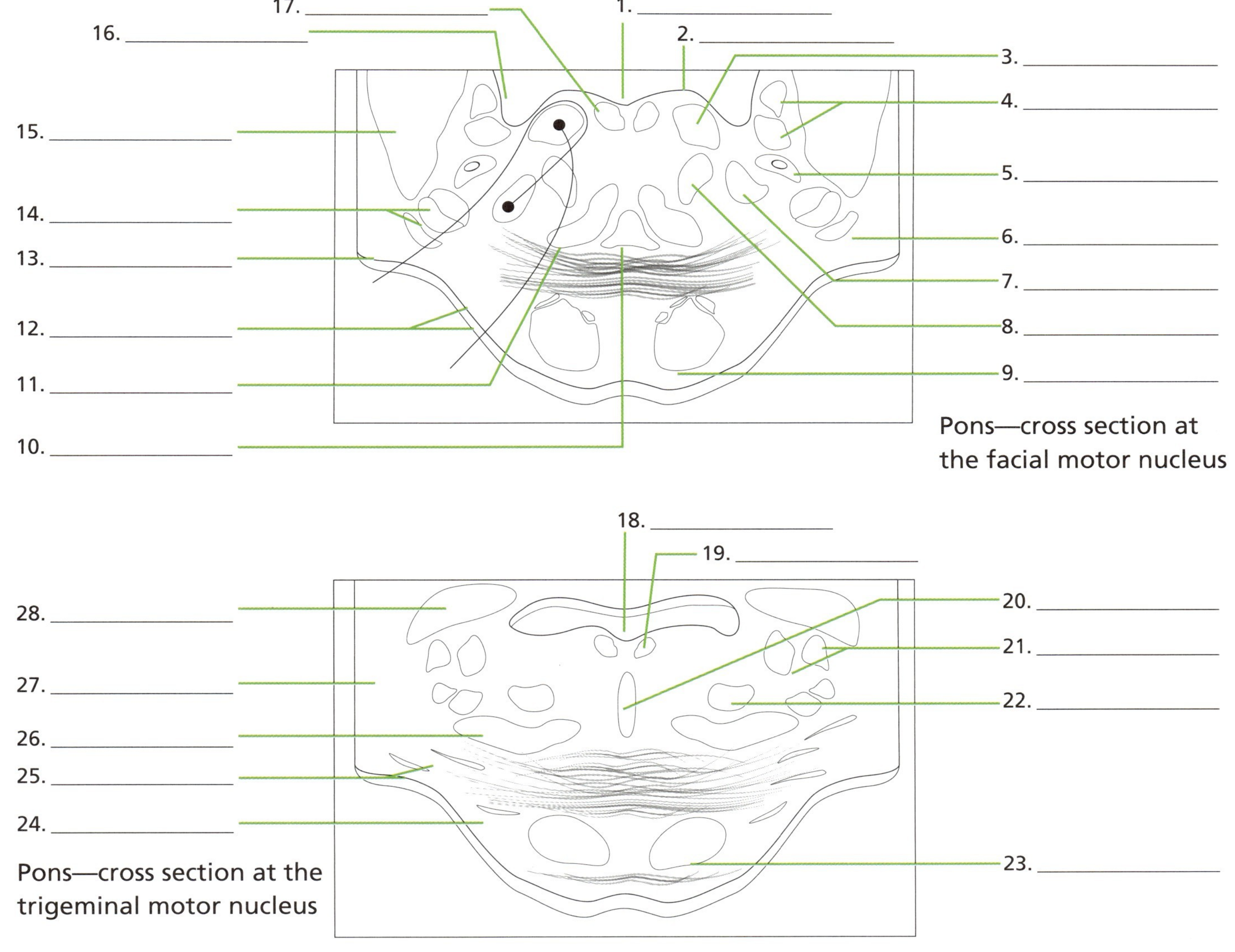

Pons—cross section at the facial motor nucleus

Pons—cross section at the trigeminal motor nucleus

Answers

1. fourth ventricle, 2. facial colliculus, 3. abducens nucleus of abducens nerve (CN VI), 4. vestibular nuclei (lateral and superior), 5. solitary nucleus and tract, 6. spinothalamic tract, 7. facial nucleus of facial nerve (CN VII), 8. central tegmental tract, 9. corticobulbar and corticospinal tracts, 10. trapezoid body, 11. medial lemniscus, 12. transverse pontine fibers and deep pontine nuclei, 13. middle cerebellar peduncle, 14. spinal nucleus and tract, 15. inferior cerebellar peduncle, 16. genu of facial nerve, 17. medial longitudinal fasciculus fibers, 18. fourth ventricle, 19. medial longitudinal fasciculus fibers, 20. raphe nuclei, 21. sensory and motor nuclei of trigeminal nerve (CN V), 22. central tegmental tract, 23. corticobulbar and corticospinal tracts, 24. deep pontine nuclei, 25. transverse pontine fibers, 26. medial lemniscus, 27. middle cerebellar peduncle, 28. superior cerebellar peduncle

Midbrain

Level of Inferior Colliculus

The inferior colliculus is a hemisphere of nerves that form the caudal part of the midbrain. It is divided into the tectum region and the cerebral peduncle (crus cerebri and tegmentum regions). The tectum region contains the inferior colliculus, a key component of the auditory pathway that is responsible for frequency recognition, signal integration, and pitch discrimination. The cerebral aqueduct is a cerebrospinal fluid (CSF)-filled conduit of the ventricular system that connects the third and fourth ventricles. It is surrounded by the gray matter of the periaqueductal gray (PAG) and separates the tegmentum from the tectum. The PAG is the interface between the forebrain and the lower brainstem and is involved in processing pain and analgesia as well as fear and anxiety. The trochlear nerve (CN IV), which innervates the superior oblique muscle of the eye, is present in the dorsal aspect of the caudal midbrain. The substantia nigra is a group of neurons that produce dopamine; it is located between the crus cerebri and the tegmentum. The lateral aspect of the tegmentum contains pathways that provide auditory input, including the lateral lemniscus and the large ascending medial lemniscus. Overall, the inferior colliculus is vital for orientation of the head in response to all types of incoming stimuli.

1. 2. 3. 4. 5. 6. 7. 8. 9. 10. 11. 12.

Level of Inferior Colliculus

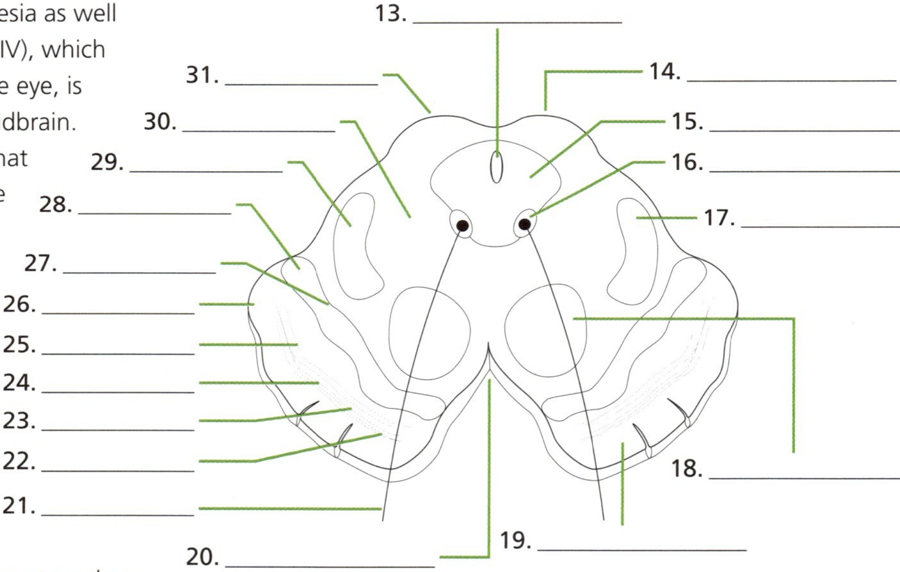

Level of Superior Colliculus

Level of Superior Colliculus

The superior colliculus is a rostral midbrain structure that is primarily involved in processing visual sensory input and projecting connections to associated motor areas. It is divided into two sections: the tectum and the cerebral peduncle. The tectum is a layered structure that processes visual sensory input, controlling responses through eye movement. It has many connections to other visual centers in the visual cortex of the occipital lobe. The cerebral peduncle includes three regions: the tegmentum, the crus cerebri, and the substantia nigra. The tegmentum contains key sensory pathways, including the reticular formation (central tegmentum area), spinothalamic and trigeminothalamic tracts (lateral tegmentum area), and the medial lemniscus (medial tegmentum area). Other notable features of the tegmentum include the red nucleus, a subcortical center that is key for motor coordination. The crus cerebri contains large bundles of descending nerve fibers that are important for movement of the eye: the occipito, parieto, and temporopontine fibers. The substantia nigra, which is located between the crus cerebri and the tegmentum, contains dopaminergic neurons that are stimulated by unexpected visual information.

Answers

1. tectum, 2. inferior colliculus, 3. cerebral aqueduct, 4. periaqueductal gray, 5. tegmentum (reticular formation), 6. trochlear nerve (CN IV), 7. crus cerebri, 8. substantia nigra, 9. superior cerebellar peduncle, 10. cerebral peduncle, 11. medial lemniscus, 12. lateral lemniscus: trigeminal thalamic fibers and spinothalamic fibers, 13. cerebral aqueduct, 14. superior colliculus, 15. periaqueductal gray, 16. oculomotor nerve (CN III), 17. spinothalamic and trigeminothalamic tracts, 18. red nucleus, 19. crus cerebri, 20. ventral tegmental area, 21. root fibers of oculomotor nerve (CN III), 22. frontopontine fibers, 23. corticobulbar fibers, 24. corticospinal fibers, 25. occipito, parieto, and temporopontine fibers, 26. cerebral peduncle, 27. pars compacta, 28. substantia nigra, 29. medial lemniscus, 30. tegmentum (reticular formation), 31. superior colliculus

Reticular Formation

Brainstem and Reticular Formation—left lateral view

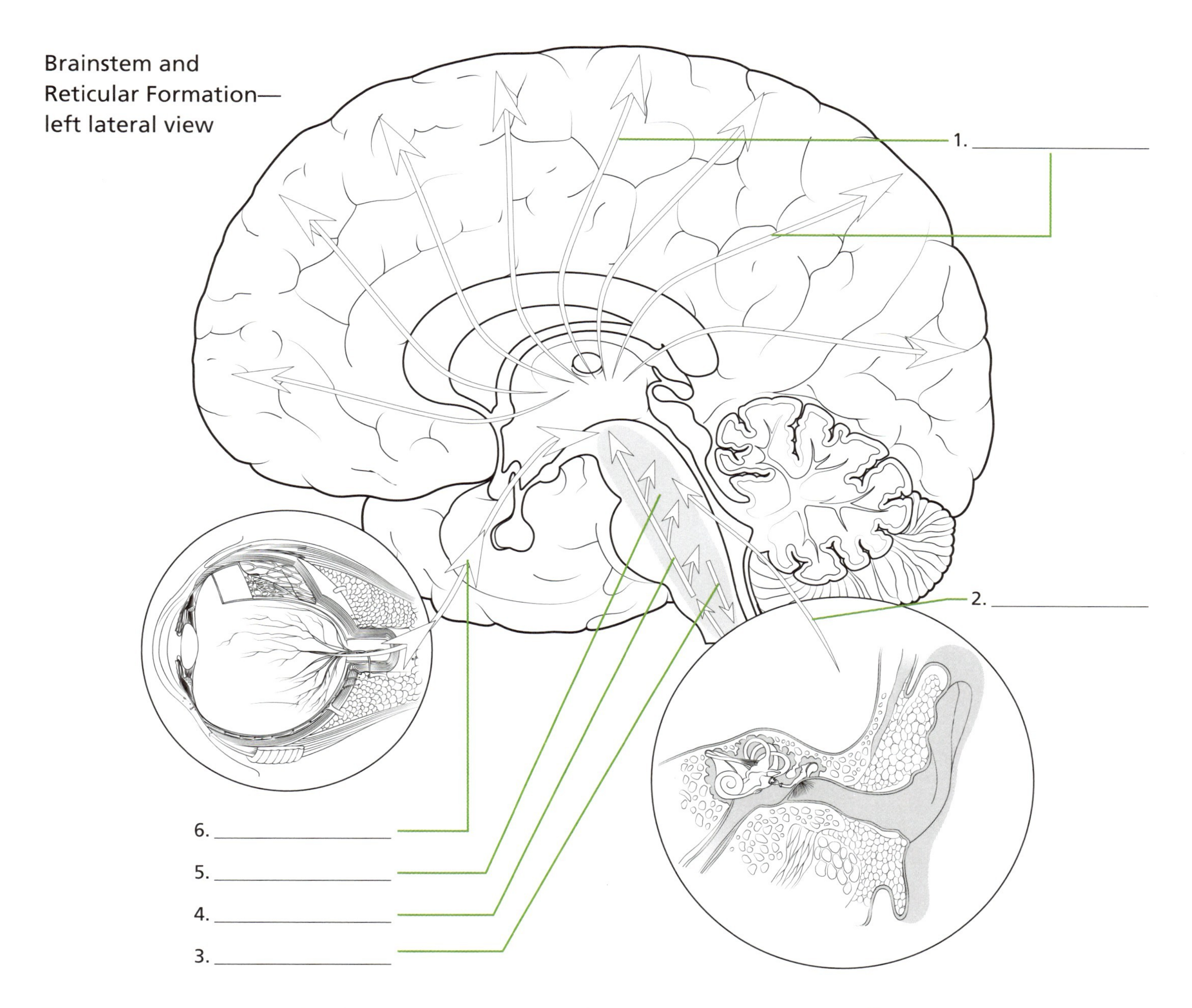

The reticular formation is a network of connected nuclei throughout the length of the brainstem. The nuclei are divided into three columns: median column (raphe nuclei), medial column (magnocellular nuclei), and lateral column (parvocellular nuclei). These columns are essential for regulation of many basic but important functions. Ascending sensory tracts are involved in sleep and consciousness through projections to the thalamus and cerebral cortex. Descending motor projections are involved in the moderation of somatic motor control (e.g., balance and posture), cardiac function (e.g., heart rate), and vasomotor centers (e.g., dilation/constriction of blood vessels). In addition, the reticular formation is used to relay visual and auditory impulses to the cerebellum and allow integration of these signals for modulating motor responses. The reticular formation is also the pathway for pain signals from the lower body to the cerebral cortex. Here, neurons in the reticular formation can filter and block some pain signals.

Answers

1. radiations to cerebral cortex, 2. auditory impulses, 3. descending motor projections to spinal cord, 4. ascending sensory tracts, 5. reticular formation, 6. visual impulses

Overview of Cranial Nerves

The 12 cranial nerves are those that emerge directly from the brain, rather than from the spinal cord. They are responsible for carrying sensory information to the brain, as well as motor and parasympathetic signals to muscles and organs, respectively. Cranial nerves primarily innervate the head and face, as well as the sensory organs on them, although some also innervate tissues in the neck and below. All cranial nerves come in pairs, one for each side of the body. They are traditionally numbered I to XII (in Roman numerals) in the order in which they emerge from the brain, starting anterior and moving posterior. Cranial nerves III through XII emerge from the brainstem and belong to the PNS, as traditionally denoted by the word "nerve." Cranial nerves I and II (the olfactory and optic nerves, respectively), however, emerge from the forebrain and are actually part of the CNS.

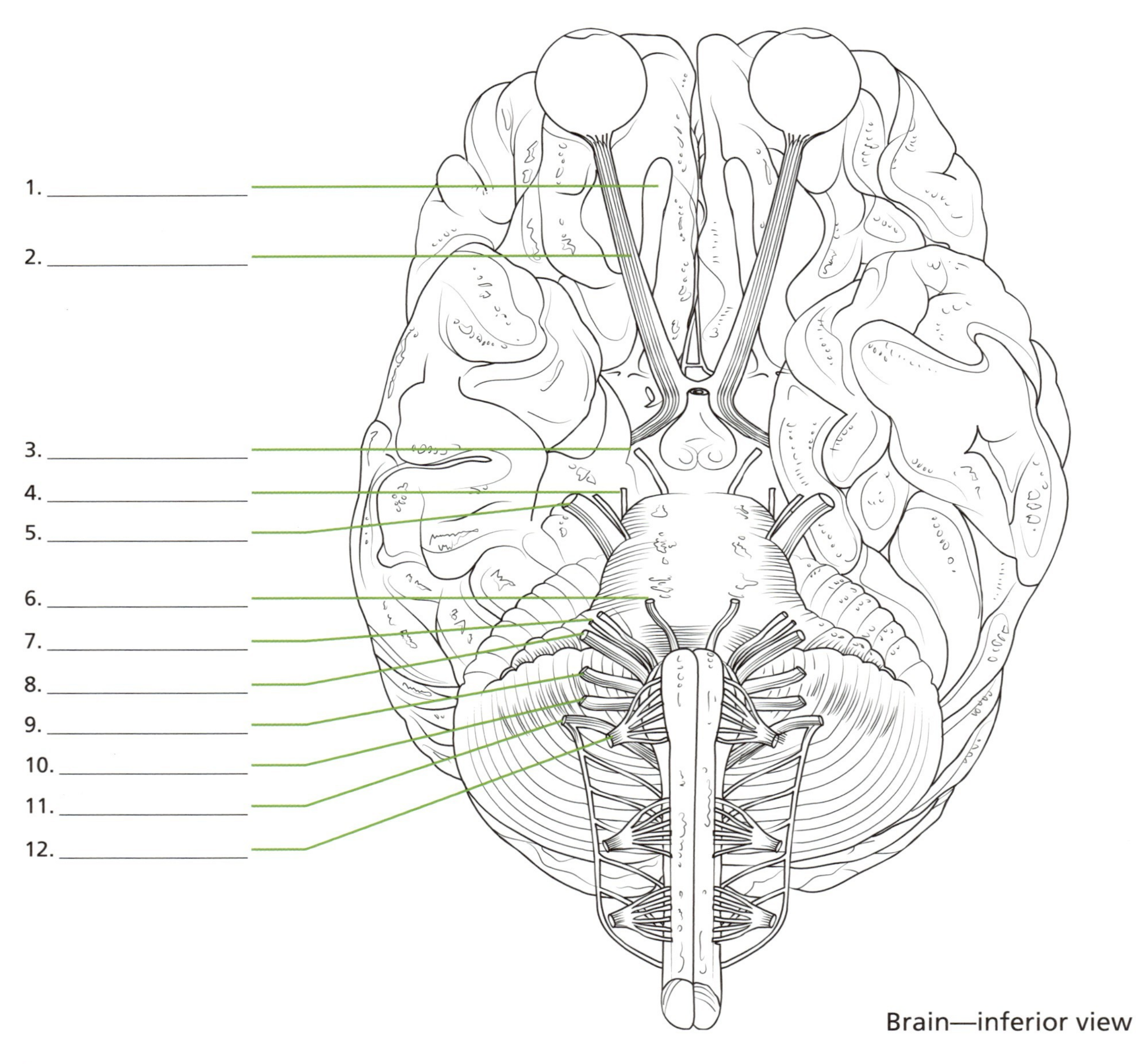

Brain—inferior view

Answers

1. olfactory bulb, 2. optic nerve (CN II), 3. oculomotor nerve (CN III), 4. trochlear nerve (CN IV), 5. trigeminal nerve (CN V), 6. abducens nerve (CN VI), 7. facial nerve (CN VII), 8. vestibulocochlear nerve (CN VIII), 9. glossopharyngeal nerve (CN IX), 10. vagus nerve (CN X), 11. accessory nerve (CN XI), 12. hypoglossal nerve (CN XII)

Cranial Nerve Nuclei

The cranial nerves, like all nerves, are bundles of axons that course together toward their eventual targets. For many of the cranial nerves (those that emerge from the brainstem), the neurons that give rise to these axons (in the case of motor output) or receive input from these axons (in the case of sensory input) are clustered together into discrete nuclei. Most of these brainstem nuclei contain neurons that are associated with a single cranial nerve. However, a significant subset of nuclei contain neurons associated with multiple cranial nerves, usually with portions of the nerves that are carrying the same type of information (such as pain). Likewise, many nerves that carry several types of information are associated with multiple nuclei. In particular, motor and sensory components of a nerve emerge from, or target, distinct nuclei. As a rule of thumb, motor nuclei lie on the ventral side of the brainstem, while sensory nuclei lie on the dorsal side. All of the nuclei except the trochlear nucleus are associated with nerves on the ipsilateral side of the body.

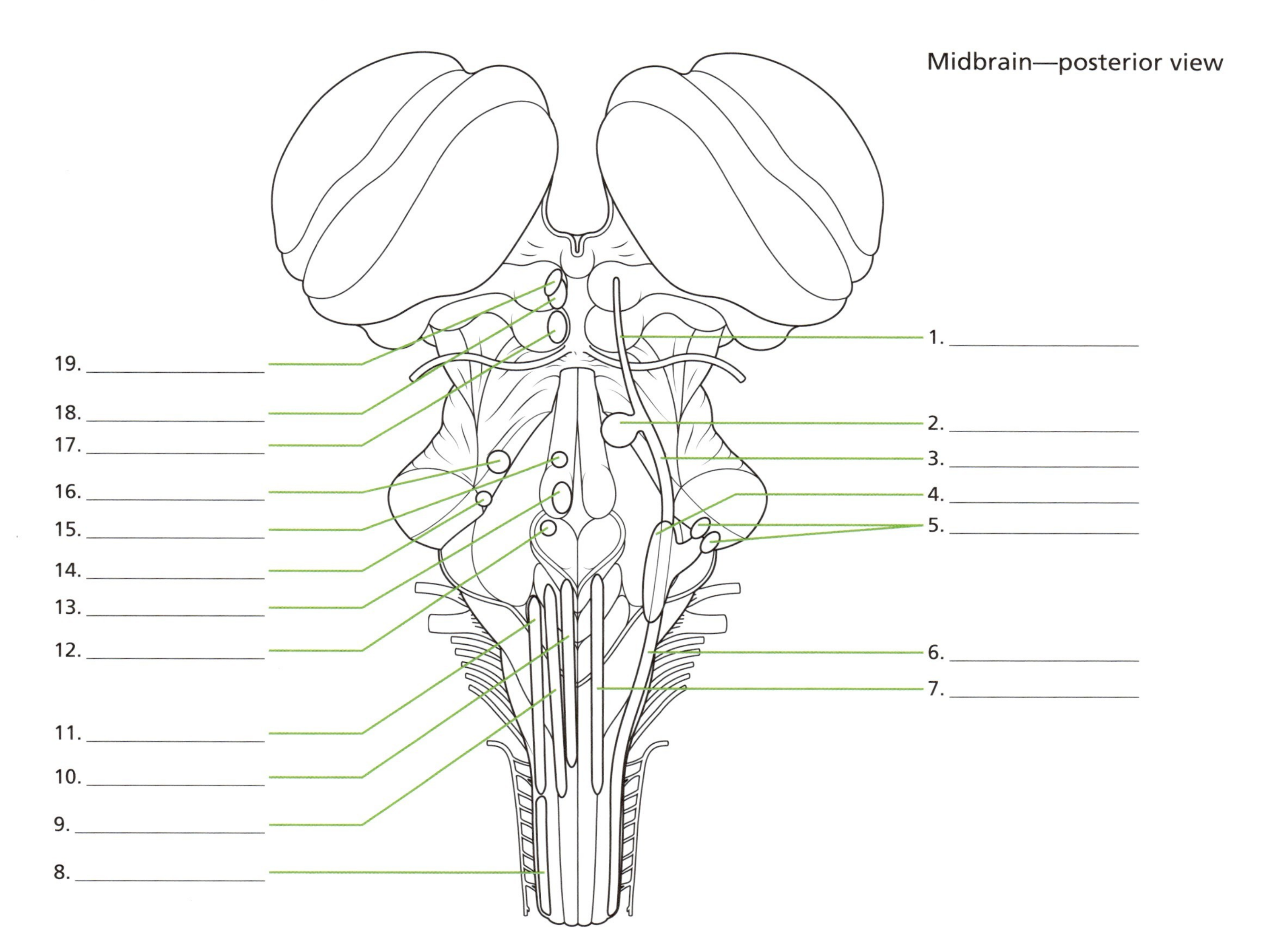

Answers

1. mesencephalic nucleus of trigeminal nerve (CN V), 2. chief sensory nucleus of trigeminal nerve (CN V), 3. spinal trigeminal nucleus (CN V, VII, IX, X), 4. vestibular nuclei (CN VIII), 5. dorsal and ventral cochlear nuclei (CN VIII), 6. nucleus solitarius, rostral portion (CN VII, IX, X), 7. nucleus solitarius, caudal portion (CN IX, X), 8. spinal accessory nucleus (CN XI), 9. dorsal motor nucleus of vagus nerve (CN X), 10. hypoglossal nucleus (CN XII), 11. nucleus ambiguus (CN IX, X), 12. inferior salivatory nucleus (CN IX), 13. abducens nucleus (CN VI), 14. facial nucleus (CN VII), 15. superior salivatory nucleus (CN VII), 16. trigeminal motor nucleus (CN V), 17. trochlear nucleus (CN IV), 18. oculomotor nucleus (CN III), 19. Edinger-Westphal nucleus (CN III)

Cranial Nerves and Skull Foramina

The cranial nerves (CN) enter and exit the skull through tunnels in the bone called foramen (plural: foramina). In some cases a single nerve traverses a foramen (as with the optic nerve through the optic canal), while in other cases several nerves do (as with the glossopharyngeal, vagus, and accessory nerves through the jugular foramen). Some nerves exit through multiple foramina (as with the trigeminal nerve and the rotundum, ovale, and spinosum foramina, as well as the superior and inferior orbital fissures). The olfactory nerve is a special case: It sends many nerve fibers through numerous tiny foramina in the cribriform plate of the ethmoid bone, collectively called the olfactory foramina.

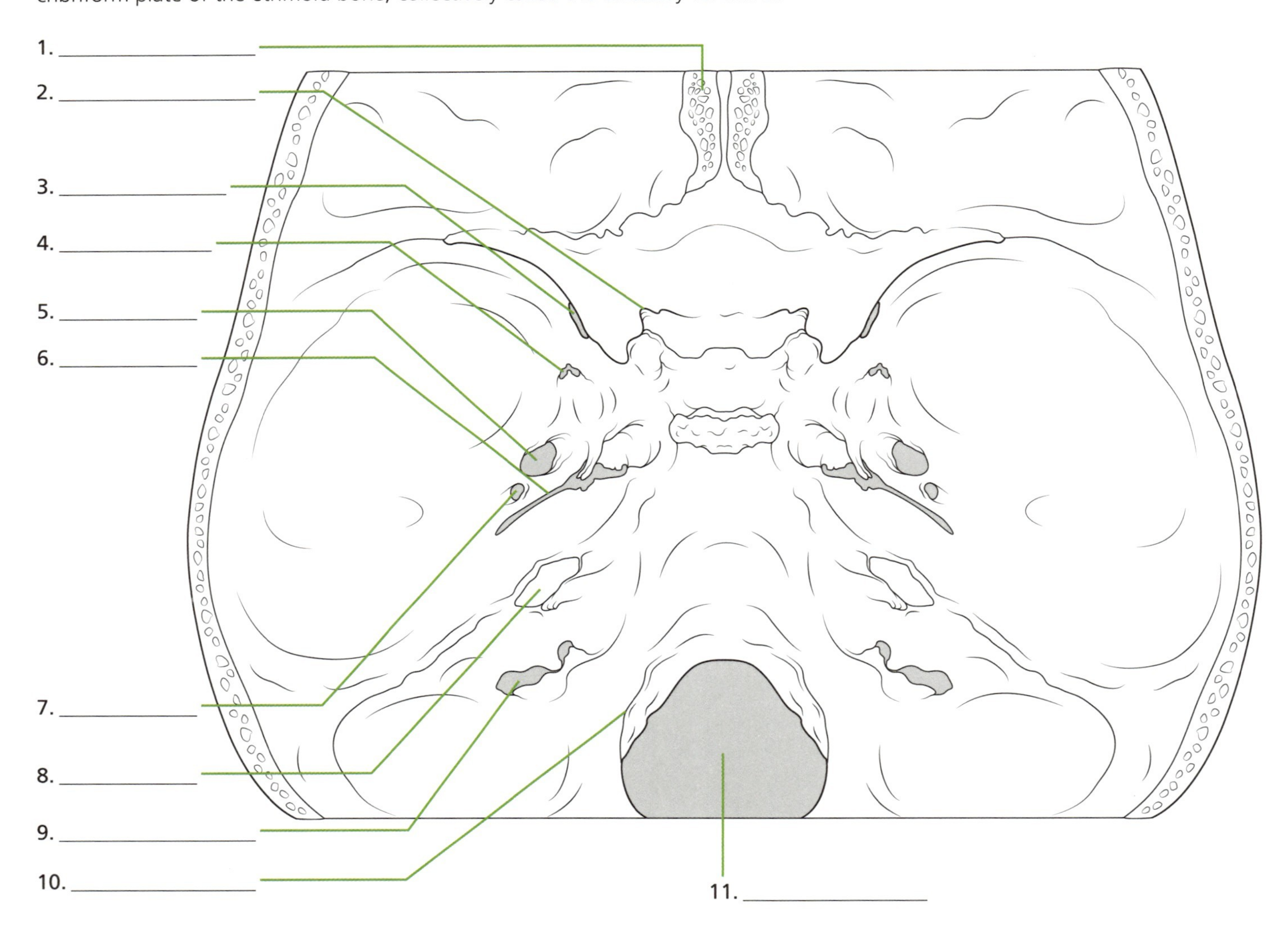

Interior of Skull—inferior surface

Answers

1. olfactory foramina (CN I), 2. optic canal (CN II), 3. superior orbital fissure (CN III, IV, V, VI), 4. foramen rotundum (CN V), 5. foramen ovale (CN V), 6. foramen lacerum (CN VII), 7. foramen spinosum (CN V), 8. internal acoustic meatus (CN VII, VIII), 9. jugular foramen (CN IX, X, XI), 10. hypoglossal canal (CN XII), 11. foramen magnum (CN XI)

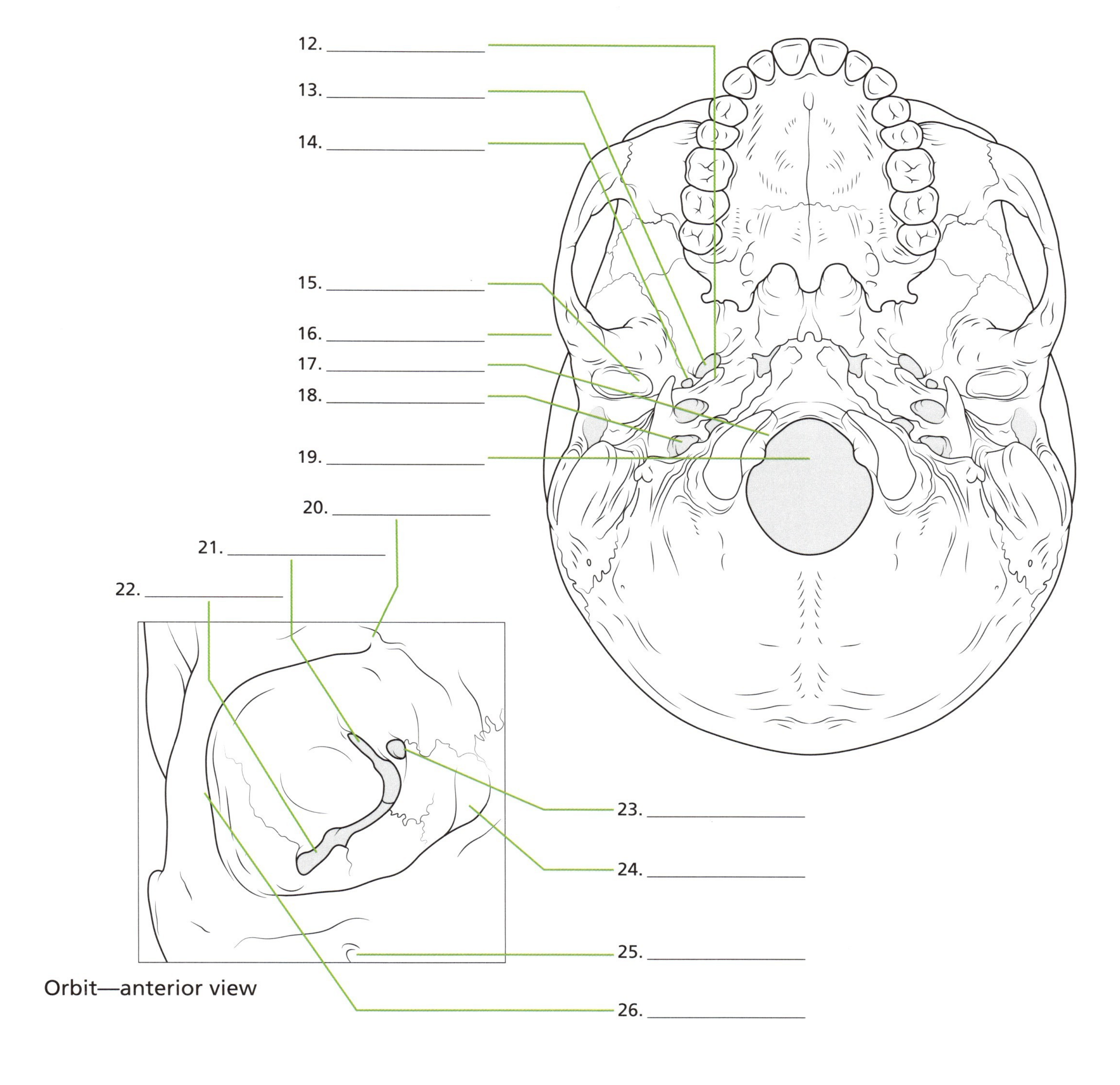

Answers

12. foramen lacerum (CN VII), 13. foramen ovale (CN V), 14. foramen spinosum (CN V), 15. external acoustic meatus (CN VII, VIII), 16. stylomastoid foramen (CN VII), 17. hypoglossal canal (CN XII), 18. jugular foramen (CN IX, X, XI), 19. foramen magnum (CN XI), 20. supraorbital notch, 21. superior orbital fissure, 22. inferior orbital fissure, 23. optic canal, 24. lacrimal fossa, 25. infraorbital foramen, 26. Whitnall's tubercle

Structure of Olfactory Epithelium

The olfactory nerve (CN I) carries all of the sensory information for the sense of smell. The process of detecting odorants begins when inhaled molecules are captured in the mucus (produced by Bowman's gland) coating the olfactory mucosa. This dissolves the odorants, which then bind to receptors on olfactory receptor neurons (ORNs), also called olfactory sensory neurons (OSNs). The axons of the ORNs form the olfactory nerve. The olfactory mucosa is also notable for containing stem cells that constantly give rise to new ORNs, making the olfactory nerve capable of regeneration. Close examination shows that the olfactory nerve is actually many small bundles of axons rather than one large bundle. These enter through the skull by way of the many small olfactory foramina in the cribriform plate of the ethmoid bone. The axons of the ORNs terminate in a specialized structure called the olfactory bulb. Each ORN contains receptors for a single odorant molecule, and all of the ORNs expressing the same type of receptor target anatomically distinct regions of the olfactory bulb called glomeruli (singular: glomerulus). Neural circuits in each glomerulus process the odorant signal and then send that information to higher-order olfactory centers of the brain via the olfactory tract.

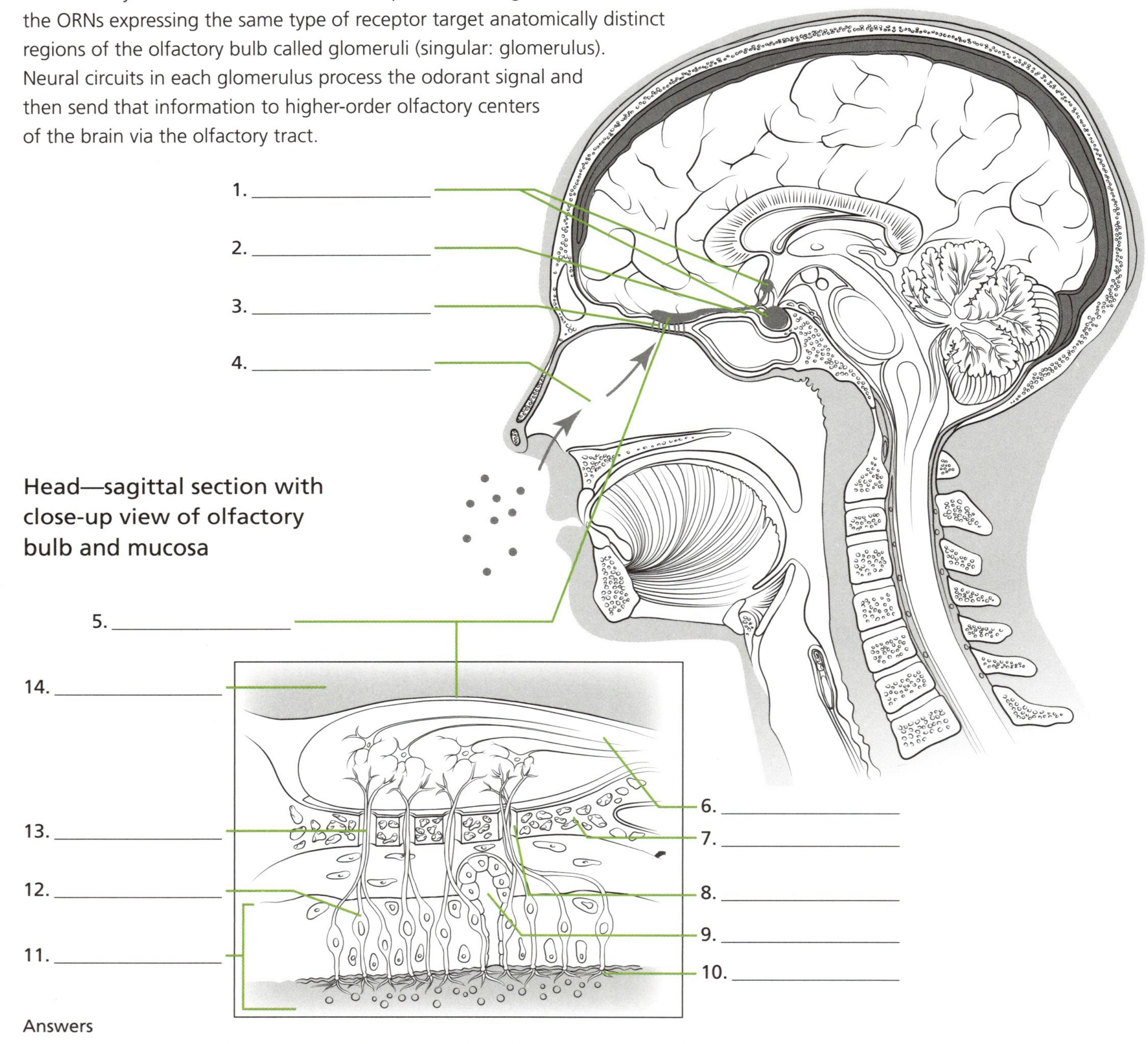

Head—sagittal section with close-up view of olfactory bulb and mucosa

Answers

1. olfactory centers in the brain, 2. sphenoid bone, 3. olfactory nerves, 4. nasal cavity, 5. olfactory bulb, 6. olfactory tract, 7. cribriform plate, 8. olfactory foramina, 9. Bowman's gland (olfactory gland), 10. cilia of olfactory sensory neurons, 11. olfactory mucosa, 12. olfactory receptor neuron, 13. olfactory nerve, 14. frontal lobe of the brain

Olfactory Nerve—Central Olfactory Pathways and Anosmia

After the initial processing of olfactory information in the glomeruli of the olfactory bulb, the resulting signal is transmitted down the olfactory tract. The olfactory tract is composed of bundles of axons from neurons in the glomeruli and runs posterior from the olfactory bulb along the inferior aspect of the frontal lobe until it splits into lateral- and medial-running branches. The targets of the lateral olfactory tract include the piriform cortex (traditionally considered the principal olfactory area of cortex), entorhinal cortex (associated with memory formation, in this case of smells), olfactory tubercle (a multisensory integration center associated with reward-driven behavior and attention), and medial amygdala (which processes pheromones and the emotional associations of smells). The axons that comprise the medial olfactory tract target the contralateral olfactory centers. It is important to note that the sense of smell does not directly synapse in the thalamus for integration. This is another reason why the sense of smell has strong associations with memories. Damage to one olfactory bulb or tract can cause a permanent, ipsilateral loss of the sense of smell (anosmia). Complete (bilateral) anosmia can be caused temporarily by inflammation of the nasal mucosa (such as by a cold or flu) or permanently as a side effect of certain drugs or infections.

Anterolateral Aspect of Brain—inferior view

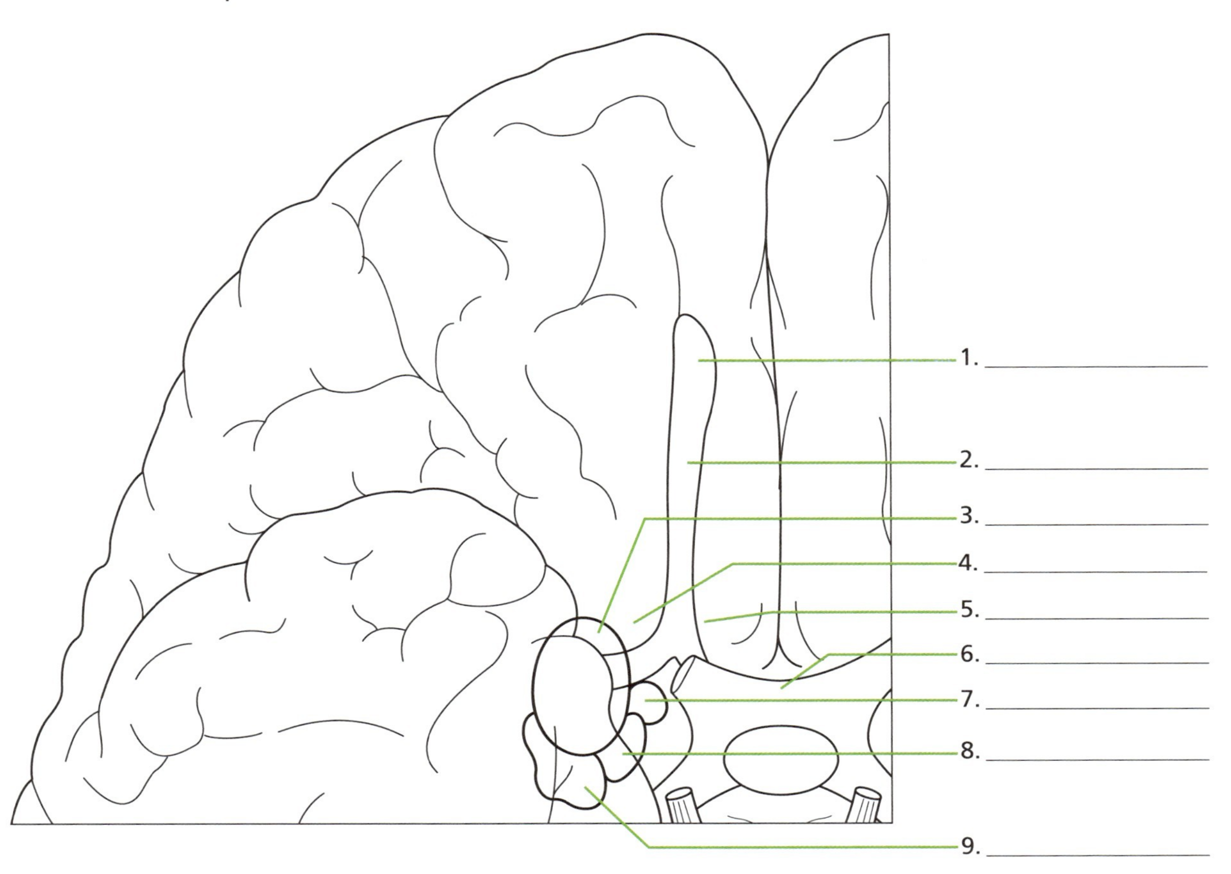

Answers

1. olfactory bulb, 2. olfactory tract, 3. amygdala, 4. lateral olfactory tract, 5. medial olfactory tract (to contralateral olfactory centers), 6. optic chiasm, 7. olfactory tubercle, 8. piriform cortex, 9. entorhinal cortex

Optic Nerve—Topography and Layered Structure of the Retina

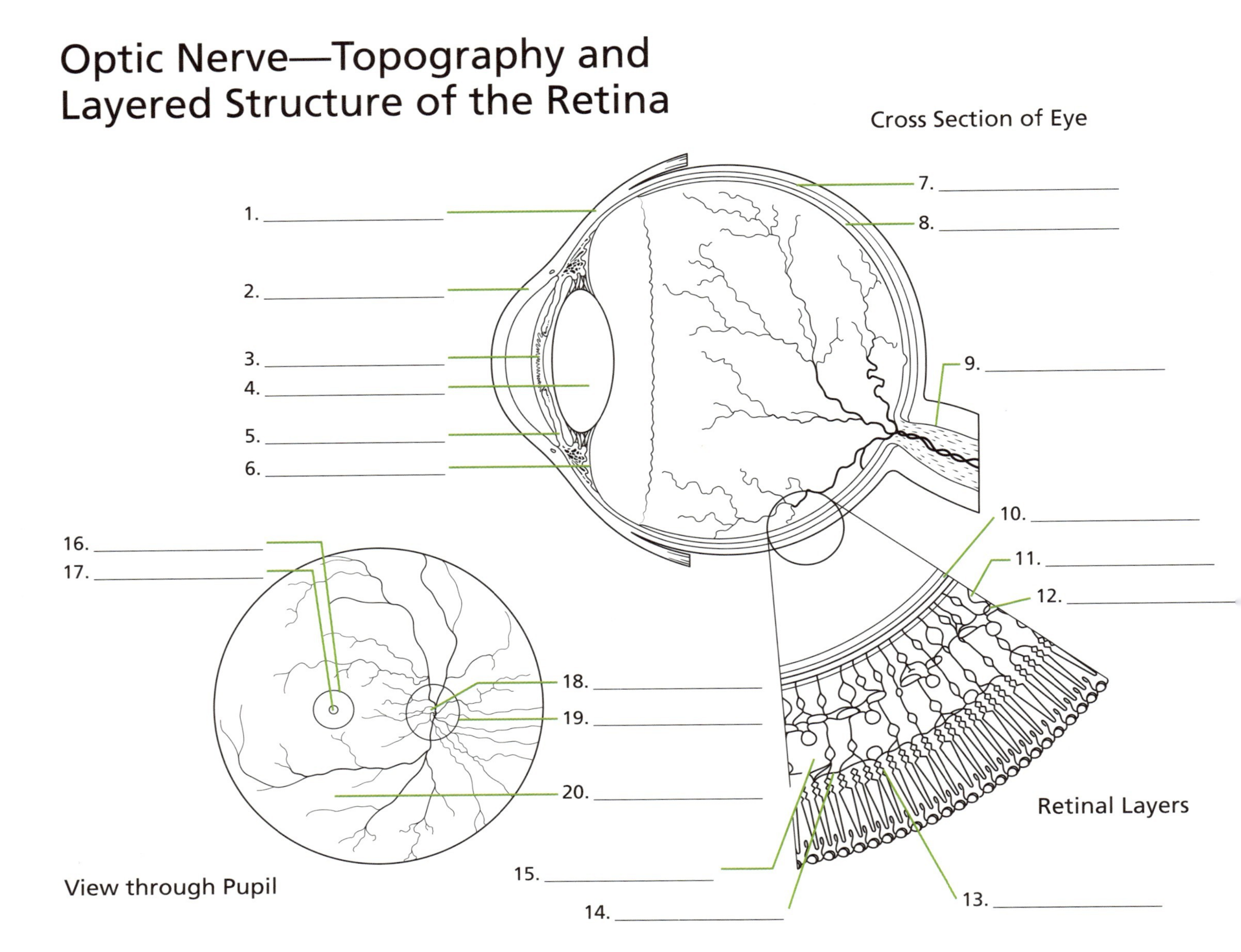

The optic nerve (CN II) carries all visual sensory information from the eye to the brain. Light enters the eye by passing through the cornea and pupil, and it is focused onto the retina by the lens. The retina is a multilayered sheet of neurons that captures photons with specialized photoreceptor neurons that reside in the outer nuclear layer. The photoreceptors convert the photons into electrical signals and transmit them to interneurons in the inner nuclear layer (via synapses in the outer plexiform layer). These interneurons begin to process the visual signal into multiple parallel streams containing subsets of information about the visual world, and they transmit that information to retinal ganglion cells (via synapses in the inner plexiform layer). The retinal ganglion cells finish encoding the parallel streams and send the information to the brain via their axons, which bundle together in the nerve fiber layer. These bundles course toward the optic disc, where they exit the eye and form the optic nerve. The optic nerve then enters the skull through the optic canal. When the retina is viewed through the pupil, a number of retinal landmarks can be seen. Foremost is the macula lutea (a yellowish circle), and at its center is the fovea centralis, which corresponds to the area of retina with the highest visual acuity.

Answers

1. sclera, 2. cornea, 3. pupil, 4. lens, 5. iris, 6. ciliary body, 7. choroid, 8. retina, 9. optic nerve, 10. nerve fiber layer, 11. ganglion cell layer, 12. inner plexiform layer, 13. outer nuclear layer, 14. outer plexiform layer, 15. inner nuclear layer, 16. macula lutea, 17. fovea centralis, 18. optic nerve head, 19. optic disc, 20. peripheral retina

Optic Nerve—Central Visual Pathways and Effects of Lesions

Visual field describes the portion of the world that reflects light onto a retina. The left portion of each retina views the right aspect of its visual field, and vice versa. The ipsilateral visual field of each eye is sent to the contralateral side of the brain. Axons from each retina carrying ipsilateral visual field information must cross over to the other hemisphere; this happens at a juncture called the optic chiasm. Posterior to the optic chiasm, retinal ganglion cell axons terminate in the lateral geniculate nucleus of the thalamus. Axons from thalamic neurons carry visual signals onward to the visual cortex via a tract called the optic radiation. Complex patterns of visual field loss result from lesions at different points in the visual pathway. An optic nerve lesion would result in total loss of visual field contributed by the source eye. Lesions that occur in the optic chiasm ablate only contralaterally projecting axons, resulting in loss of the temporal visual field from each eye (bitemporal hemianopia). A lesion in the optic tract causes deficits in the contralateral visual field of each eye (homonymous hemianopia). Damage to the optic radiation or in visual cortex causes more specific visual field losses, depending on the exact location.

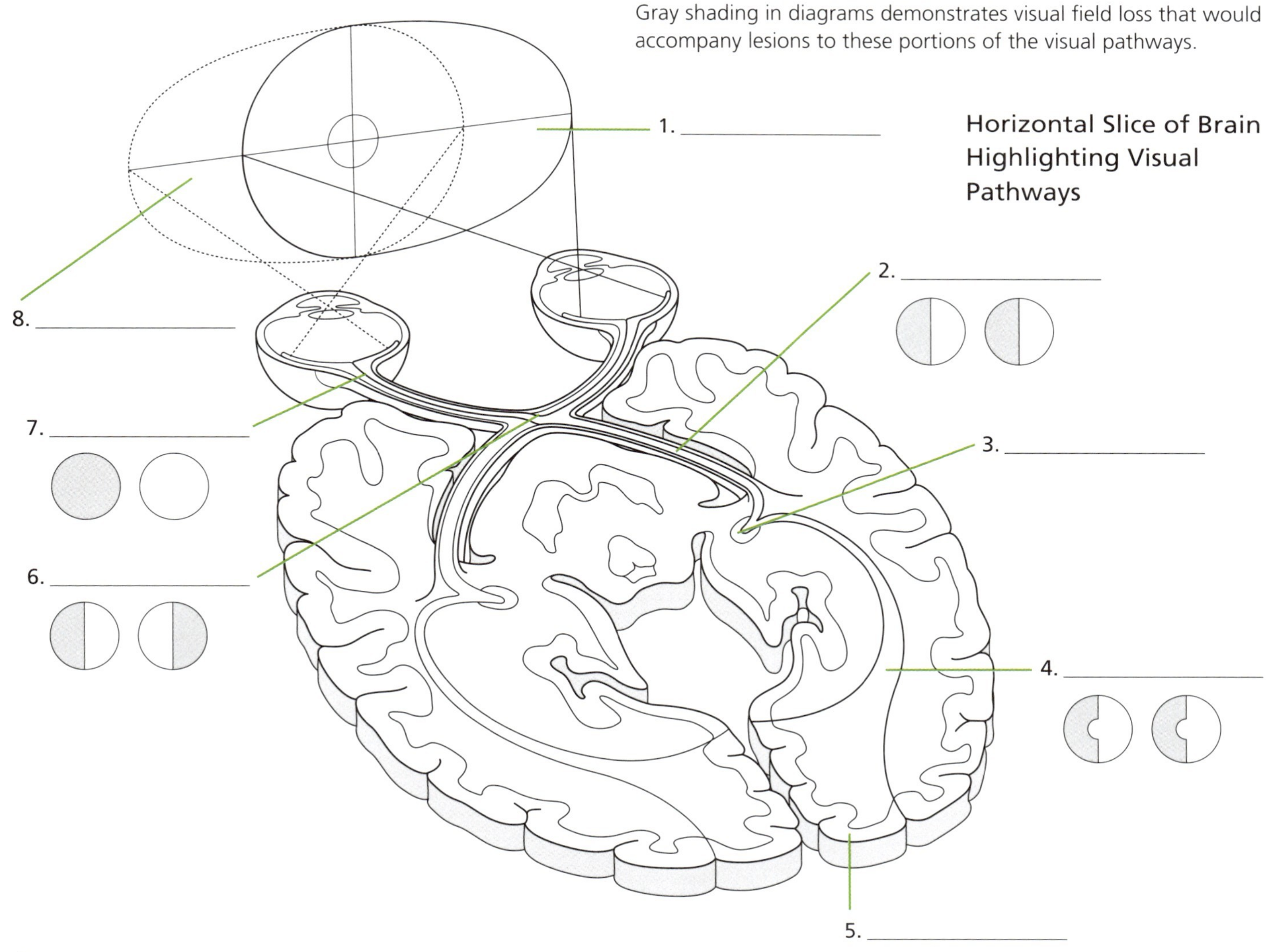

Gray shading in diagrams demonstrates visual field loss that would accompany lesions to these portions of the visual pathways.

Horizontal Slice of Brain Highlighting Visual Pathways

Answers

1. right visual field, 2. optic tract, 3. lateral geniculate nucleus of the thalamus, 4. optic radiation, 5. primary visual cortex, 6. optic chiasm, 7. optic nerve, 8. left visual field

Nerves to Eye Muscle Nuclei—Course of Nerves

Three nerves are dedicated to controlling movements of the eyes: the oculomotor, the trochlear, and the abducens (CN III, IV, and VI, respectively). The oculomotor and trochlear nerves emerge from the midbrain, while the abducens nerve emerges from the pons. All three nerves exit the skull and enter the orbit of the eye through the superior orbital fissure. The oculomotor nerve innervates most of the muscles responsible for eye movements: the superior rectus, inferior rectus, medial rectus, and inferior oblique. It also has a parasympathetic branch that forms the ciliary ganglion. Neurons from this ganglion innervate the pupillary sphincter (which constricts the pupil) and the ciliary muscle (which changes the shape of the lens to focus light on the retina). Finally, the oculomotor nerve also innervates the levator palpebrae superioris, which lifts the superior eyelid. The trochlear and abducens nerves each control a single muscle for eye movements: the superior oblique and the lateral rectus, respectively. The trochlear nerve is the only cranial nerve that innervates a muscle on the contralateral side of the body. It is so named because the tendon that connects the superior oblique to the eye passes through a cartilaginous loop attached to the orbit, called the trochlea.

Orbit and Brainstem—sagittal section

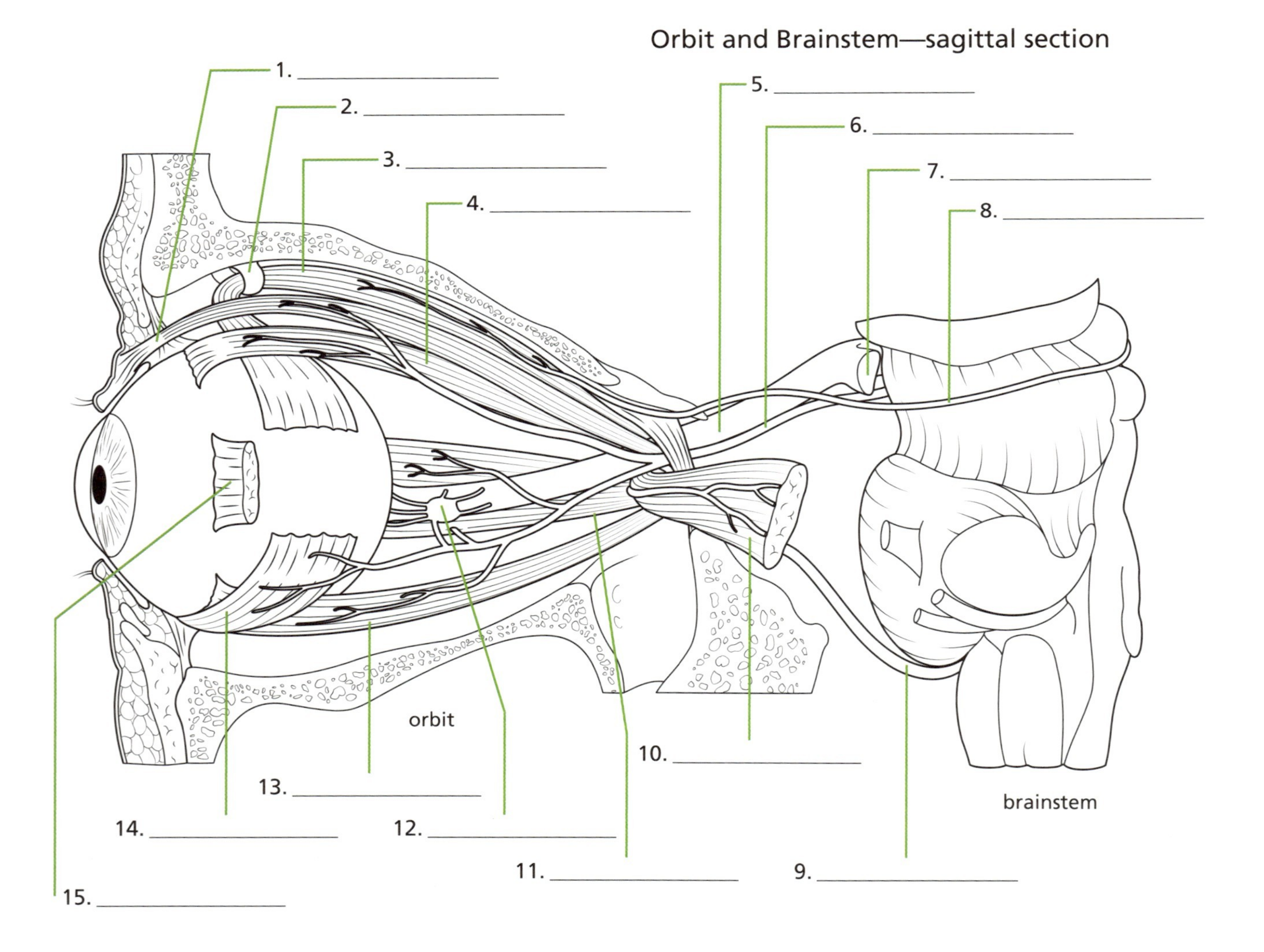

Answers

1. levator palpebrae superioris, 2. trochlea, 3. superior oblique, 4. superior rectus, 5. optic nerve, 6. oculomotor nerve (CN III), 7. optic chiasm, 8. trochlear nerve (CN IV), 9. abducens nerve (CN VI), 10. lateral rectus (cut), 11. medial rectus, 12. ciliary ganglion, 13. inferior rectus, 14. inferior oblique, 15. lateral rectus (cut)

Nerves to Eye Muscle Nuclei—Eye Movements and Effects of Lesions

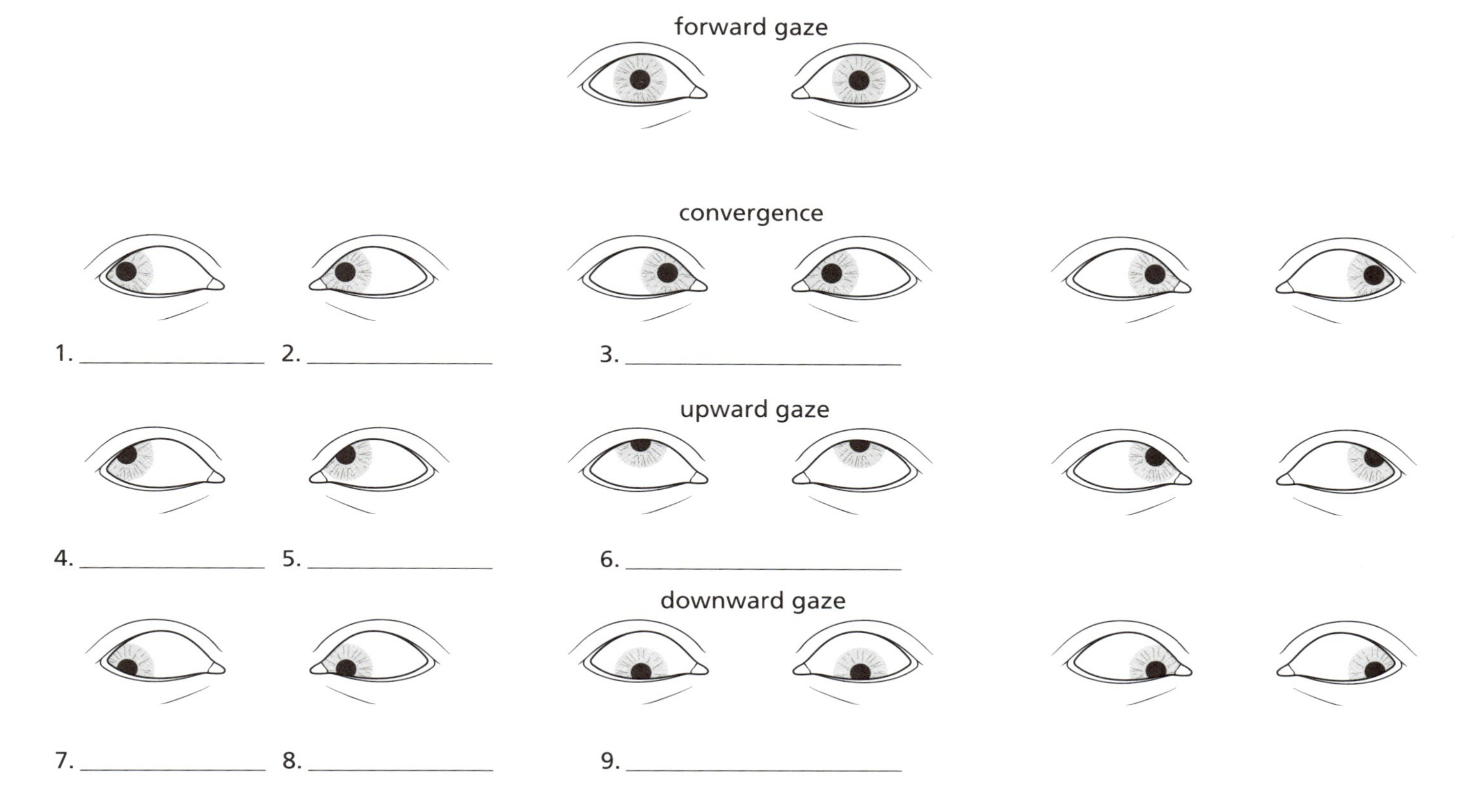

Label the muscle used to perform each eye movement and the corresponding cranial nerve that innervates that muscle.

The extraocular muscles work in a coordinated fashion to direct a person's gaze, either actively or reflexively. The four rectus muscles are responsible for moving the eyes in the four cardinal directions—up, down, left, and right—while the oblique muscles rotate the eyes and also contribute to up and down movements. Assessment of the function of cranial nerves III, IV, and VI is done with the H-test—asking a subject's eyes to follow an imaginary "H" that is drawn in the air. To gaze to the right, the lateral rectus of the right eye and the medial rectus of the left eye contract. Since these muscles are innervated by the abducens and oculomotor nerves, respectively, abnormalities in the function of one nerve can be identified here. Likewise, gazing down and to the right requires contraction of the inferior rectus (controlled by the oculomotor nerve) of the right eye and the superior oblique (controlled by the trochlear nerve) of the left eye, again allowing for abnormalities in the functioning of one nerve to be identified. Completing the H-test allows an examiner to identify the general localization of potential lesions responsible for any abnormalities that are discovered.

Answers

1. lateral rectus (CN VI), 2. medial rectus (CN III), 3. medial rectus (CN III), 4. superior rectus (CN III), 5. inferior oblique (CN III), 6. superior rectus (CN III) and inferior oblique (CN III), 7. inferior rectus (CN III), 8. superior oblique (CN IV), 9. inferior rectus (CN III) and superior oblique (CN IV)

Trigeminal Nerve—Sensory Distribution

The trigeminal nerve (CN V) carries sensory signals from the face and motor signals to the muscles of mastication. Prior to exiting the skull, the trigeminal nerve enlarges at the trigeminal ganglion, which contains the somas of the trigeminal nerve's sensory neurons. From the anterior aspect of the trigeminal ganglion sprout the three branches of the trigeminal nerve: the ophthalmic, maxillary, and mandibular nerves. The ophthalmic nerve (V_1) exits the skull through the superior orbital fissure and carries somatosensory signals from the nose, cornea, upper eyelids, forehead, frontal portion of the scalp, and frontal sinus. The maxillary nerve (V_2) exits the skull through the foramen rotundum and carries somatosensory signals from the palate; upper teeth and gums; upper lip; nostrils; cheeks; lower eyelids; and the ethmoid, maxillary, and sphenoid sinuses. The mandibular nerve (V_3) exits the skull through the foramen ovale and carries somatosensory signals from the chin, jaw, lower teeth and gums, lower lip, and temporal portions of the scalp. The three nerves that branch from the trigeminal nerve continue to branch further themselves. Several of these pass through foramina in the facial bones on the way to their targets: from V_1 the supraorbital nerve (supraorbital foramen), from V_2 the infraorbital nerve (infraorbital foramen), and from V_3 the mental nerve (mental foramen).

Head—sagittal section

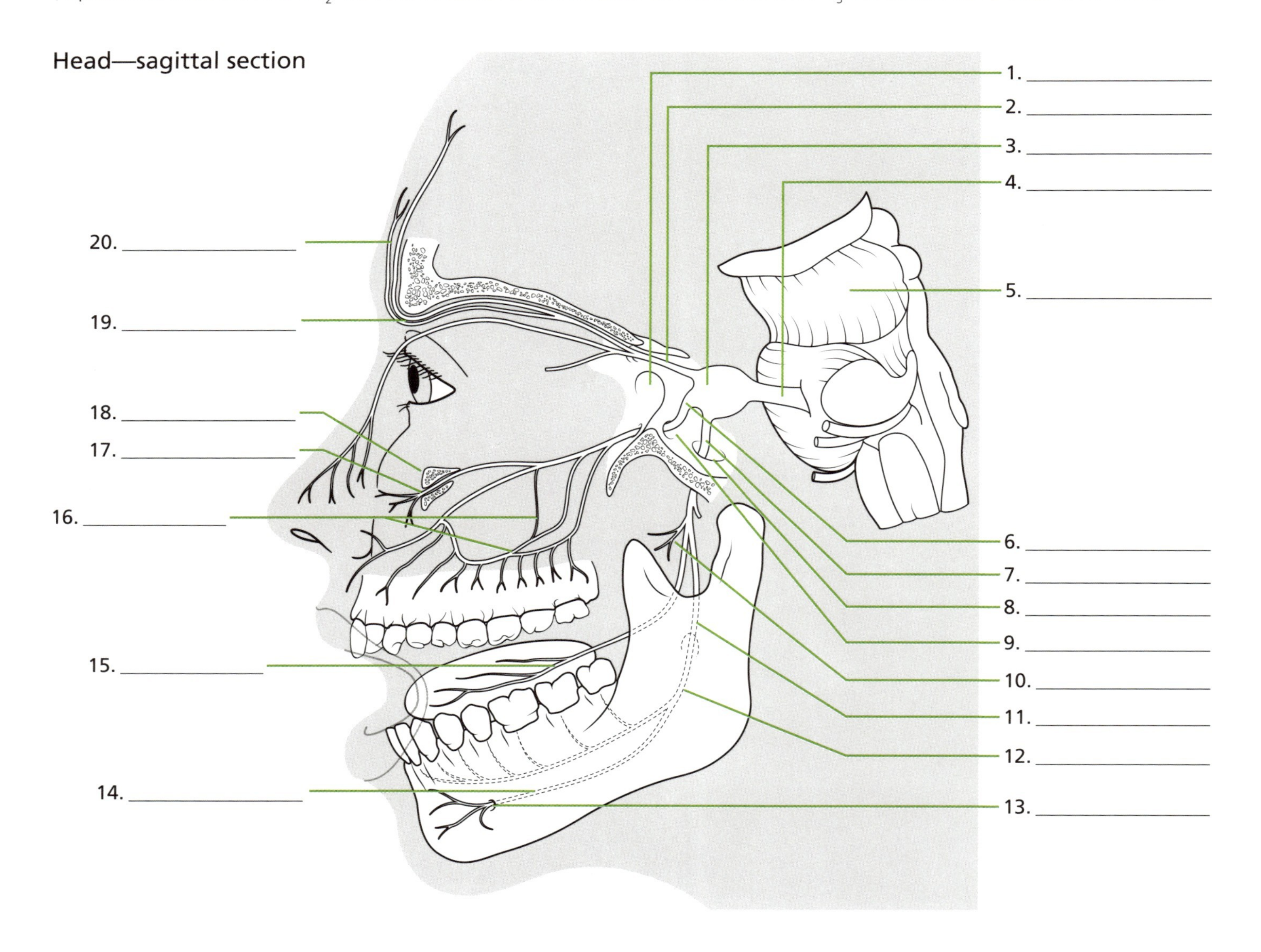

Answers

1. superior orbital fissure, 2. ophthalmic nerve (V_1), 3. trigeminal ganglion, 4. trigeminal nerve (CN V), 5. pons, 6. maxillary nerve (V_2), 7. mandibular nerve (V_3), 8. foramen ovale, 9. foramen rotundum, 10. anterior branch to muscles of mastication, 11. mandibular foramen, 12. inferior alveolar nerve, 13. mental foramen, 14. mental nerve, 15. lingual nerve, 16. superior alveolar nerves, 17. infraorbital nerve, 18. infraorbital foramen, 19. supraorbital foramen, 20. supraorbital nerve

Trigeminal Nerve—Motor Distribution, Muscles of Mastication, and Corneal Reflex

The trigeminal nerve (CN V) innervates a number of muscles through its mandibular branch (V_3), including the tensor tympani in the ear, which reduces the noise produced from chewing; the tensor veli palatini in the soft palate, which prevents food from entering the airway when swallowing; and the mylohyoid muscle and the anterior belly of the digastric muscle, which elevate the hyoid bone (used in part for speaking and swallowing). The mandibular branch also innervates the muscles of mastication: the masseter, lateral and medial pterygoid muscles, and the temporalis. These four muscles anchor the mandible to the skull and are responsible for the complex range of motion produced as part of biting and chewing.

The trigeminal nerve also plays a role in the corneal (blink) reflex, which can be a useful component of the neurological exam. When the cornea is stimulated (as with a cotton swab), sensory fibers from the ophthalmic branch of the trigeminal nerve relay that information to the chief sensory trigeminal nucleus. Interneurons relay that information bilaterally to the facial motor nucleus, activating motor efferents that synapse onto the orbicularis oculi muscles and causing both eyelids to blink.

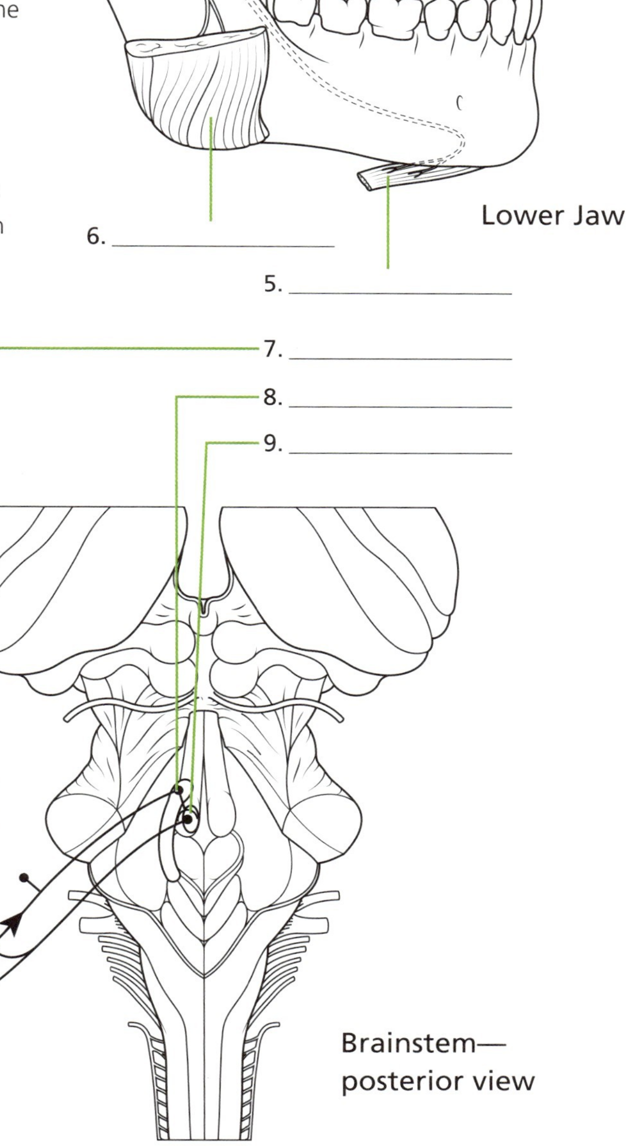

Answers

1. anterior trunk of the mandibular nerve, 2. temporalis muscle, 3. lateral pterygoid muscle, 4. medial pterygoid muscle, 5. anterior belly of the digastric muscle, 6. masseter muscle, 7. orbicularis oculi, 8. spinal trigeminal nucleus, 9. facial motor nucleus, 10. ophthalmic nerve, 11. cornea

Facial Nerve—Motor Distribution and Muscles of Facial Expression

The facial nerve (CN VII) carries motor, sensory, and parasympathetic signals. It emerges from the pons and exits the skull by first entering the internal acoustic meatus, then the facial canal (which branches off from the meatus), and exiting from the stylomastoid foramen. The facial nerve branches extensively, with five branches arrayed like the fingers of a hand sending motor efferents anterior to the facial muscles. From superior to inferior, they are (and innervate these muscles): the temporal branch (frontalis, orbicularis oris), the zygomatic branch (orbicularis oris, zygomaticus), the buccal branch (levator labii superioris, buccinator, orbicularis oris, risorius), the mandibular branch (mentalis, depressor labii inferioris, depressor anguli oris), and the cervical branch (platysma). The posterior auricular nerve also branches from the facial nerve but sends motor efferents posterior to the auricularis posterior and the occipitalis. The facial nerve also carries taste information from the anterior two-thirds of the tongue to the brain.

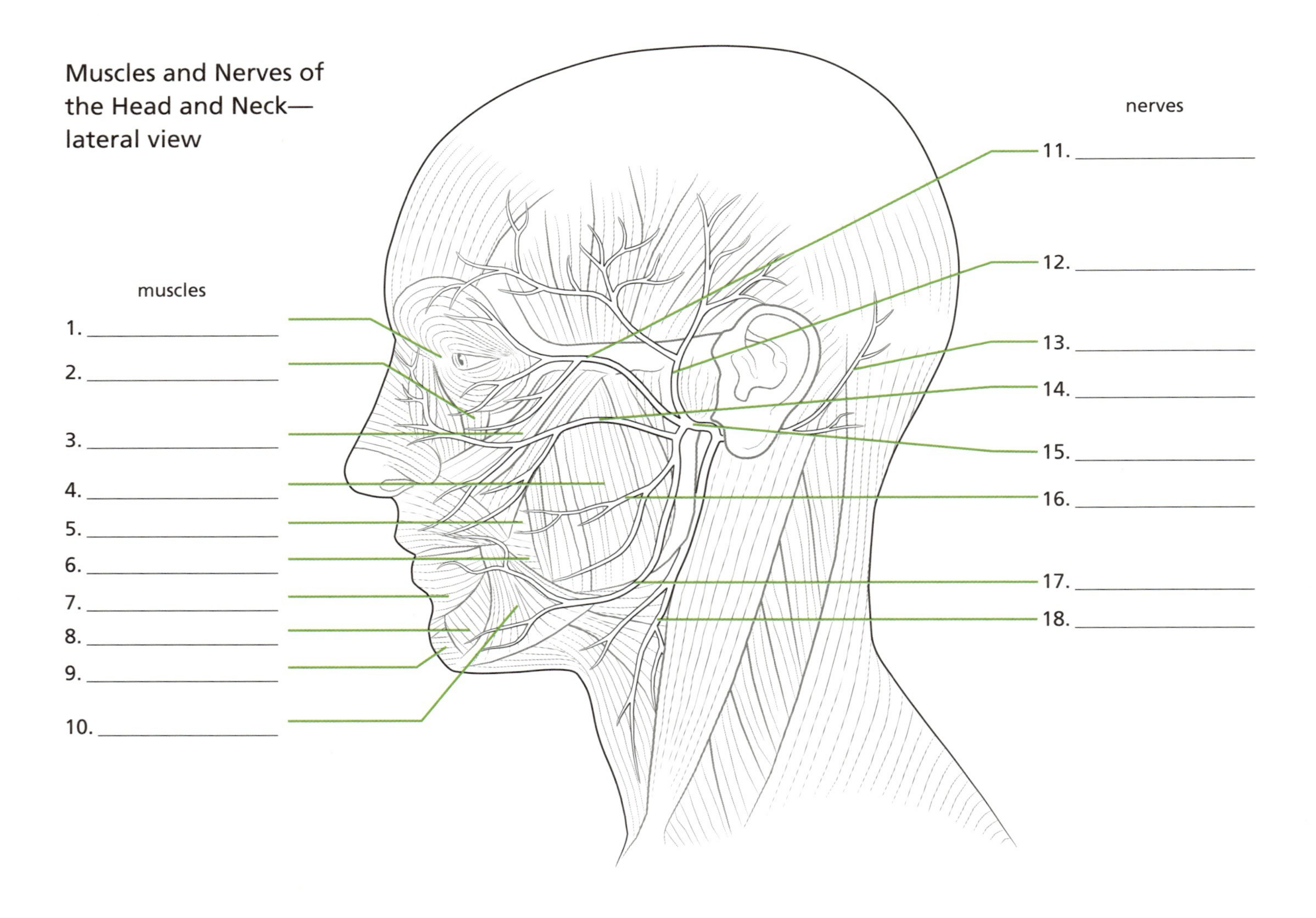

Answers

1. orbicularis oculi, 2. levator labii superioris, 3. zygomaticus, 4. masseter, 5. buccinator, 6. risorius, 7. orbicularis oris, 8. depressor labii inferioris, 9. mentalis, 10. depressor anguli oris, 11. zygomatic branch of facial nerve, 12. temporal branch of facial nerve, 13. posterior auricular nerve, 14. upper buccal branch of facial nerve, 15. trunk of facial nerve, 16. lower buccal branch of facial nerve, 17. mandibular branch of facial nerve, 18. cervical branch of facial nerve

Facial Nerve—Control of Salivary Glands and Bell's Palsy

The facial nerve provides parasympathetic innervation to the sublingual and submandibular glands (which produce saliva), the lacrimal gland (which produces tears), and the palatine gland (which produces mucus) and sinuses of the skull. Stimulation of the facial nerve increases production from these glands and membranes.

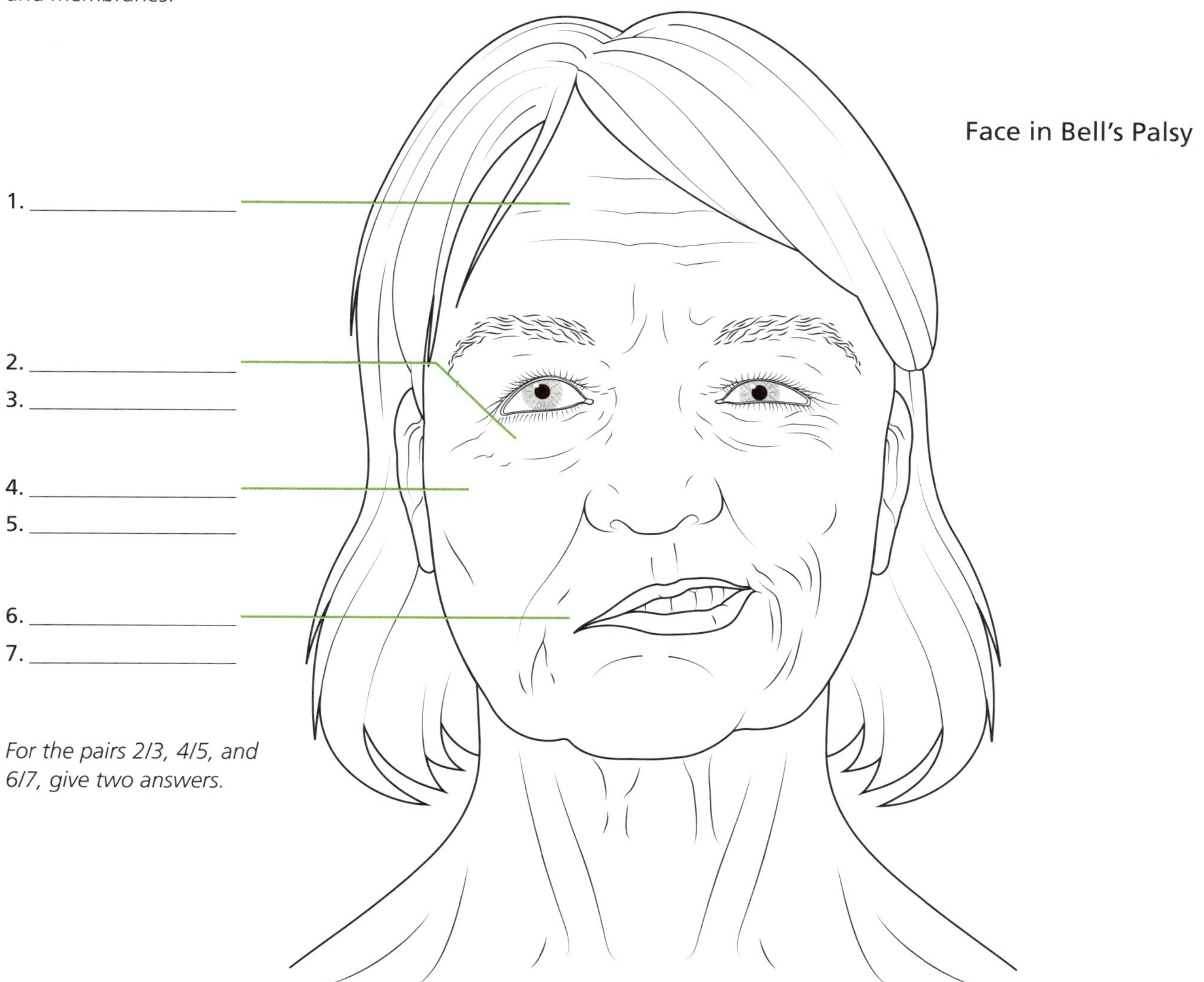

Damage to the facial nerve or its motor nucleus can cause facial nerve paralysis, the most common form of which is called Bell's palsy. This sudden-onset condition can arise at any age, but most commonly affects mature adults, and is characterized by a droop on one half of the face, which causes an inability to smile, wrinkle the brow, or move cheek muscles on that side only. The lower eyelid also droops and the eye cannot be closed; thus, care must be taken to prevent damage to the eye due to dryness. Most individuals begin to recover from this palsy over several months. The cause of Bell's palsy is unknown, although improvement associated with steroid treatment suggests that inflammation of the facial nerve is involved.

Answers

1. inability to wrinkle brow, 2. drooping eyelid, 3. inability to close eye, 4. inability to puff cheeks, 5. no cheek muscle tone, 6. drooping mouth, 7. inability to smile or pucker

Vestibulocochlear Nerve—Structure of the Inner Ear and Central Connections

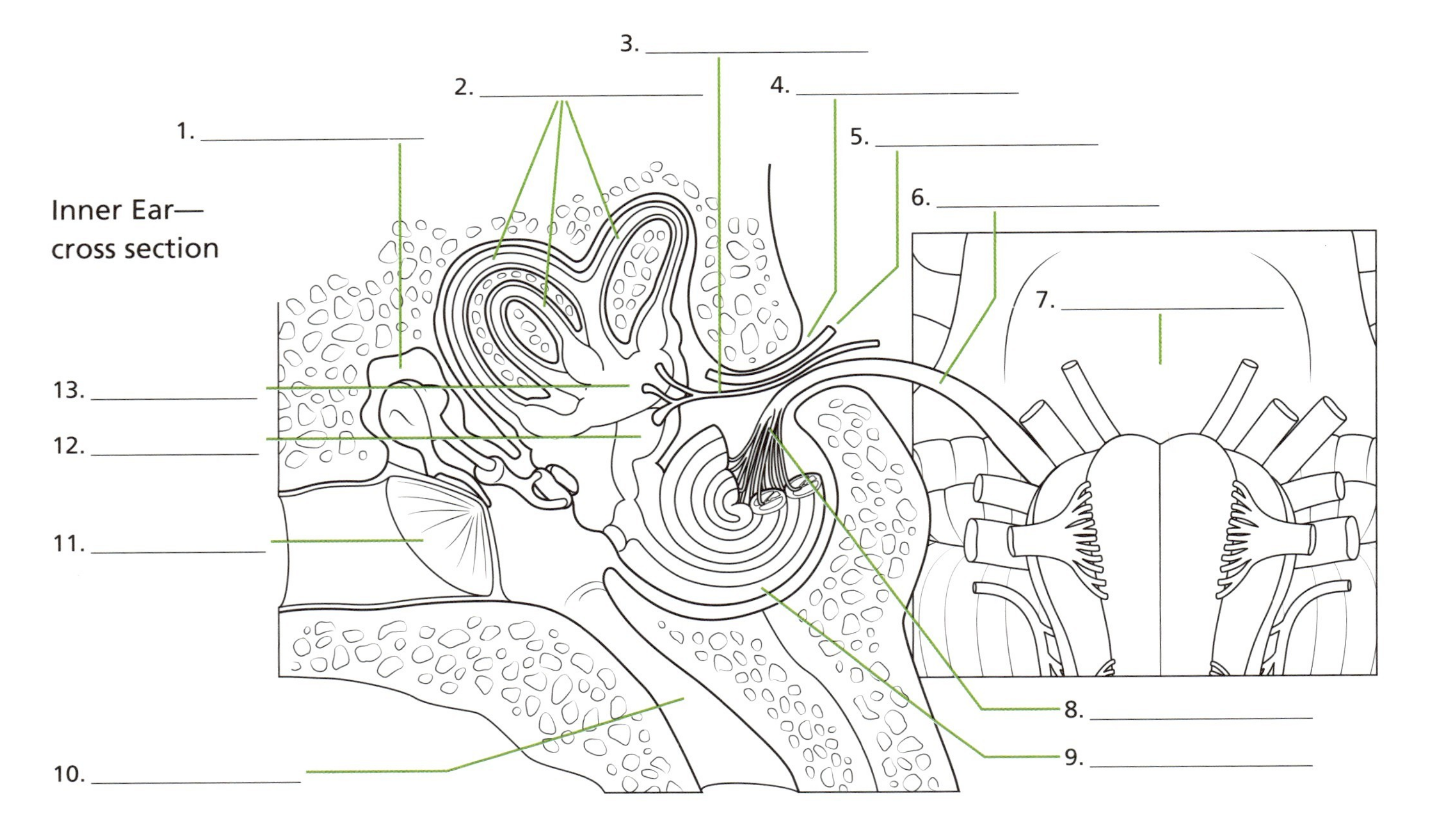

Inner Ear—cross section

The vestibulocochlear nerve (CN VIII) carries sensory signals about auditory and balance information that originate in the inner ear. It emerges from the pons and exits the skull through the internal acoustic meatus, after which it separates into the vestibular and cochlear branches. Both branches receive input from specialized sensory neurons called hair cells, which detect movement of the endolymph fluid that fills the chambers in which they reside. The vestibular branch gathers information on balance and movement in the semicircular canals, saccule, and utricle. The three semicircular canals detect rotational head movement in the pitch, roll, and yaw axes; the utricle detects linear acceleration in the horizontal plane; and the saccule detects linear acceleration in the vertical plane. The cochlear branch gathers auditory information from the cochlea, a complex spiral-shaped structure that detects progressively lower audible frequencies along its length. Neurons in the cochlear branch project to two cochlear nuclei, while neurons in the vestibular branch project to four vestibular nuclei. All of these nuclei are located in the brainstem at the junction of the pons and medulla. The vestibular branch also contains neurons that project to the inferior cerebellar peduncle, which connects to the vestibulocerebellum.

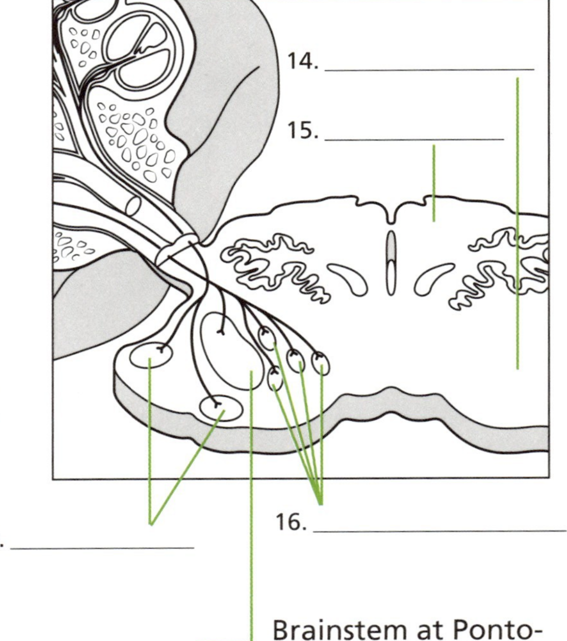

Brainstem at Ponto-medullary Junction—cross section

Answers

1. tympanic cavity, 2. semicircular canals, 3. vestibular branch of CN VIII, 4. internal acoustic meatus, 5. facial nerve (cut), 6. vestibulocochlear nerve (CN VIII), 7. pons, 8. cochlear branch of CN VIII, 9. cochlea, 10. auditory tube, 11. tympanic membrane, 12. saccule, 13. utricle, 14. brainstem, 15. medulla oblongata, 16. vestibular nuclei (medial, rostral, caudal, and lateral), 17. inferior cerebellar peduncle, 18. cochlear nuclei (dorsal and ventral)

Vestibulocochlear Nerve—Vestibulo-ocular Reflex

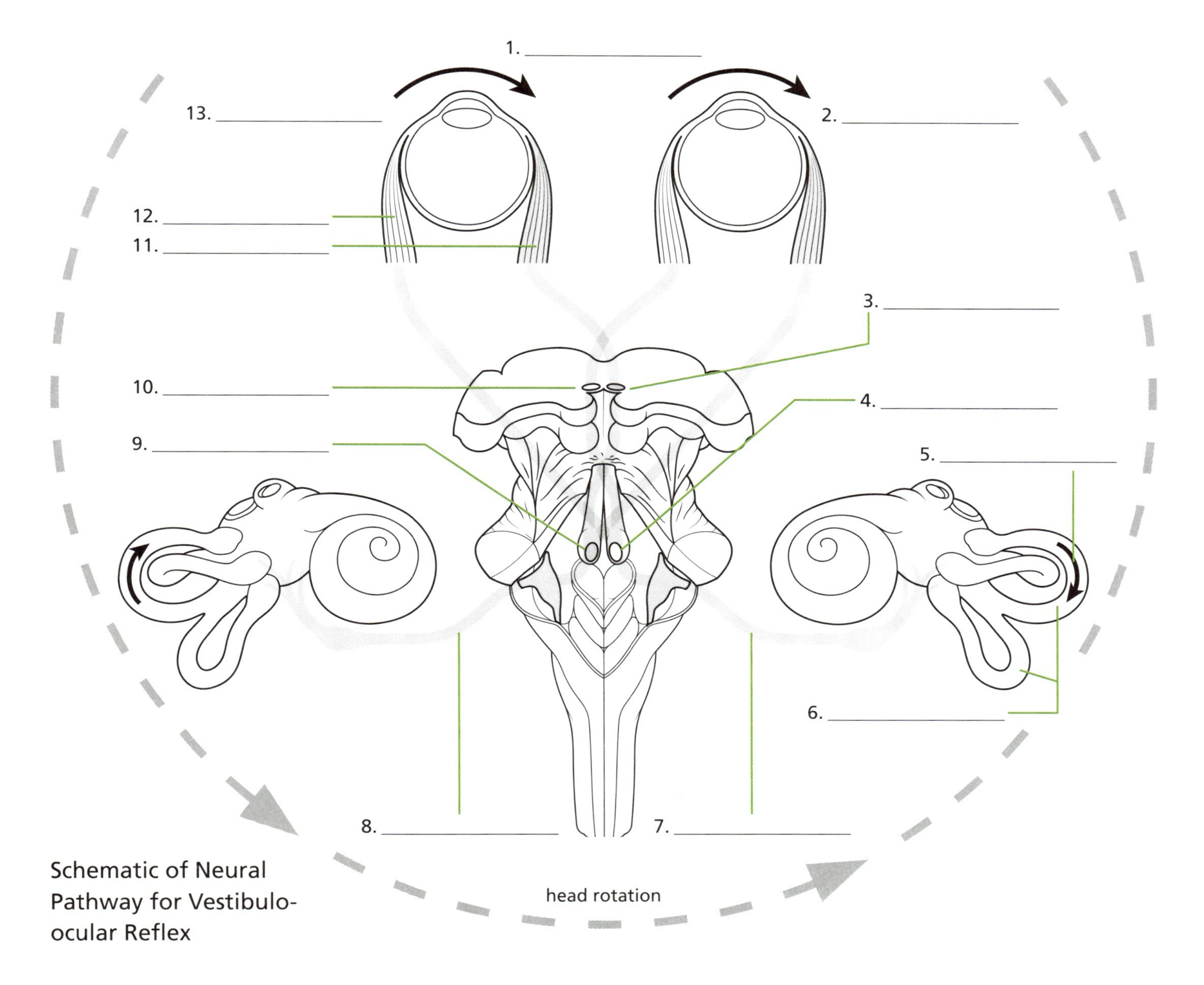

Schematic of Neural Pathway for Vestibulo-ocular Reflex

The vestibulo-ocular reflex (VOR) is an eye movement that occurs reflexively in response to a triggering head movement in the opposite direction. This allows the eyes to remain fixated on a target even as the position of the head is changed, and it stabilizes the image that falls on the retina. The VOR is initiated by movement of the head, which causes movement of endolymph in the semicircular canals. This fluid moves in opposite directions in the right and left canals, causing excitation of hair cells in one and inhibition in the other. The vestibular sensory neurons downstream of the hair cells then synapse onto interneurons in the vestibular nuclei. These project onto motor neurons and interneurons in the abducens nucleus. The abducens motor neurons control the lateral rectus muscles; the one downstream of the excited hair cells will contract, and the other will relax. Simultaneously, the abducens nuclei interneurons send signals to the contralateral oculomotor nucleus via the medial longitudinal fasciculus (two collections of bundled axonal fibers that cross in the brainstem). They synapse onto motor neurons in the oculomotor nucleus, which then contract or relax the medial rectus muscles. The VOR is an extremely fast reflex that keeps the visual image very stable as we look around.

Answers

1. compensatory eye movement, 2. inhibition of left medial rectus and right lateral rectus, 3. oculomotor nucleus, 4. abducens nucleus, 5. direction of endolymph movement, 6. semicircular canals, 7. inhibitory pathway, 8. excitatory pathway, 9. vestibular nucleus, 10. medial longitudinal fasciculus, 11. medial rectus, 12. lateral rectus, 13. excitation of left lateral rectus and right medial rectus

Glossopharyngeal Nerve

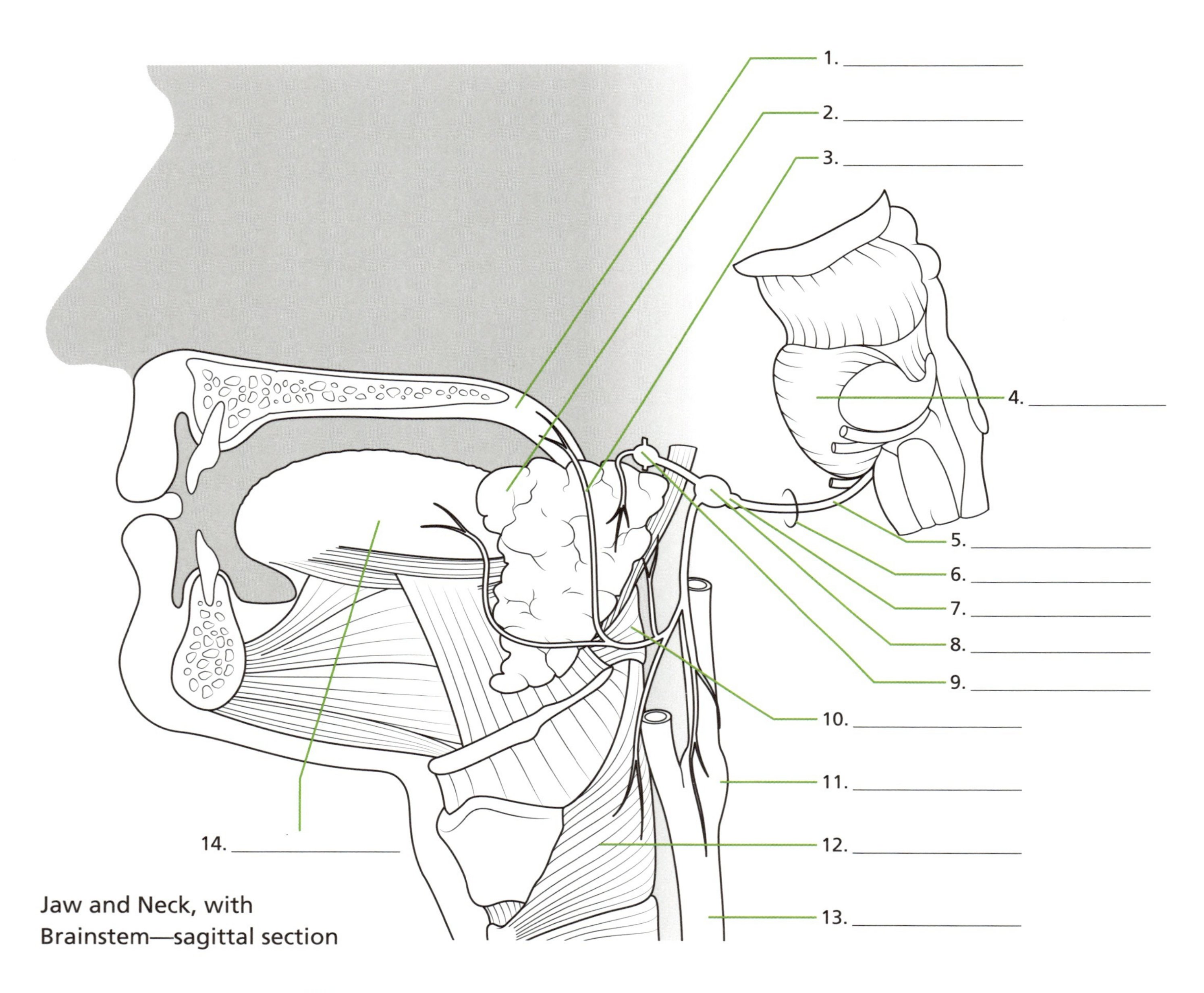

Jaw and Neck, with Brainstem—sagittal section

The glossopharyngeal nerve (CN IX) carries sensory, motor, and parasympathetic signals. It emerges from the medulla oblongata and exits the skull through the jugular foramen. There are several ganglionic enlargements of the nerve: the superior and inferior ganglia contain somatosensory and visceral sensory neurons, respectively, while the otic ganglion controls salivation via parasympathetic innervation of the parotid gland. The glossopharyngeal nerve provides motor innervation to the stylopharyngeus muscle (important for swallowing) and receives sensory information from the middle ear, palatine tonsil, and pharynx. It also carries sensory and taste signals from the posterior third of the tongue, as well as sensory information on blood oxygen levels and blood pressure from the carotid body and carotid sinus, respectively.

Answers

1. palatine tonsil, 2. parotid gland, 3. parasympathetic fibers, 4. pons, 5. glossopharyngeal nerve (CN IX), 6. jugular foramen, 7. superior ganglion, 8. inferior ganglion, 9. otic ganglion, 10. stylopharyngeus muscle, 11. carotid sinus, 12. pharyngeal muscles, 13. common carotid artery, 14. tongue

Vagus Nerve—Sensory and Motor Function

The vagus nerve (CN X) is named for its wandering path through the body. It emerges from the medulla oblongata and exits the skull through the jugular foramen. Two ganglionic enlargements of the vagus nerve (superior and inferior) contain somatosensory and visceral sensory neurons, respectively. It descends parallel to the carotid artery, branching frequently to innervate numerous muscles and organs. The vagus nerve is composed primarily of visceral sensory and motor axons that connect the body's organs to the brain and vice versa. It also sends motor efferents to some skeletal muscles, notably controlling the pharynx and larynx, and thus is important for swallowing, breathing, and speech. The vagus nerve plays a critical role in sensing the internal environment of the body and regulating it. Its functions are many and diverse, such as lowering the heart rate, constricting the bronchi of the lungs, reducing blood pressure, and moving food out of the stomach and through the small intestine, to name just a few.

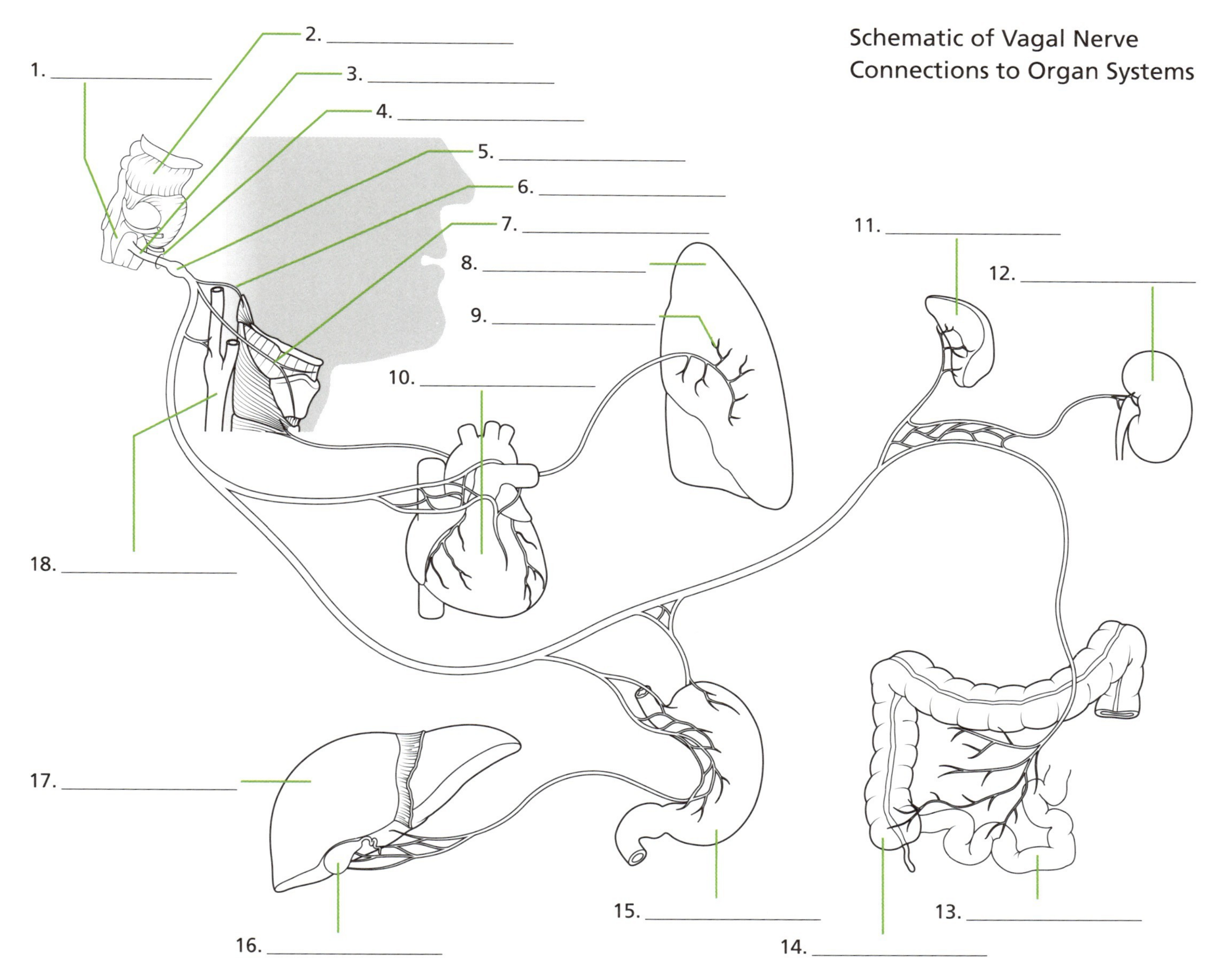

Schematic of Vagal Nerve Connections to Organ Systems

Answers

1. medulla oblongata, 2. pons, 3. vagus nerve (CN X), 4. jugular foramen, 5. ganglia (superior and inferior), 6. pharyngeal nerve branches, 7. laryngeal branches, 8. lung, 9. pulmonary plexus, 10. heart, 11. spleen, 12. kidney, 13. small intestine, 14. colon (proximal portion), 15. stomach, 16. gallbladder, 17. liver, 18. carotid sinus and carotid body

Accessory Nerve

The accessory nerve (CN XI) is considered a cranial nerve because, when it was first discovered, it was thought to emerge from the brain. However, it was later found that the accessory nerve actually originates from the spinal cord. What was previously thought to be the origination point of the cranial nerve is called the cranial root, and today this portion is known to be part of the vagus nerve. The actual origination point of the accessory nerve is called the spinal root, and because of this the nerve is also sometimes called the spinal accessory nerve. From its true origin, the accessory nerve enters the skull through the foramen magnum and then exits almost immediately through the jugular foramen. The accessory nerve provides motor innervation to the sternocleidomastoid and trapezius muscles; these muscles tilt and rotate the head or lift the shoulders and move the arms away from the body, respectively.

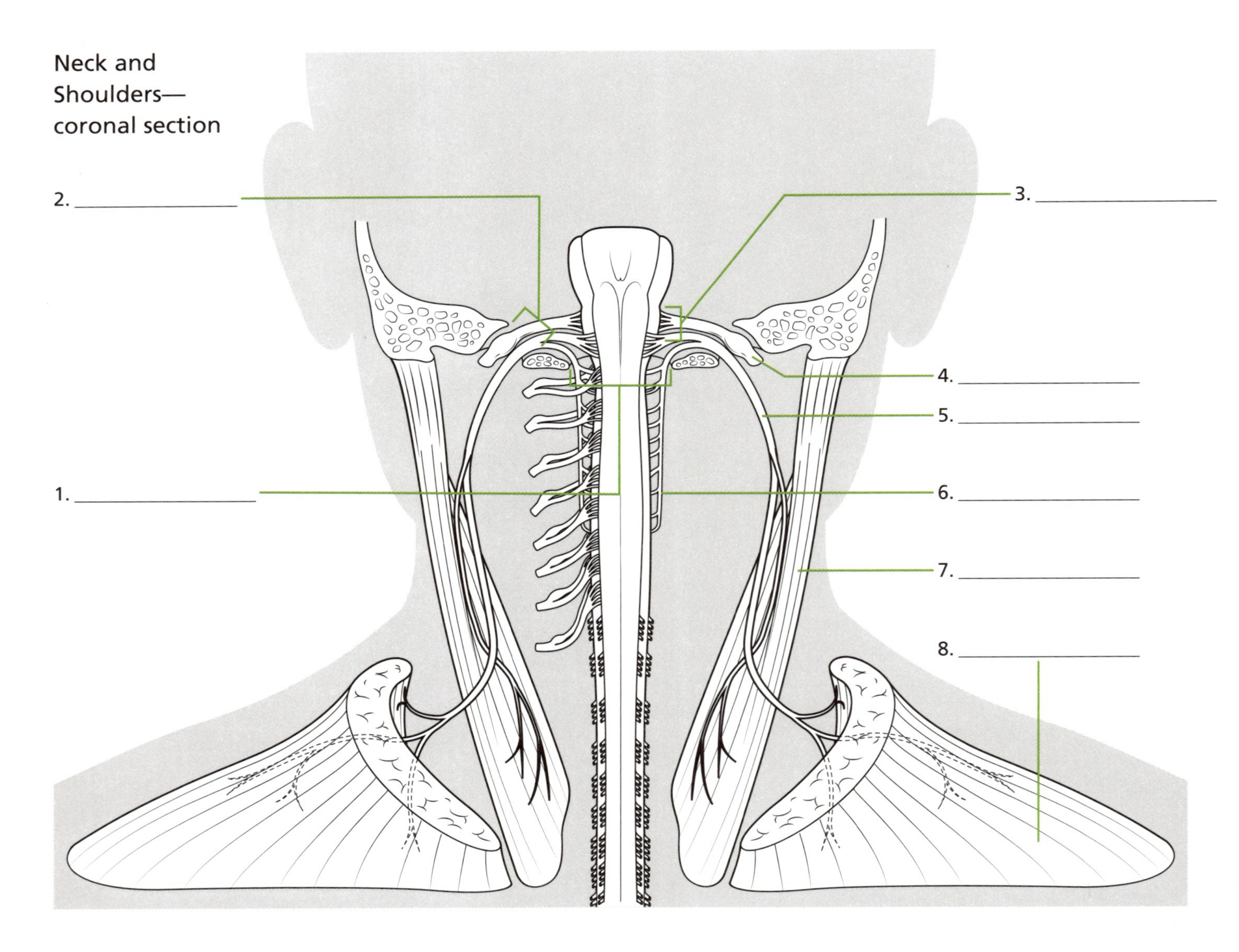

Neck and Shoulders—coronal section

Answers

1. foramen magnum, 2. jugular foramen, 3. cranial root of accessory nerve, 4. vagus nerve (CN X), 5. accessory nerve (CN XI), 6. spinal root of accessory nerve, 7. sternocleidomastoid muscle, 8. trapezius muscle

Hypoglossal Nerve

The hypoglossal nerve (CN XII) controls the muscles of the tongue. It emerges from the medulla oblongata and exits the skull through the hypoglossal canal. It branches to innervate the genioglossus, hyoglossus, and styloglossus muscles, which are three of the four extrinsic tongue muscles (muscles attached to the tongue). It also innervates all four of the intrinsic tongue muscles (the tongue itself). As such, it is responsible for both moving the position of the tongue (via the extrinsic muscles) and changing the shape of the tongue (via the intrinsic muscles), including involuntary and voluntary functions (such as habitual swallowing and speech production, respectively).

Lower Jaw, with Brainstem—sagittal section

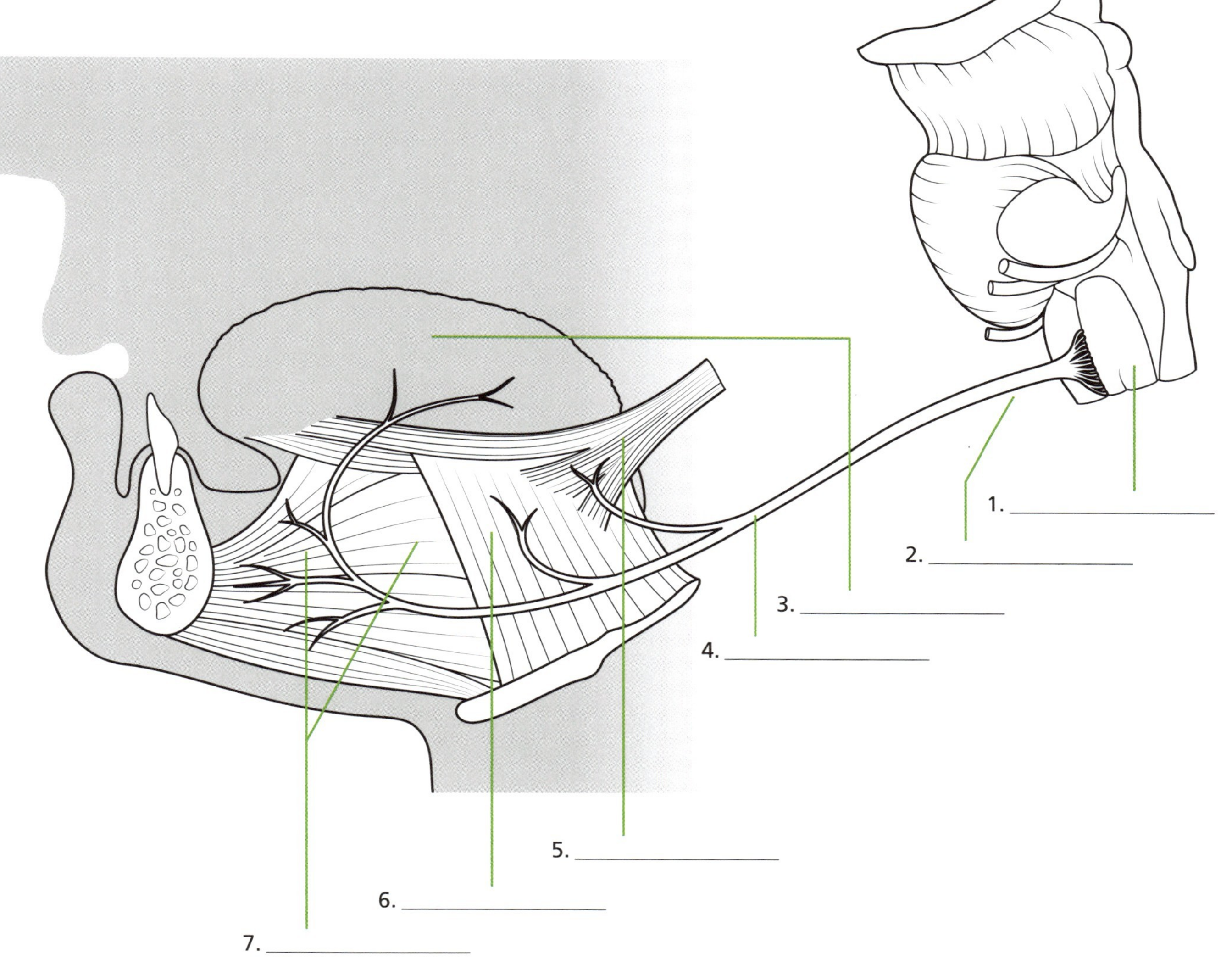

Answers

1. medulla oblongata, 2. hypoglossal canal, 3. intrinsic muscles of the tongue, 4. hypoglossal nerve (CN XII), 5. styloglossus, 6. hyoglossus, 7. genioglossus

Cranial Nerve Functional Columns in the Brainstem

Cranial nerve nuclei can be organized into vertical columns within the brainstem based upon the functions associated with the neurons within those nuclei. There are three functional columns each for motor and sensory nuclei and seven associated functions (one sensory column is associated with two functions). The most medial motor column contains general somatic efferent (GSE) fibers, which innervate skeletal muscles. The most lateral motor column contains special visceral efferent (SVE) fibers. These also innervate skeletal muscles, but only those derived from the branchial arches. In between these columns are the general visceral efferent (GVE) fibers, which innervate smooth muscles and glands. The most medial sensory column contains the special and general visceral afferent (SVA and GVA) fibers. The SVA fibers receive smell and taste sensory information, while the GVA fibers receive parasympathetic information. The most lateral sensory column contains special somatic afferent (SSA) fibers, which receive visual, auditory, and vestibular sensory information. In between these columns are the general somatic afferent (GSA) fibers, which receive sensory information from the skin, muscles, and joints.

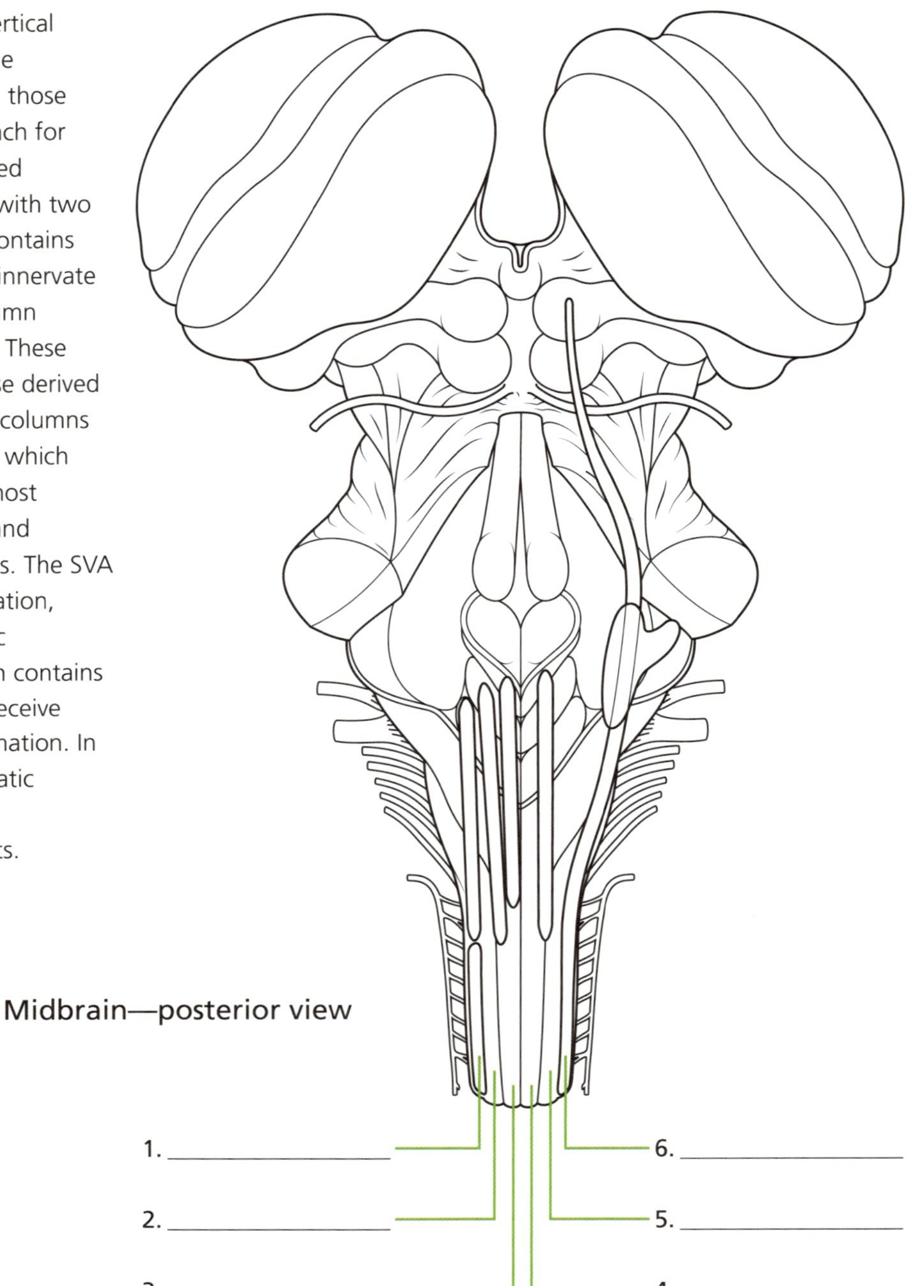

Answers

1. branchial motor column (SVE), 2. parasympathetic column (GVE), 3. somatic motor column (GSE), 4. visceral sensory column (SVA and GVA), 5. general somatic sensory column (GSA), 6. special somatic sensory column (SSA)

External Features of the Cerebellum—Anterior Lobe

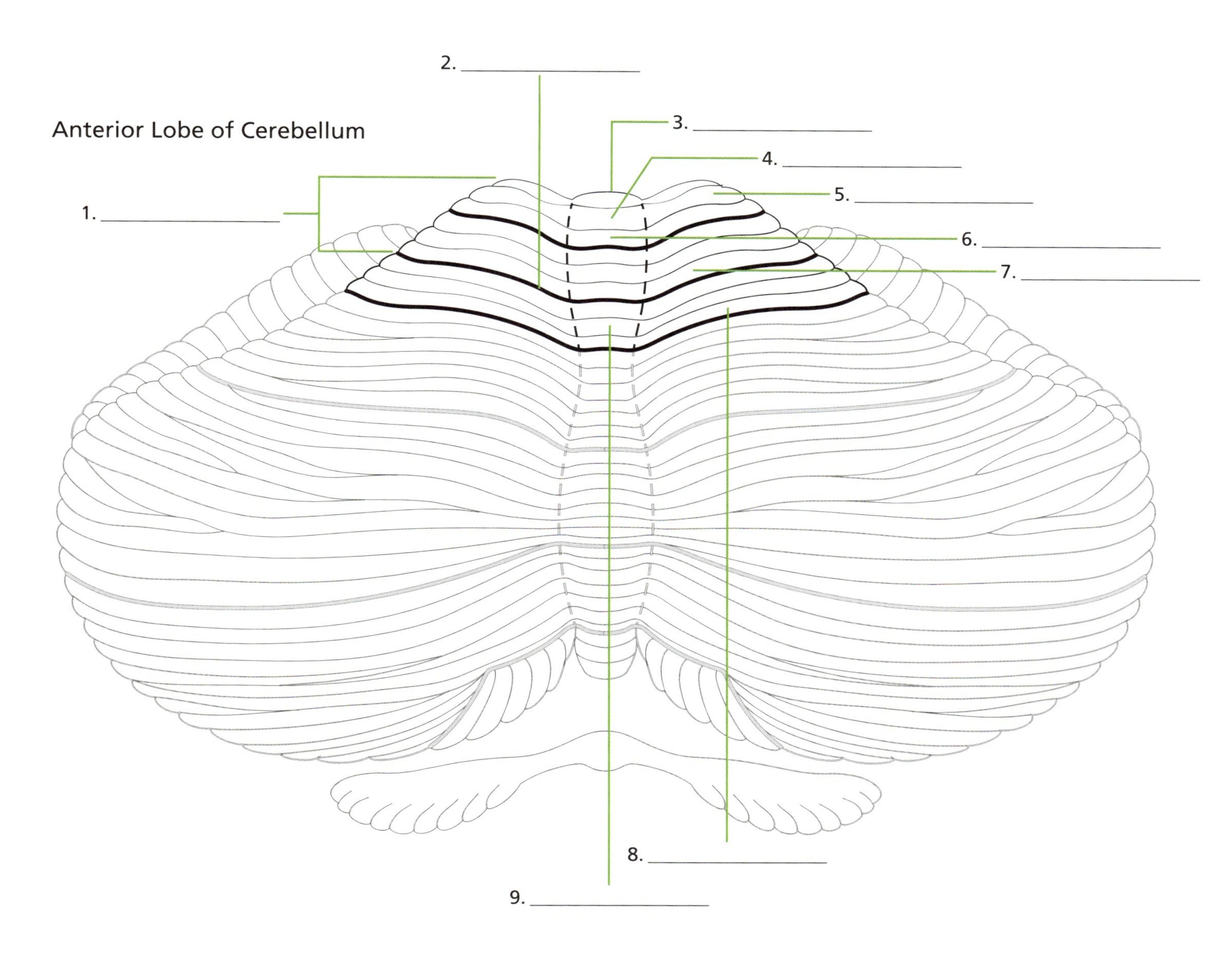

The anterior lobe receives proprioceptive input from the spinal cord, which allows for the sensation of stimuli within the body regarding position, motion, and equilibrium. The anterior lobe lies caudal to the primary fissure. Along with the rest of the cerebellum, it is divided into subdivisions known as lobules. The most anterior of the lobules is the lingula, which is attached to the superior surface of the medullary velum. Posterior to the lingula are lobules II and III, which form the central lobule. Lobules IV and V form the culmen, which lies adjacent to the primary fissure on the rostral side.

Answers

1. anterior lobe, 2. primary fissure, 3. lingula, 4. central lobule, 5. ala, 6. culmen, 7. quadrangular lobule, 8. lobulus simplex, 9. declive

External Features of the Cerebellum—Posterior Lobe

The posterior lobe is situated rostrally to the flocculonodular lobe and caudally to the anterior lobe, separated by the posterolateral and primary fissures, respectively. It is responsible for the coordination of fine motor skills and muscle movement. This is the largest of the three cerebellar lobes and receives input mainly from the brainstem and cortex. The posterior lobe contains six out of the ten lobules of the cerebellum. These lobules are responsible for coordinating and modifying movement in different areas of the body. The most anterior region of the posterior lobe is the simple lobule, lobule VI. The folium (VII A) and tuber (VII B) make up the ansiform lobule and are responsible for bilateral movements of the upper and lower limbs. The folium is anterior to, and the tuber is posterior to, the horizontal fissure. The superior and inferior semilunar lobules are paired areas that control unilateral arm movements, with the superior semilunar lobule on the anterior side of the horizontal fissure and the inferior semilunar lobule on the posterior side. The pyramid is posterior to the inferior semilunar lobule and is involved in muscle contraction of the trunk. The posteriormost lobules are the uvula (VIII) and tonsil (IX), which line the posterolateral fissure on the anterior and posterior sides, respectively.

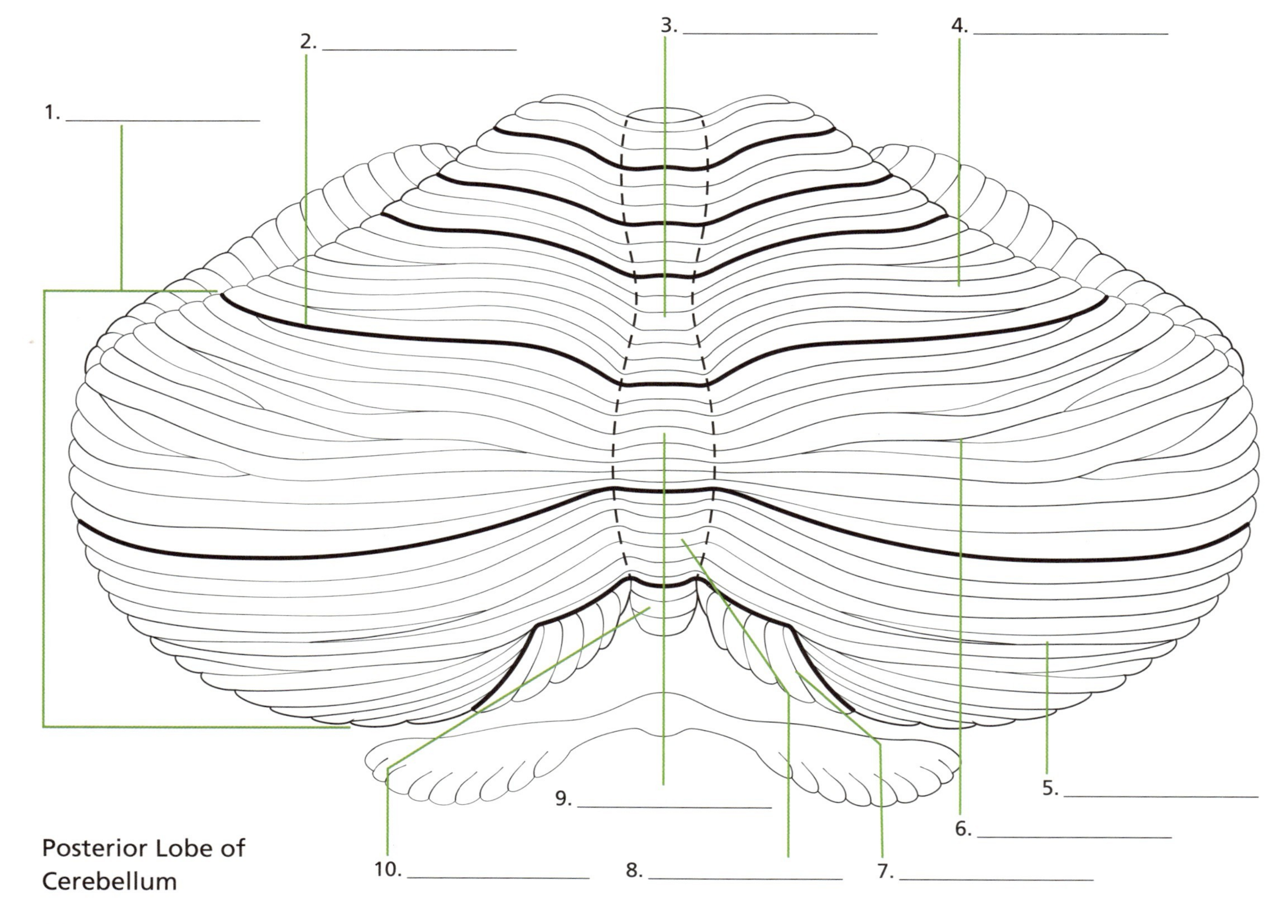

Posterior Lobe of Cerebellum

Answers

1. posterior lobe, 2. horizontal fissure, 3. folium, 4. superior semilunar lobule, 5. biventral lobule, 6. inferior semilunar lobule, 7. tonsil, 8. pyramid, 9. tuber, 10. uvula

External Features of the Cerebellum—Flocculonodular Lobe

The flocculonodular is the smallest of the three cerebellar lobes. It receives information from the vestibular system in the inner ear for awareness of the placement and location of the body. It is composed of the flocculus and nodule, which are involved in the maintenance of equilibrium, learning of basic motor skills, and influence of eye movement. The flocculus also regulates balance and helps to stabilize gaze during head rotation. The nodule is the only lobule in the flocculonodular lobe, lobule X. The flocculonodular lobe is separated from the remainder of the cerebellum by the posterolateral, or uvulonodular, fissure. This lobe contains vestibular nuclei, which are located in the fourth ventricle in the medulla and pons. They project to the spinal cord, thalamus, and cortex to control eye movements, posture, and perception of movement.

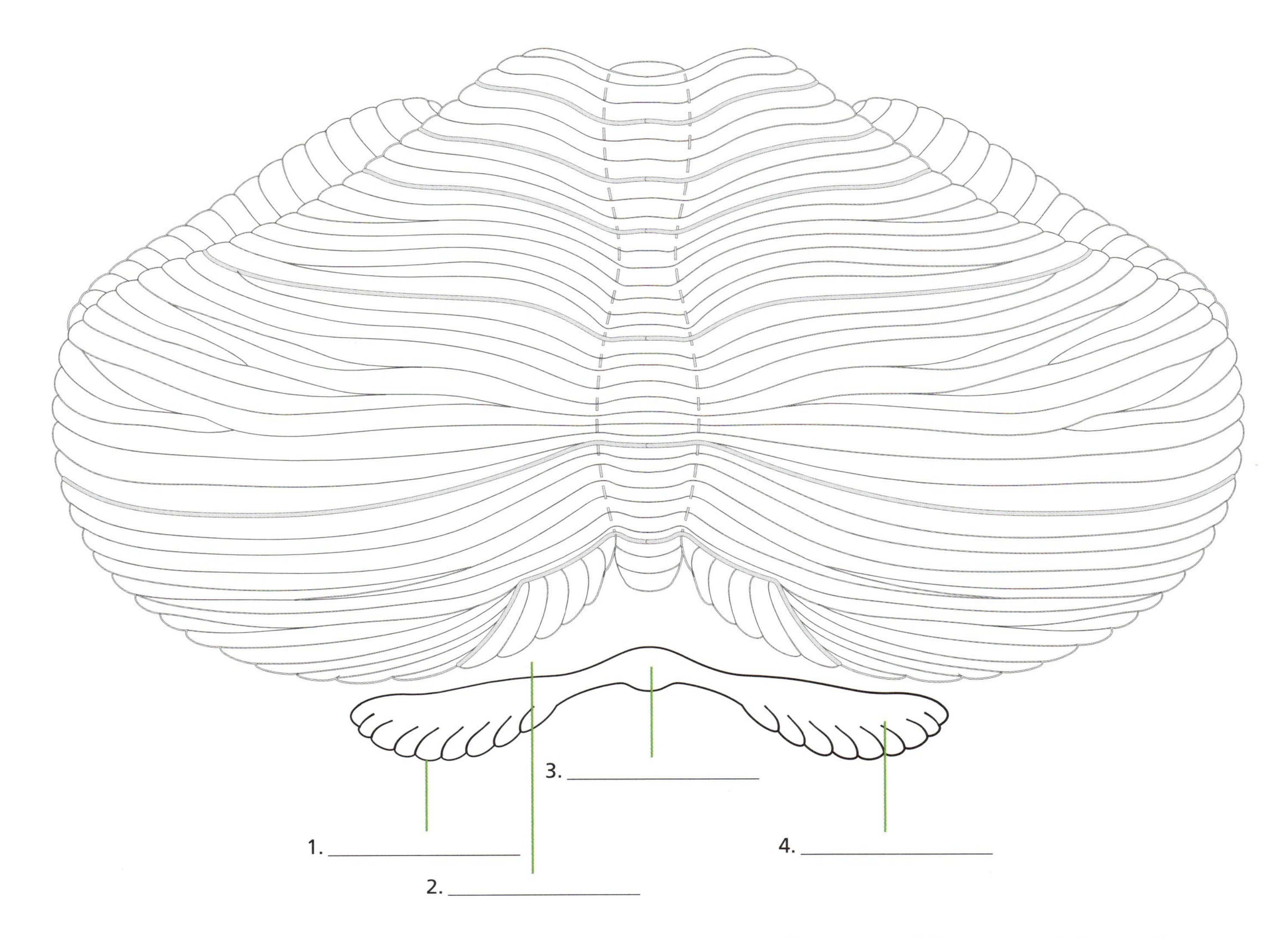

Flocculonodular Lobe of Cerebellum

Answers

1. flocculonodular lobe, 2. uvulonodular fissure, 3. nodule, 4. flocculus

Functional Components of the Cerebellum

The cerebellum is made up of three functional components known as the vestibulocerebellum, spinocerebellum, and cerebrocerebellum, also known as the pontocerebellum. The functional divisions are characterized by the way a lesion in the respective areas would affect the brain. The vestibulocerebellum is a functional division of the cerebellum that includes the flocculonodular lobe and connections with lateral vestibular nuclei. This region of the cerebellum is involved with vestibular reflexes and postural maintenance. The vestibulocerebellum is responsible for balance, and lesions of this region result in symptoms such as ipsilateral jerk nystagmus, head tilt, and circling gait. The spinocerebellum is a functional division of the cerebellum that includes the vermis and intermediate zones of the cerebellar cortex. Lesions of the spinocerebellum result in impaired gait indicative of abnormal spinal control of walking. The cerebrocerebellum is the largest functional subdivision of the human cerebellum and contains the lateral hemispheres and dentate nuclei. It is involved with the planning and timing of movements, as well as cognitive functions of the cerebellum. Lesions of the cerebrocerebellum result in inaccuracy of reaching and clumsiness of hand movements, or impaired voluntary control of body parts.

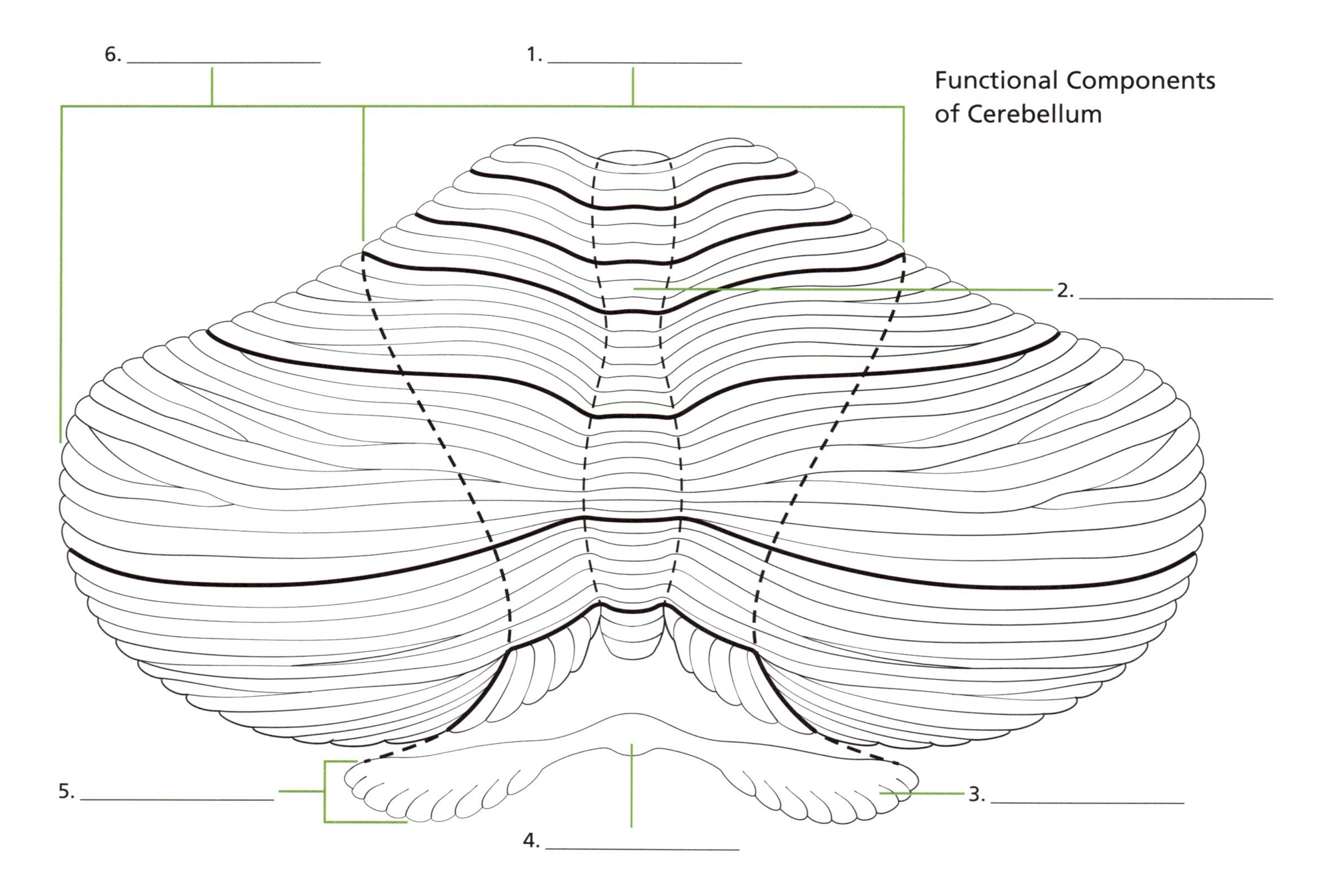

Functional Components of Cerebellum

Answers

1. spinocerebellum, 2. vermis, 3. flocculus, 4. nodule, 5. vestibulocerebellum, 6. cerebrocerebellum

Cerebellar Cortical Circuitry

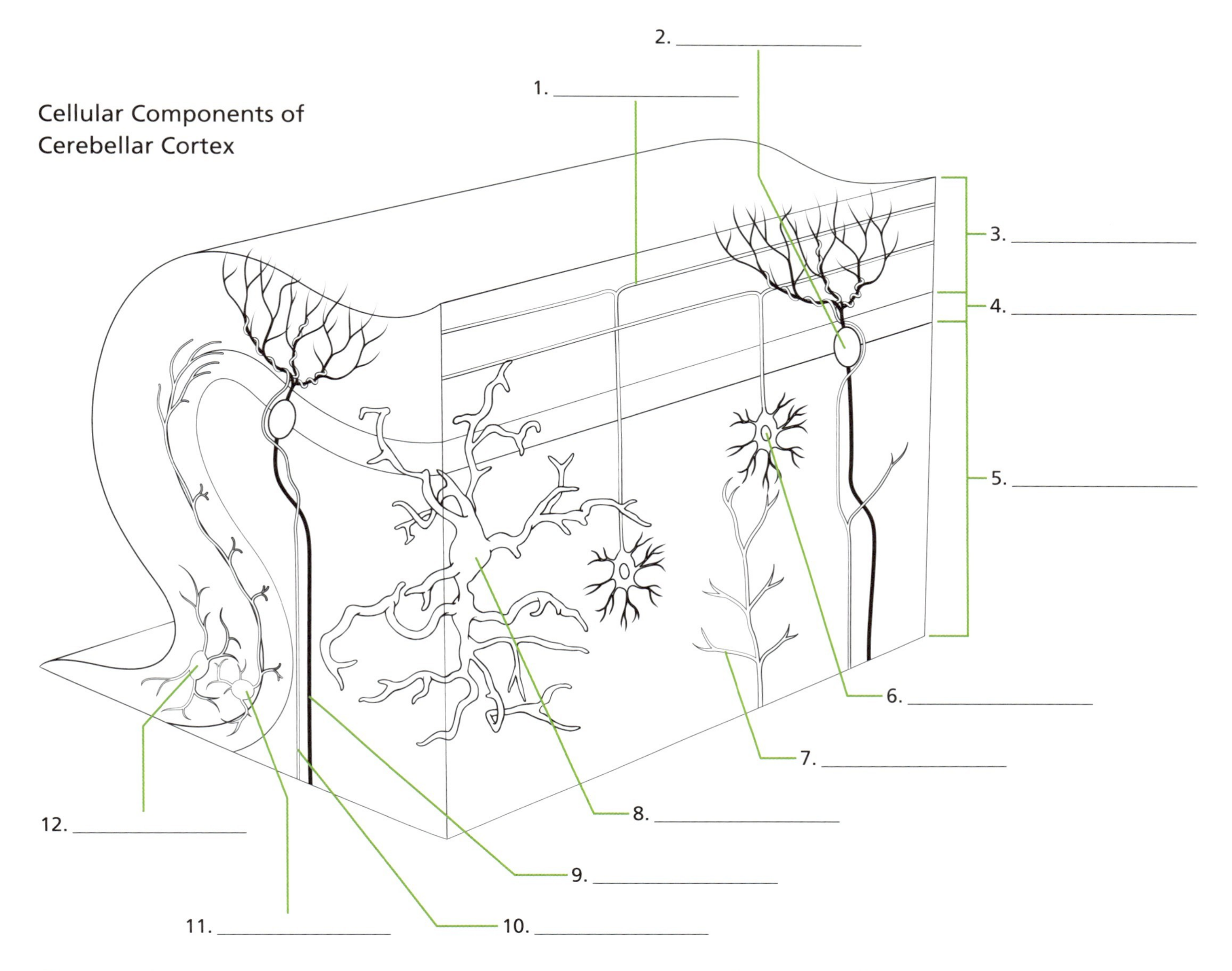

The cerebellar cortex contains almost all of the neurons in the cerebellum and is made up of three layers: the innermost granule layer, the Purkinje layer, and the outer molecular layer. The granule layer is made up of small, densely packed neurons called granule cells and Golgi neurons, which receive input from mossy fibers. Mossy fibers originate in the pontine nuclei, spinal cord, and brainstem and make excitatory projections onto cerebellar nuclei and granule cells. Mossy fibers are essential in the modulation of simple spikes in Purkinje cells. The middle layer, consisting of Purkinje cells, is only one cell layer thick. Purkinje cells receive information about sensory conditions and are the sole source of output from the cerebellar cortex. They make inhibitory connections onto cerebellar nuclei. Climbing fibers, however, make excitatory projections onto cerebellar nuclei, leading to signal transmission to the brainstem and spinal cord. Their axons climb and wrap around the dendrites of Purkinje cells, which they also excite. The molecular layer contains stellate, basket, and unipolar brush cells. Stellate and basket cells are excited by parallel fibers and form feed-forward inhibitory circuits by synapsing on Purkinje cells. Unipolar brush cells interact with the mossy fiber and granule cell system.

Answers

1. parallel fiber, 2. Purkinje cell, 3. molecular layer, 4. Purkinje cell layer, 5. granular layer, 6. granule cell, 7. mossy fiber, 8. Golgi cell, 9. climbing fiber, 10. Purkinje cell axon, 11. basket cell, 12. stellate cell

Cerebellar Connections—Vestibulocerebellum

The vestibulocerebellum is a functional division of the cerebellum that includes the flocculonodular lobe and connections with lateral vestibular nuclei, which are found below the vermis. This region of the cerebellum is involved with vestibular reflexes and postural maintenance. The inferior cerebellar peduncle, also known as the restiform body, contains afferent fibers from the medulla and efferent fibers to the vestibular nuclei in the inner ear. The vestibulocerebellum is responsible for vestibular function, and lesions of this region result in symptoms such as ipsilateral jerk nystagmus, head tilt, and circling gait.

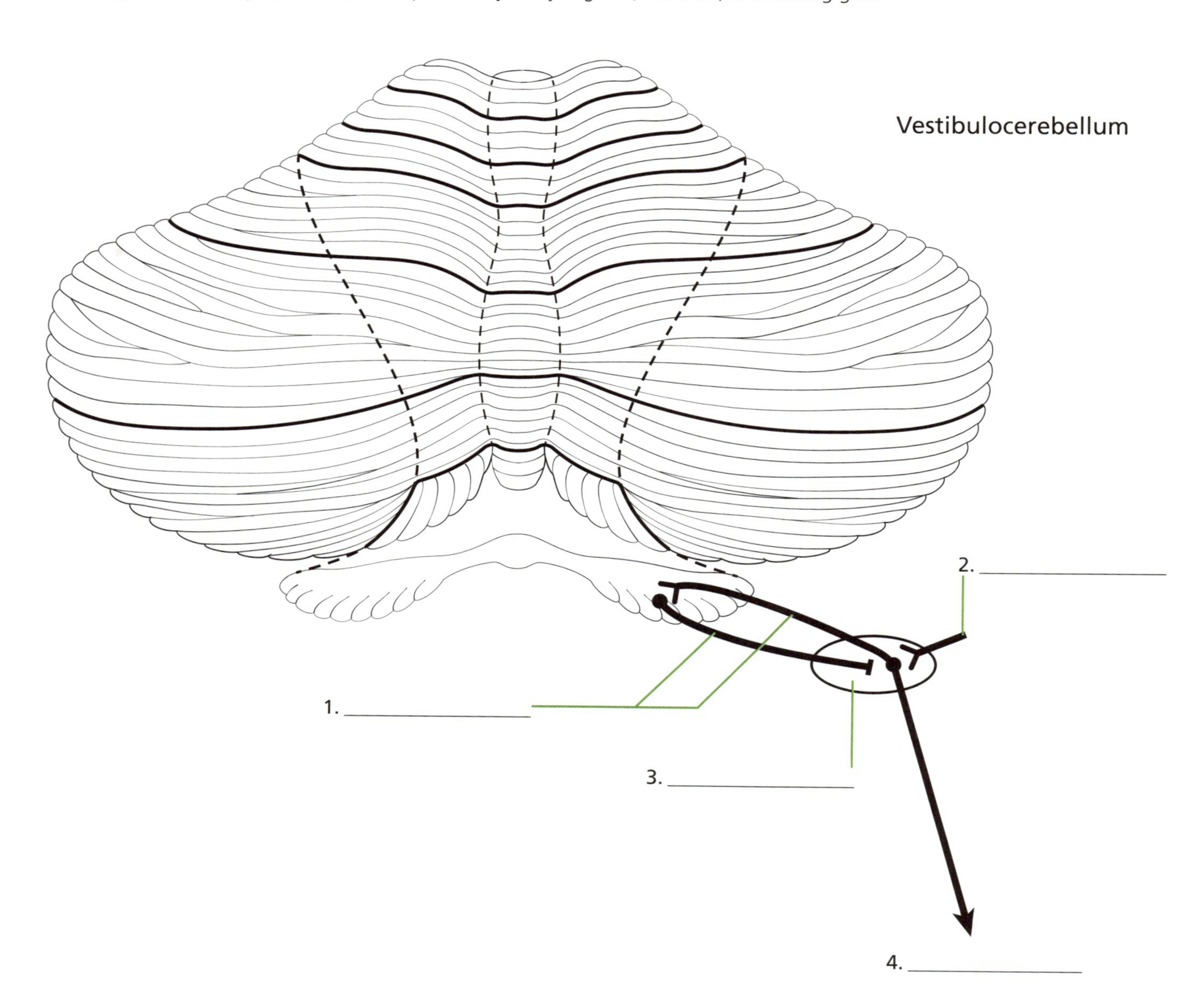

Answers

1. inferior cerebellar peduncle, 2. vestibular labyrinth, 3. vestibular nuclei, 4. vestibulospinal tracts

Cerebellar Connections—Spinocerebellum

The spinocerebellum is a functional division of the cerebellum that includes the vermis, intermediate zones of the cerebellar cortex, and fastigial and interposed nuclei. Fastigial nuclei receive input from the vestibular system and aid in the interpretation of body movement through space, and interposed nuclei are involved with coordinating agonist/antagonist muscle pairs. The spinocerebellum includes the superior cerebellar peduncle, also known as the brachium conjunctivum, which contains efferent fibers from the cerebellar nuclei and afferent fibers from the spinocerebellar tract. Output from this region projects to rubrospinal, vestibulospinal, and reticulospinal tracts. Sensory input is integrated with motor commands to produce adaptive motor coordination. Lesions of the spinocerebellum result in impaired gait indicative of abnormal spinal control of walking.

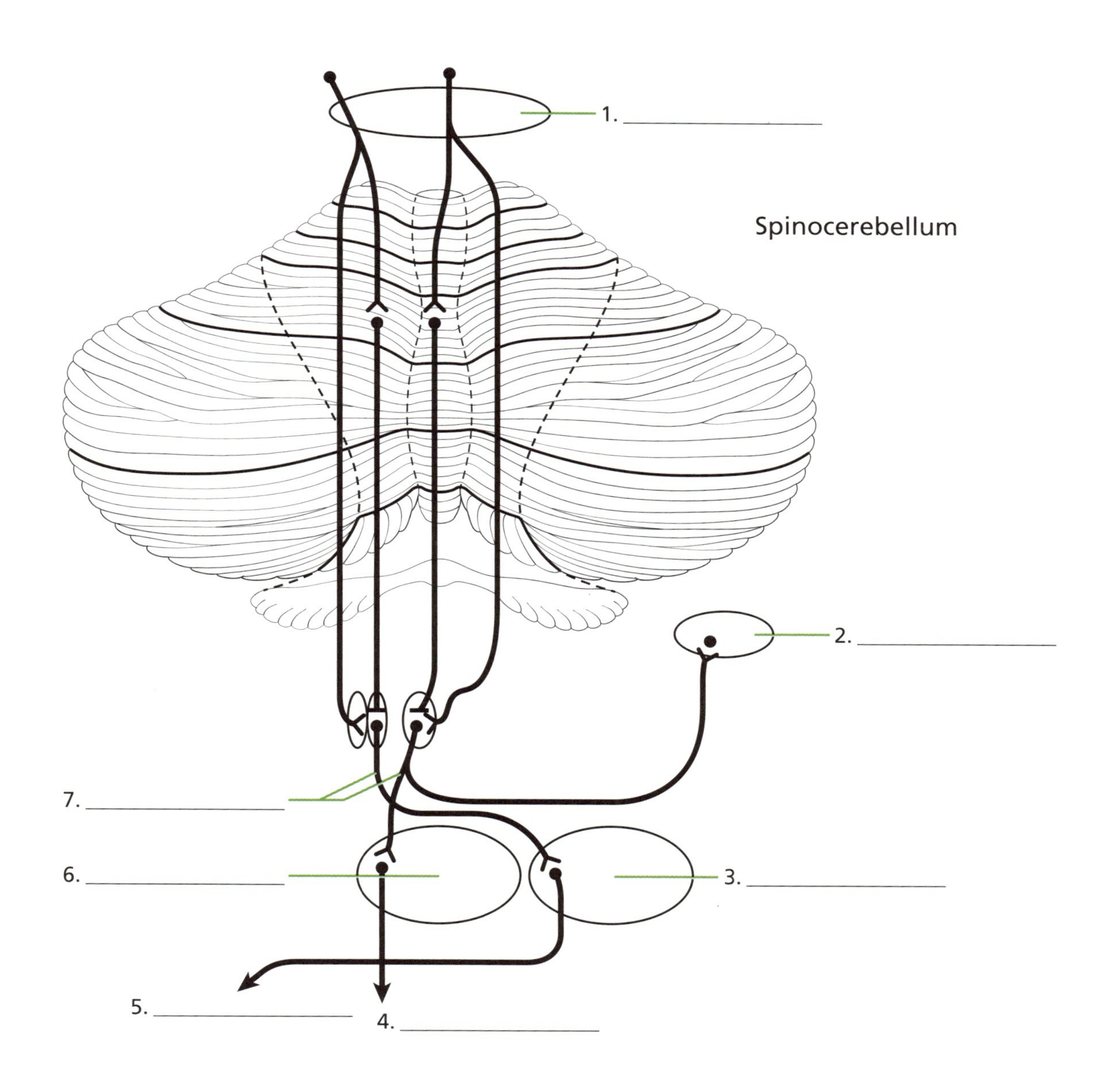

Answers

1. inferior cerebellar peduncle, 2. vestibular nuclei, 3. red nucleus, 4. medial descending tracts, 5. lateral descending tracts, 6. reticular formation, 7. superior cerebellar peduncle

Cerebellar Connections—Cerebrocerebellum

The cerebrocerebellum is the largest functional subdivision of the human cerebellum. It is bounded by the spinocerebellum and the vestibulocerebellum and contains the lateral hemispheres and dentate nuclei. The cerebrocerebellum, particularly the middle cerebellar peduncle/brachium pontis, shares extensive afferent connections with the cerebral cortex via the pontine nuclei and efferent connections via the ventrolateral (VL) thalamus. It is involved with the planning and timing of movements, as well as cognitive functions of the cerebellum. Lesions of the cerebrocerebellum result in inaccuracy of reaching and clumsiness of hand movements, or impaired voluntary control of body parts.

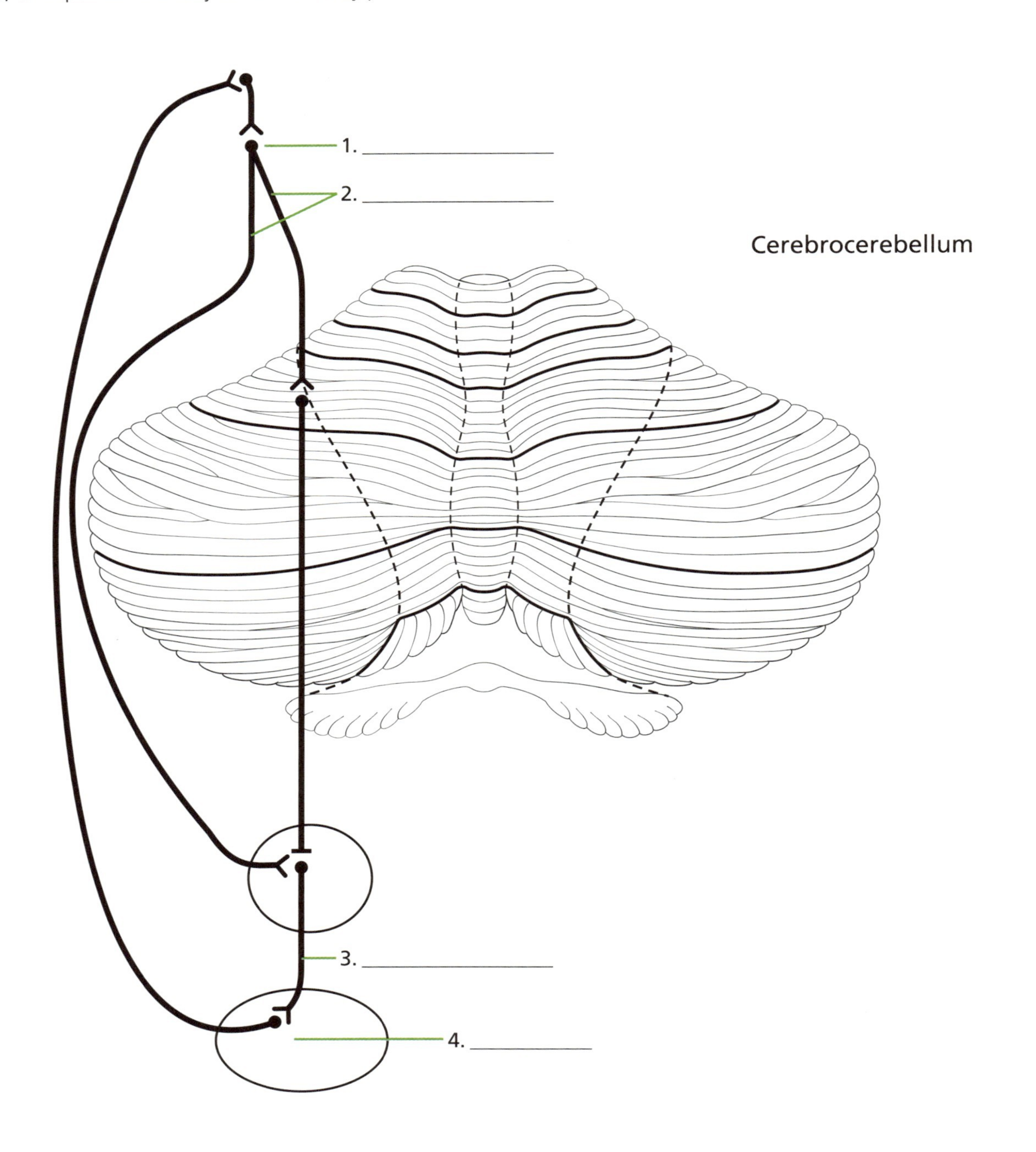

Answers

1. pontine nuclei, 2. middle cerebellar peduncle, 3. superior cerebellar peduncle, 4. ventrolateral thalamus

Overview of the Thalamus

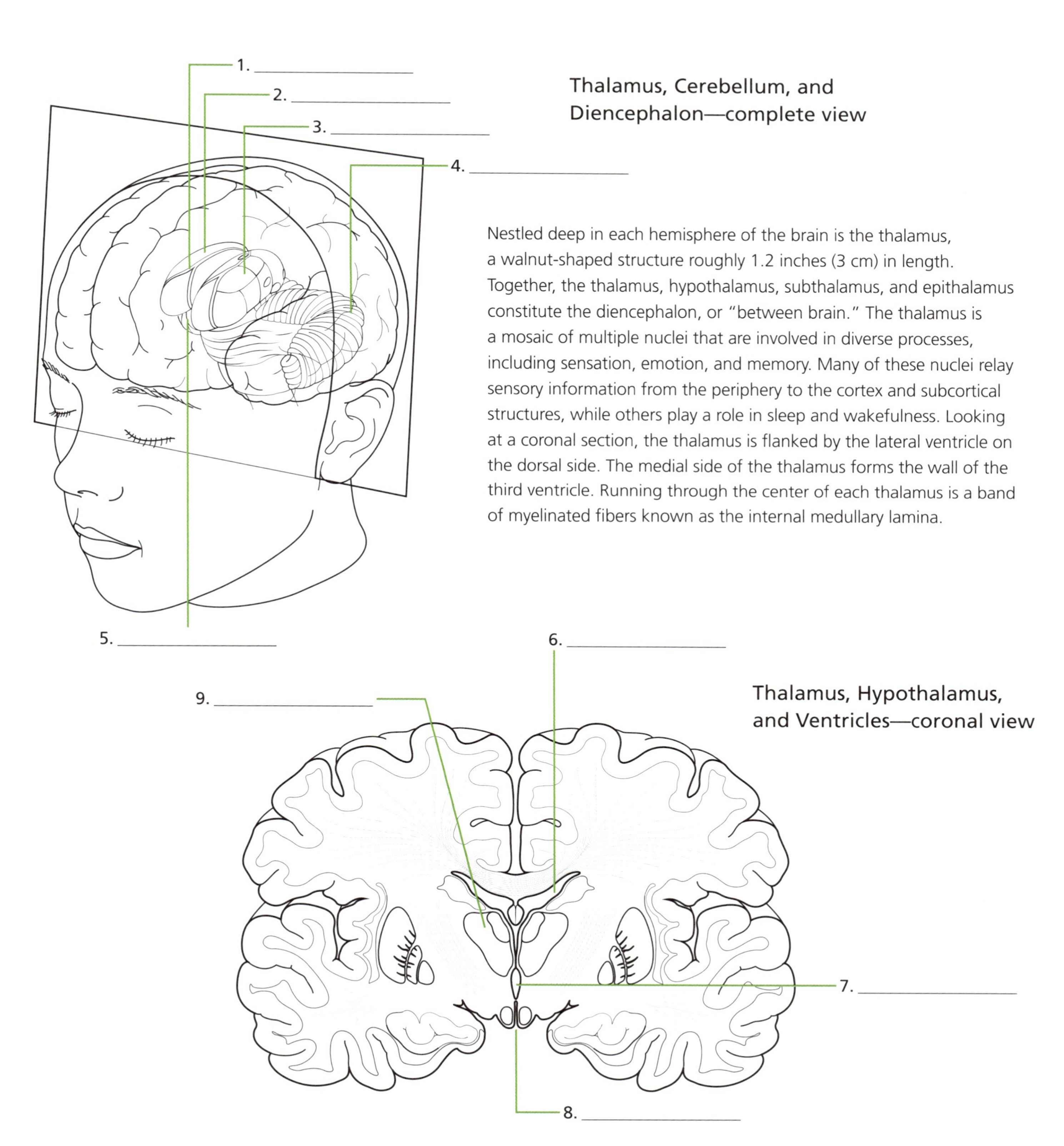

Nestled deep in each hemisphere of the brain is the thalamus, a walnut-shaped structure roughly 1.2 inches (3 cm) in length. Together, the thalamus, hypothalamus, subthalamus, and epithalamus constitute the diencephalon, or "between brain." The thalamus is a mosaic of multiple nuclei that are involved in diverse processes, including sensation, emotion, and memory. Many of these nuclei relay sensory information from the periphery to the cortex and subcortical structures, while others play a role in sleep and wakefulness. Looking at a coronal section, the thalamus is flanked by the lateral ventricle on the dorsal side. The medial side of the thalamus forms the wall of the third ventricle. Running through the center of each thalamus is a band of myelinated fibers known as the internal medullary lamina.

Answers

1. internal medullary lamina, 2. right thalamus, 3. left thalamus, 4. cerebellum, 5. hypothalamus, 6. lateral ventricle, 7. third ventricle, 8. hypothalamus, 9. thalamus

Topology of the Thalamus

Composed of numerous nuclei with diverse functions, the thalamus is a complex, multifaceted structure. Anatomically, it can be divided into three regions: anterior, medial, and lateral. The anterior nucleus is the most anterior, found between the fork of the internal medullary lamina. Within the internal medullary lamina are small groups of cells known as the intralaminar nuclei, which include the centromedian nucleus. The lateral group is composed of the ventral anterior, ventral lateral, ventral posterior, lateral geniculate, lateral posterior, and medial geniculate nuclei. The pulvinar, the largest thalamic nucleus in humans, is also part of the lateral group and is the most posterior. The medial group includes the mediodorsal nucleus. The interthalamic adhesion is a group of nerve cells and fibers between the medial surfaces of the thalamus, but it does not contain any direct connections between hemispheres.

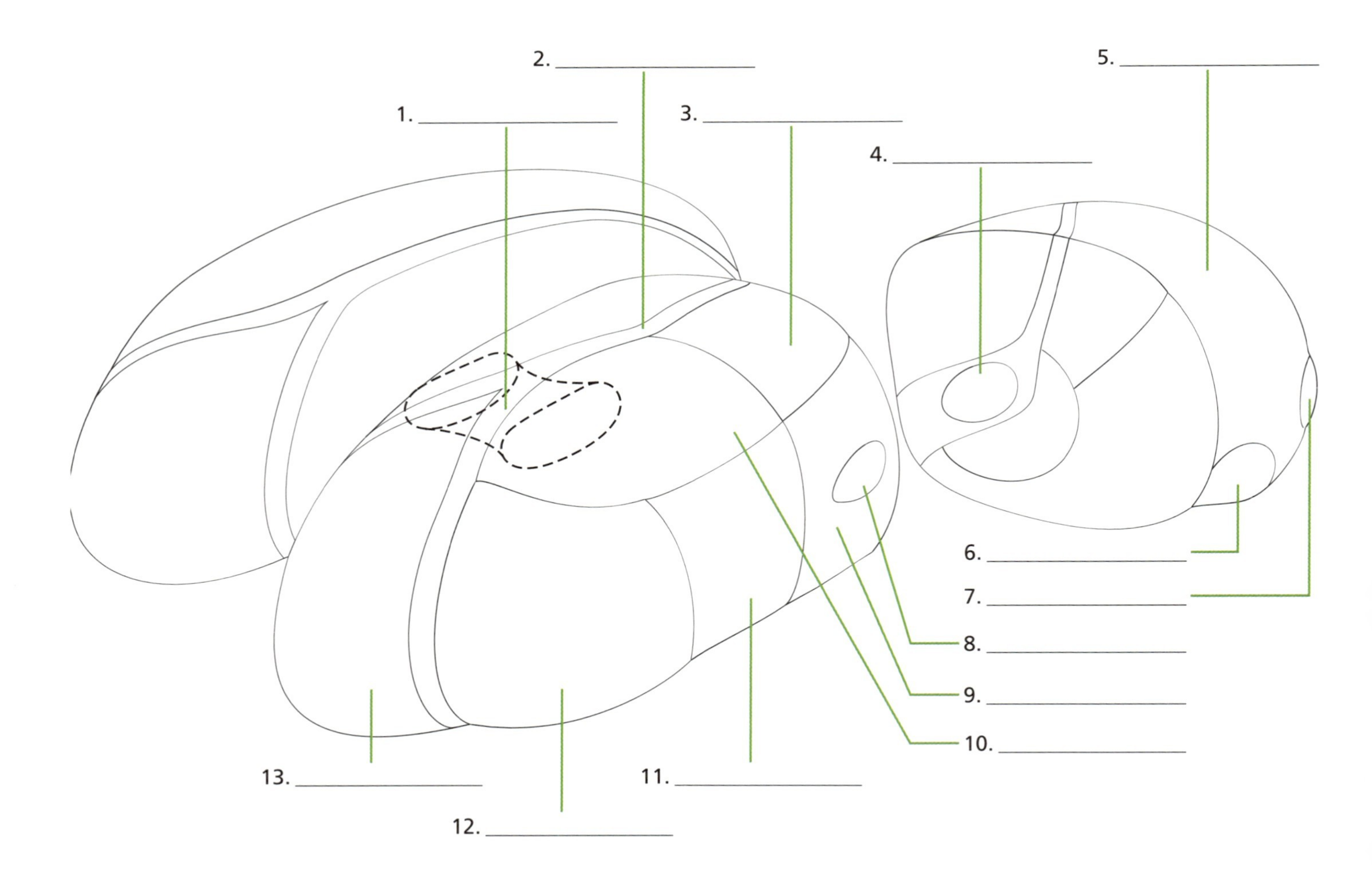

Nuclei of the Thalamus

Answers

1. interthalamic adhesion, 2. intralaminar nuclei, 3. centromedian nucleus, 4. lateral posterior nucleus, 5. pulvinar nucleus, 6. medial geniculate nucleus, 7. lateral geniculate nucleus, 8. ventral posteromedial nucleus, 9. ventral posterolateral nucleus, 10. mediodorsal nucleus, 11. ventral lateral nucleus, 12. ventral anterior nucleus, 13. anterior nucleus

Inputs to the Thalamic Nuclei

The anterior nucleus receives input from the mammillothalamic tract and the fornix, and is thereby associated with memory formation. The medial nuclei are reciprocally interconnected with the prefrontal cortex and receive inputs from the amygdala, temporal lobe, and substantia nigra. With the exception of olfaction, each sensory system in the brain sends information from the periphery to the thalamus. For the auditory system, inputs originating in the cochlea of the ear travel through the lateral lemniscus of the medulla and the inferior colliculus to the medial geniculate nucleus. Similarly, sensory information from the spinal cord is passed through the medial lemniscus of the medulla and on to the ventral posterolateral nucleus of the thalamus. In the visual system, information from the retina is conveyed directly to the lateral geniculate nucleus. The pulvinar is the secondary visual nucleus in the thalamus and is highly interconnected with the extrastriate visual cortex as well as the inferior parietal lobule. The ventral anterior nucleus receives input from basal ganglia structures.

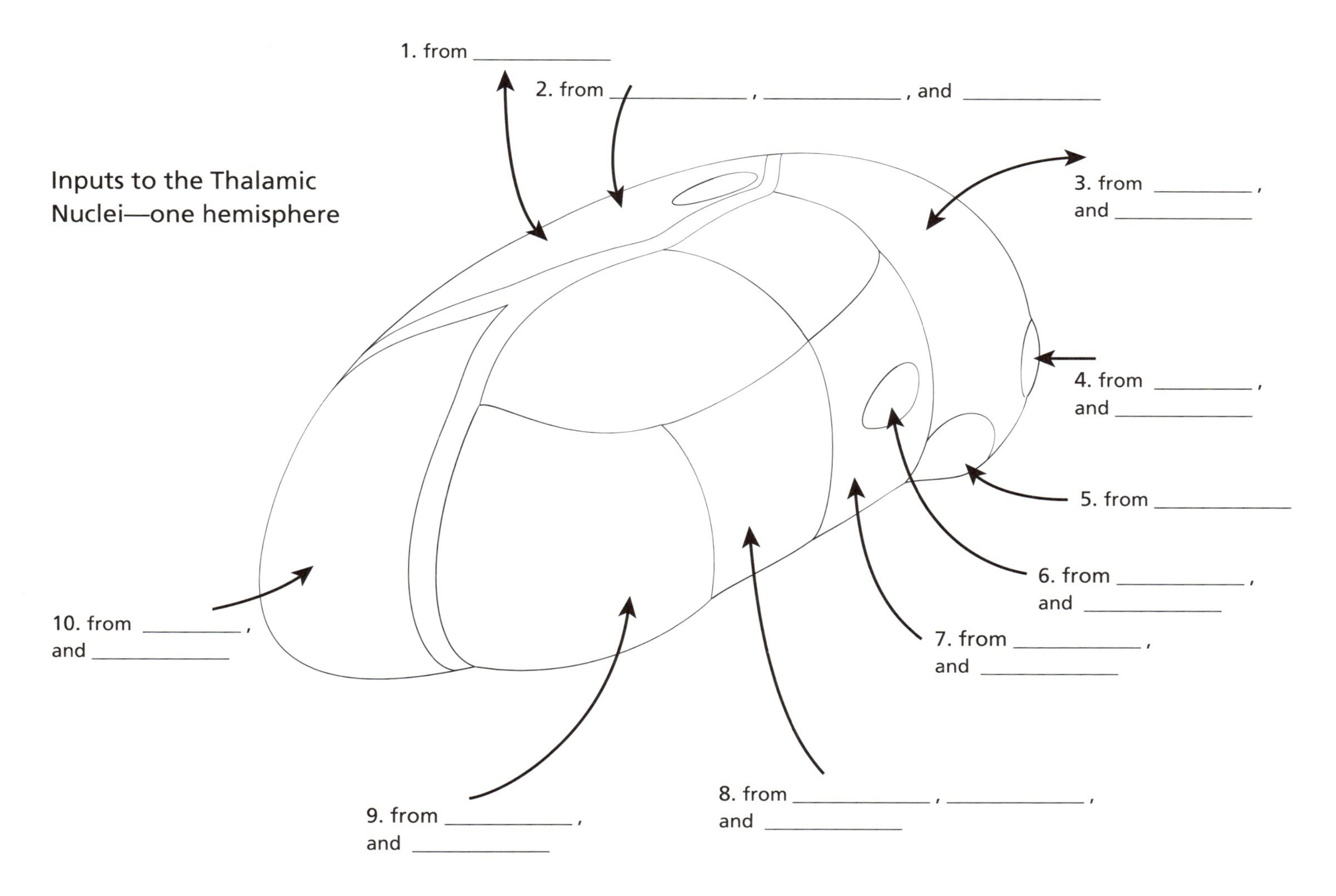

Answers

1. prefrontal cortex, 2. amygdala, temporal lobe, and substantia nigra, 3. extrastriate visual cortex and inferior parietal lobule, 4. inferior colliculus and lateral lemniscus, 5. optic tract, 6. trigeminothalamic tracts and taste pathways, 7. medial lemniscus and spinothalamic tracts, 8. dentate nucleus, globus pallidus, and substantia nigra, 9. globus pallidus and substantia nigra, 10. mammillothalamic tract and fornix

Outputs from the Thalamic Nuclei

The thalamus provides the major source of input to the cerebral cortex in multiple parallel streams. The mediodorsal nucleus is the principal thalamic nucleus for the prefrontal cortex and is heavily interconnected with it. While the ventral anterior nucleus sends information to the premotor cortex for planning of movement, the ventral lateral nucleus sends information directly to the supplementary motor area and primary motor cortex. After receiving input from the spinal cord, the ventral posterior nucleus sends this information to the somatosensory cortex. Relaying information from the eyes, the lateral geniculate nucleus projects to the primary visual cortex at the back of the brain. Likewise, the medial geniculate nucleus sends auditory information to the primary auditory cortex. With this wide array of outputs to the cortex and other structures, as well as the ability of individual cells in the thalamus to beat rhythmically, the thalamus is largely responsible for generating synchronized brain rhythms at various frequencies. While the exact function of these different rhythms is unknown, higher-frequency rhythms are thought to underlie phenomena such as arousal and attention.

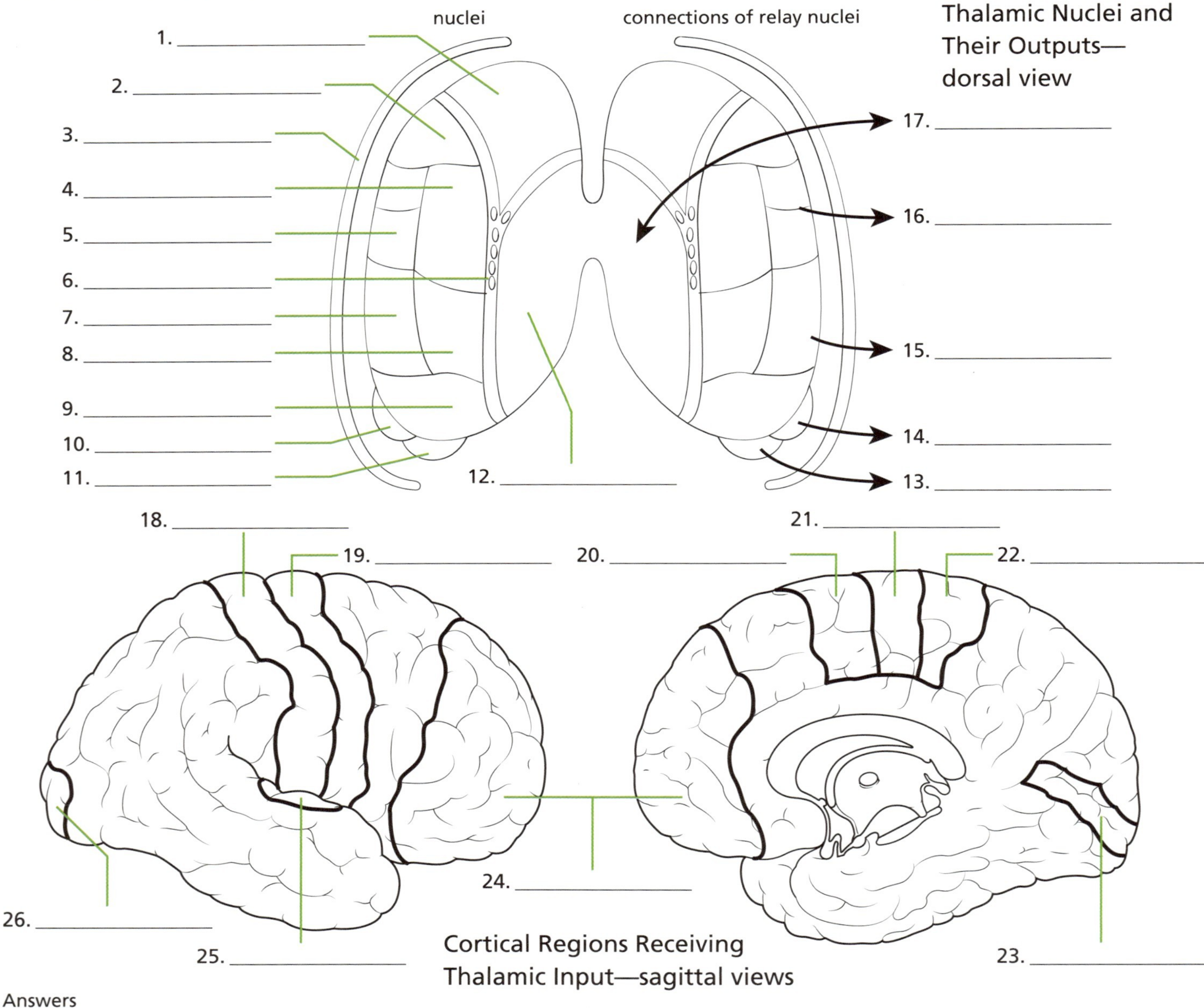

Answers

1. anterior nucleus, 2. ventral anterior nucleus, 3. reticular nucleus, 4. lateral dorsal nucleus, 5. ventral lateral nucleus, 6. intralaminar nuclei, 7. ventral posterior nucleus, 8. posterior dorsal nucleus, 9. pulvinar, 10. lateral geniculate nucleus, 11. medial geniculate nucleus, 12. mediodorsal nucleus, 13. to primary auditory cortex, 14. to primary visual cortex, 15. to somatosensory cortex, 16. to supplementary motor area and motor cortex, 17. to prefrontal cortex, 18. somatosensory cortex, 19. primary motor cortex, 20. supplementary motor area, 21. motor cortex, 22. somatosensory cortex, 23. primary visual cortex, 24. prefrontal cortex, 25. auditory cortex, 26. primary visual cortex

Structure and Function of the Subthalamus

Below the thalamus are two structures, the subthalamus and the mammillary bodies of the hypothalamus. Often called the prethalamus, the subthalamus is a region that contains multiple structures, including the subthalamic nucleus, zona incerta, subthalamic fasciculus, and ansa lenticularis. The subthalamic nucleus (STN) serves as part of the motor system and shapes motor outputs such as reaching and walking. It has reciprocal connections with the various parts of the basal ganglia, including the globus pallidus (GP) and the substantia nigra. Because of the STN's involvement in such key motor pathways, Parkinson's disease is often marked by dysfunction in this nucleus. The subthalamic fasciculus and ansa lenticularis are white matter tracts that connect the STN to the GP, and the GP to the ventral thalamus, respectively. The zona incerta (ZI) projects very widely to various cortical and subcortical structures. Although its precise function is unknown, the ZI is thought to be involved in limbic-motor integration.

Thalamus, Subthalamus, and Basal Ganglia—coronal view

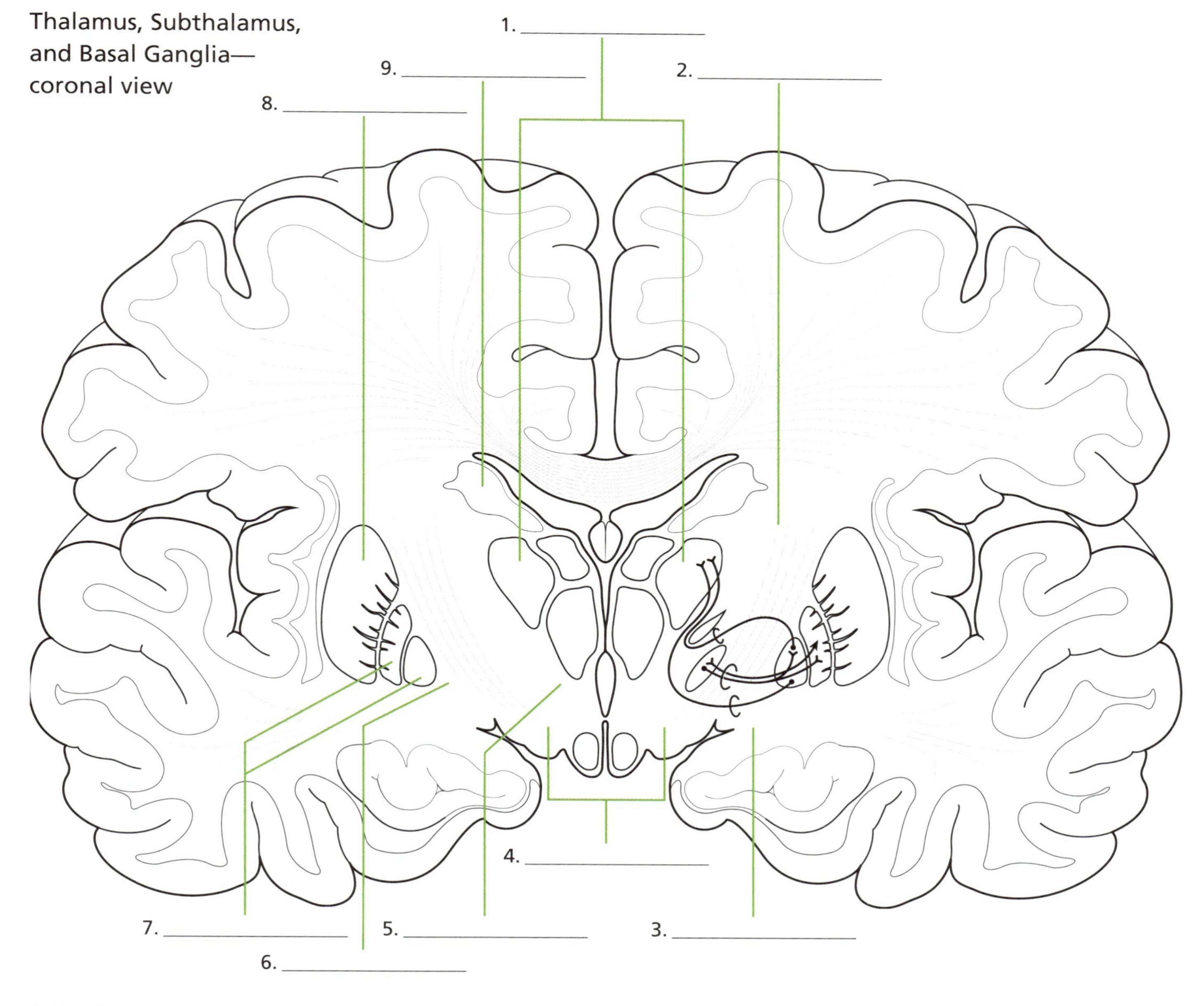

Answers

1. thalamus, 2. internal capsule, 3. ansa lenticularis, 4. subthalamic nucleus, 5. zona incerta, 6. subthalamic fasciculus, 7. globus pallidus, 8. putamen, 9. caudate nucleus

Structure and Function of the Epithalamus

The epithalamus is located just posterior to the thalamus and contains the habenula and the pineal body. The habenula is a collection of nuclei that has been associated with a diverse array of processes, including pain perception, reward learning, and sleep-wake cycles. With a significant influence on dopaminergic and serotonergic neuromodulatory systems, the habenula has been implicated in disorders such as Parkinson's disease and depression. It is often divided into medial and lateral portions, based on its gene expression and connectivity patterns. The lateral portion is connected to regions such as the basal ganglia and nucleus accumbens and is involved in reward processing and emotive decision-making, while the medial portion plays an important role in processes such as stress and memory. The pineal body (or gland) is a 0.4-inch (1 cm)-long pinecone-shaped endocrine gland that lies just ventral to the habenula. It is the only brain structure that is unpaired—it does not have a left and right half—and has therefore provoked much speculation from philosophers such as René Descartes as to its role in unifying the brain and body. Although there is no evidence that it is the "seat of the soul," the pineal body does control sleep patterns and circadian rhythms through the production and release of melatonin.

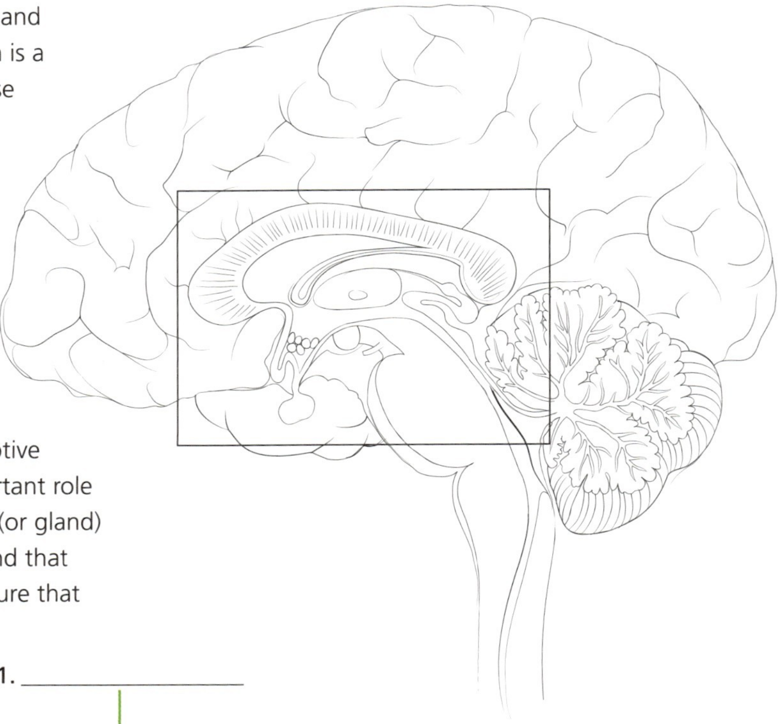

Epithalamus and Surrounding Structures—sagittal view

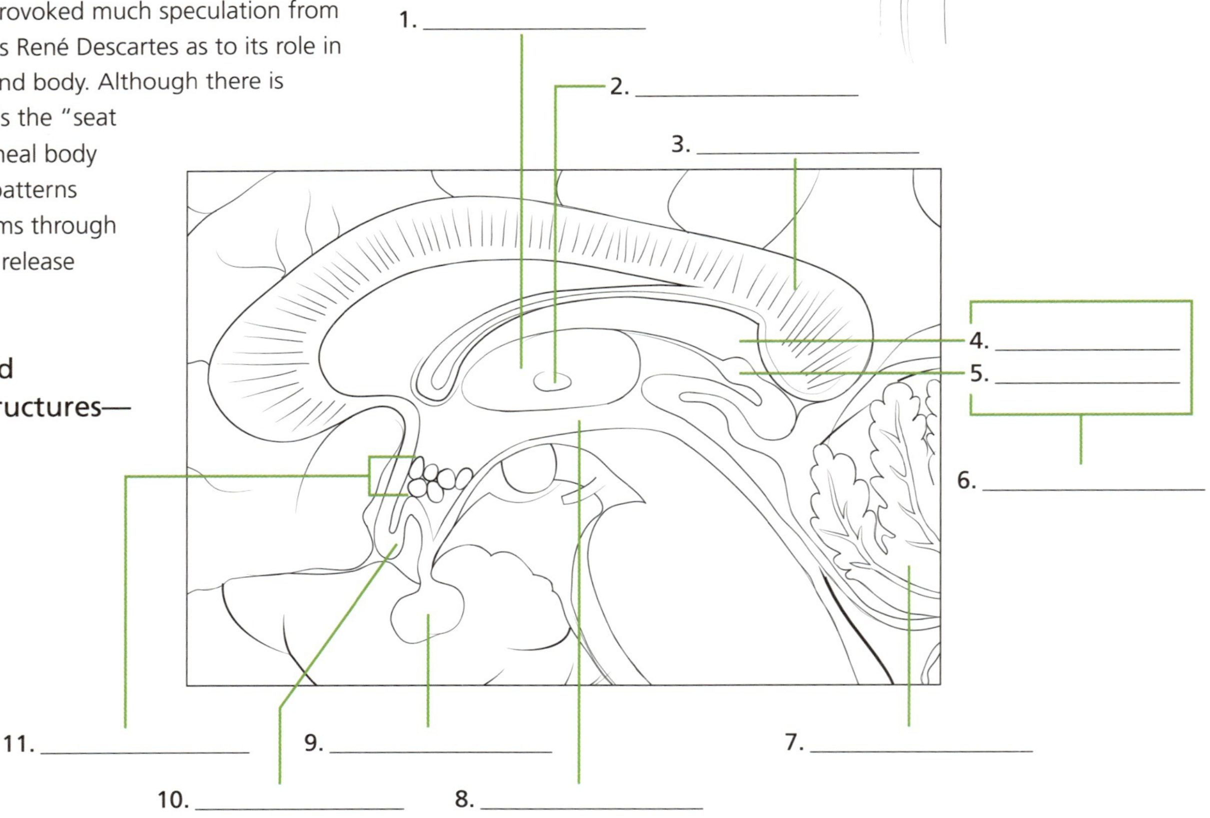

1. ____________
2. ____________
3. ____________
4. ____________
5. ____________
6. ____________
7. ____________
8. ____________
9. ____________
10. ____________
11. ____________

Answers

1. thalamus, 2. interthalamic adhesion, 3. corpus callosum, 4. habenula, 5. pineal body, 6. epithalamus, 7. cerebellum, 8. subthalamus, 9. pituitary gland, 10. optic chiasm, 11. hypothalamus

Nuclei of the Pretectum

A crucial part of the subcortical visual system, the pretectum is a midbrain structure composed of seven different nuclei. Through its reciprocal connections to the retina and efferents to the thalamus and superior colliculus, the pretectum plays a role in behavioral adaptations to light, such as the pupillary light reflex, optokinetic reflex, and circadian rhythms. Its seven highly interconnected nuclei are small and often difficult to define, although they do have different projection patterns. As part of ascending visual pathways, the olivary, anterior, and posterior pretectal nuclei project to the dorsal thalamus. The olivary pretectal nucleus is involved in pupillary constriction via its projections to the Edinger-Westphal nucleus. In contrast, descending projections from the anterior, posterior, and medial pretectal nuclei send information to the vestibulo-ocular system.

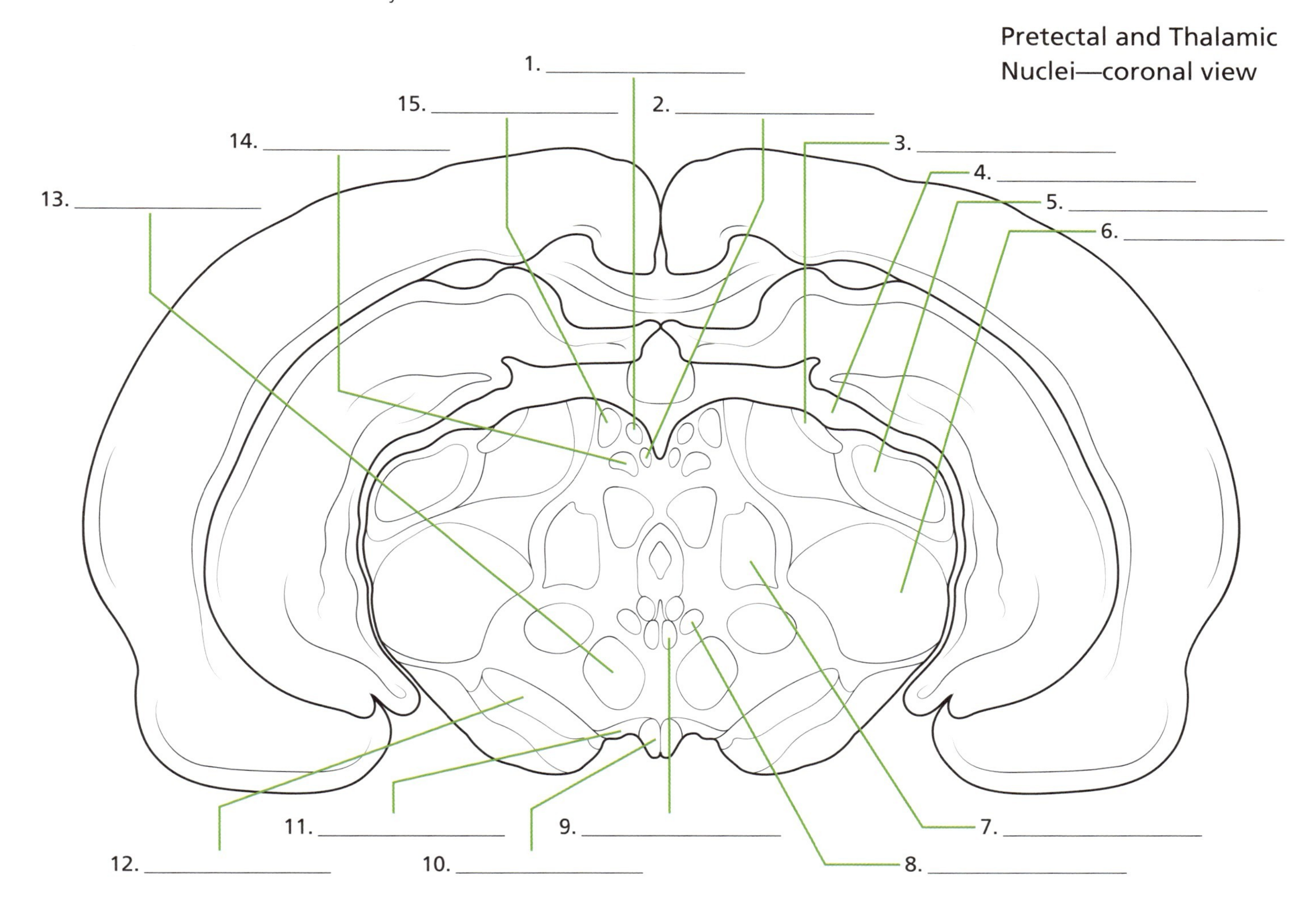

Answers

1. olivary pretectal nucleus, 2. medial pretectal nucleus, 3. pulvinar, 4. optic tract, 5. lateral geniculate nucleus, 6. anterior pretectal nucleus, 7. nucleus of posterior commissure, 8. interstitial nucleus of Cajal, 9. Edinger-Westphal nucleus, 10. interpeduncular nucleus, 11. ventral tegmental area, 12. substantia nigra, 13. red nucleus, 14. posterior pretectal nucleus, 15. nucleus of the optic tract

Nuclei of the Hypothalamus

The hypothalamus is an almond-sized region of the brain located above the brainstem, just below the thalamus. Part of the limbic system, the hypothalamus is involved in many different functions, including facilitating communication between the nervous and endocrine systems; controlling hunger, thirst, and body temperature; and regulating sleep, fatigue, and the body's circadian rhythms.

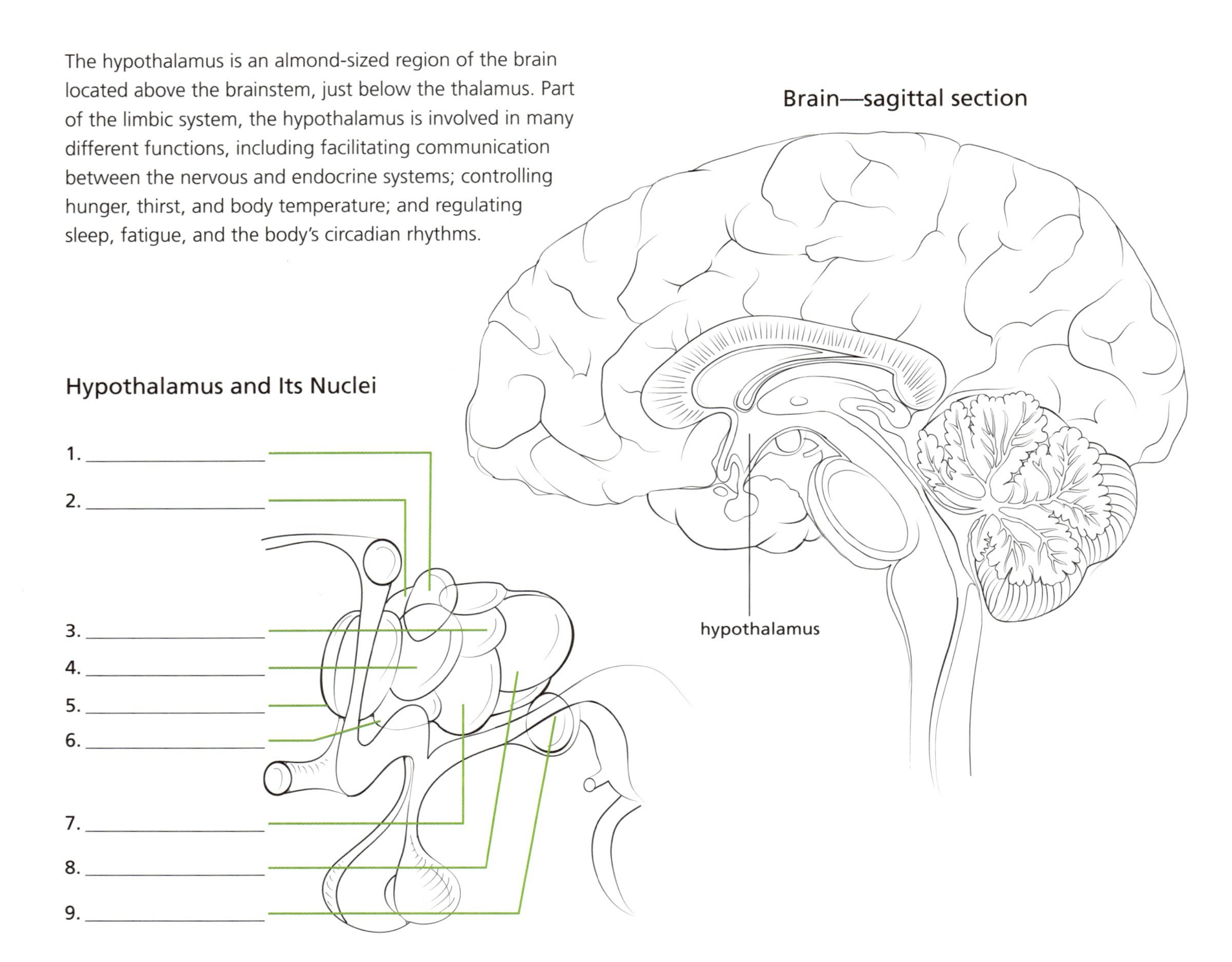

These diverse functions are carried out by various nuclei that compose the hypothalamus. While some nuclei are anatomically distinct, others are not. The nuclei are divided into three major regions. The anterior region includes the anterior hypothalamus and preoptic area (both involved in thermoregulation), the suprachiasmatic nucleus (which modulates circadian rhythm), and the paraventricular nucleus (which releases a number of metabolic hormones). The posterior region includes the posterior hypothalamus (thermoregulation) and mammillary bodies (involved in episodic memory). In the center is the lateral hypothalamus, which primarily contains orexin neuropeptides (which modulate feeding and wakefulness), and the dorsomedial and ventromedial hypothalamic nuclei (which are involved in hunger and satiety, respectively).

Answers

1. paraventricular nucleus, 2. lateral hypothalamus, 3. dorsomedial hypothalamus, 4. anterior hypothalamus, 5. preoptic area, 6. suprachiasmatic nucleus, 7. ventromedial hypothalamus, 8. posterior hypothalamus, 9. mammillary bodies

Mammillary Bodies

The mammillary bodies are a pair of small round structures located in the posterior region of the hypothalamus. Due to their unique location in the anterior arches of the fornix, the mammillary bodies are involved in episodic memory, or memories of autobiographical events, such as locations, times, and their associated emotions.

The mammillary bodies receive this information from the hippocampi, involved in consolidation of short-term memories for long-term storage, and the amygdalae, which process our emotional reactions to events. Damage to the mammillary bodies is associated with anterograde amnesia, or the inability to create new memories. Atrophy of the mammillary bodies is seen in Alzheimer's disease, schizophrenia, and heart failure, but the role of this structure in disease is not entirely clear.

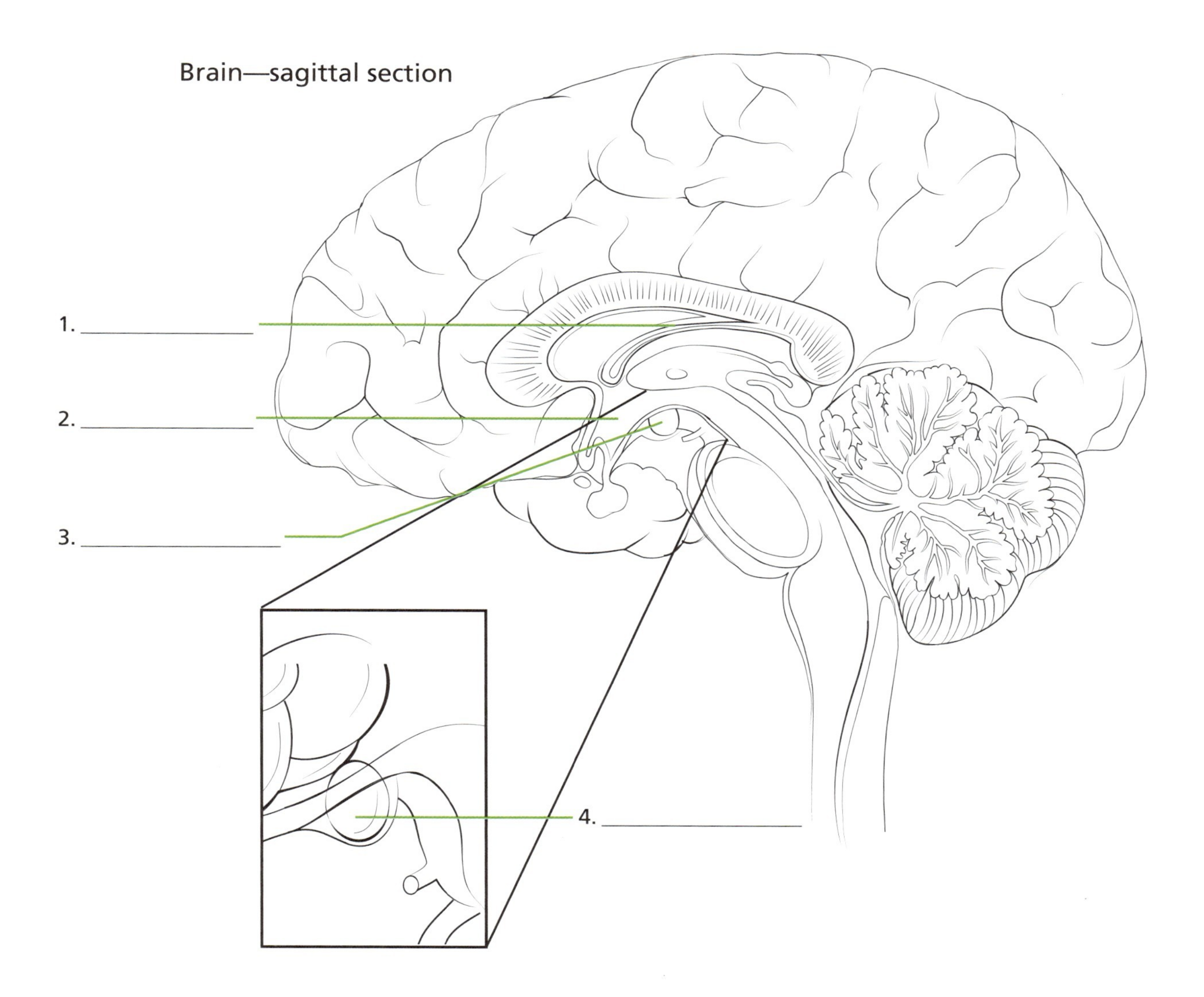

Answers

1. fornix, 2. hypothalamus, 3. mammillary bodies, 4. mammillary bodies

Hypothalamic Stimulation and Its Effects on Appetite and Rage

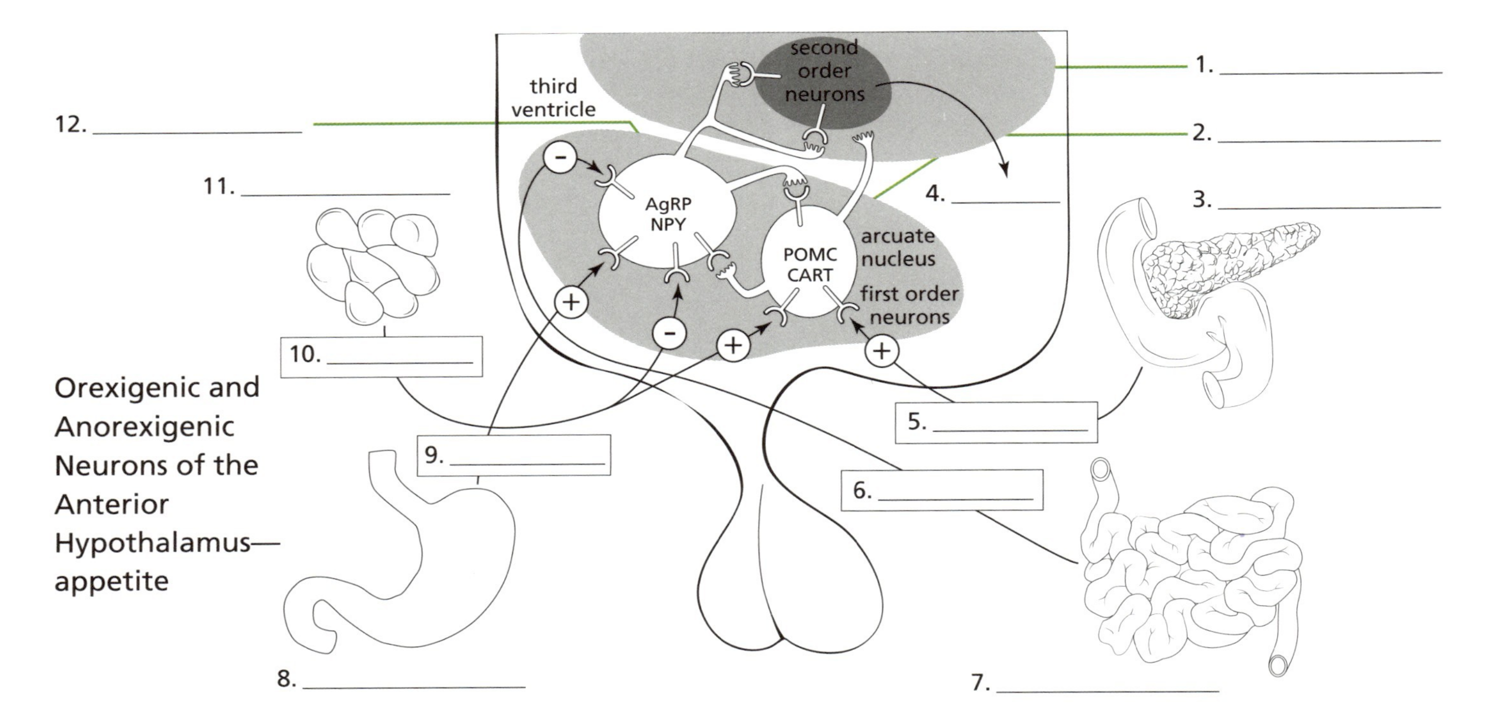

Orexigenic and Anorexigenic Neurons of the Anterior Hypothalamus—appetite

Feeding behavior is regulated by connections between orexigenic (hunger-promoting) and anorexigenic (satiety-promoting) neurons in the anterior hypothalamus with other regions of the brain. Orexigenic neurons, such as those that produce agouti-related protein (AgRP) and neuropeptide Y (NPY), work to increase appetite and decrease energy expenditure. Orexigenic neurons are activated by ghrelin (produced in the gastrointestinal tract) and inhibited by leptin (made by adipose cells) and peptide YY (PYY, from the small intestine). Alternatively, anorexigenic neurons, such as those that produce pro-opiomelanocortin (POMC) and cocaine- and amphetamine-regulated transcript (CART), are activated by leptin and insulin (produced by the pancreas) and reduce appetite. These neurons communicate with one another, as well as with second-order neurons in the paraventricular nucleus, to project to the nucleus of the solitary tract (NTS) and either promote or diminish feeding behavior.

Stimulation of the lateral hypothalamus is associated with rage and aggressive behavior; lesions to the lateral hypothalamus eliminate aggressive behavior. The ventromedial hypothalamus counters the action of the lateral hypothalamus by reducing rage reactions. Lesions to the ventromedial hypothalamus, then, trigger unchecked aggression and savage behavior.

Hypothalamus—coronal slice

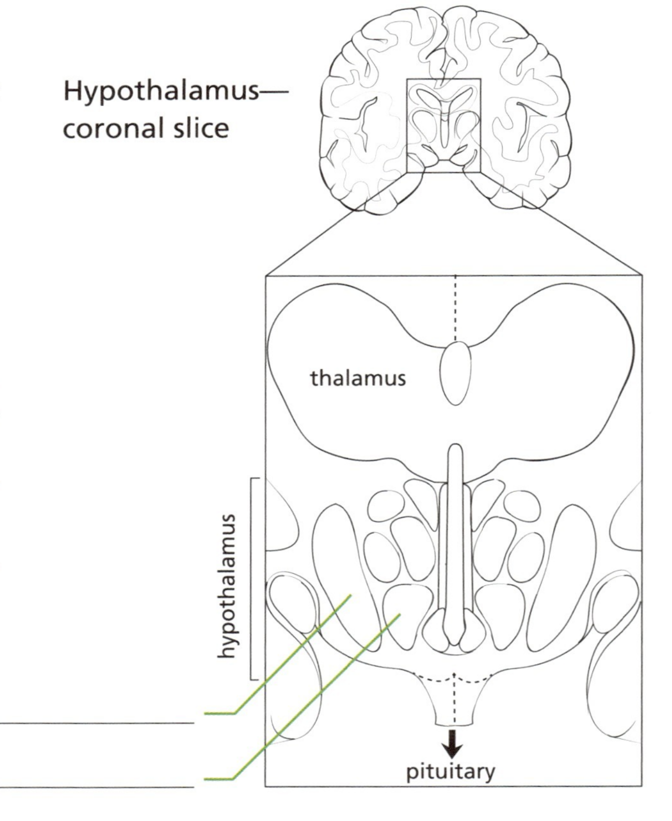

Answers

1. paraventricular nucleus, 2. anorexigenic neuron, 3. pancreas, 4. nucleus of the solitary tract, 5. insulin, 6. peptide YY, 7. small intestine, 8. gastrointestinal tract, 9. ghrelin, 10. leptin, 11. adipose tissue, 12. orexigenic neuron, 13. lateral hypothalamus, 14. ventromedial hypothalamus

Inputs to the Hypothalamus

Given the many functions of the hypothalamus, it's not surprising that its structure, despite its small size, has many different inputs. One of these inputs is the olfactory bulb, which is thought to convey information that affects feeding behavior and reproductive activity. Another input, the photosensitive ganglion cells from the retina, sends information to the suprachiasmatic nucleus regarding light intensity in an effort to synchronize our bodies to a 24-hour circadian rhythm. The hypothalamus also receives input from circumventricular organs, or brain structures characterized by extensive vasculature and lack of a blood–brain barrier, as mechanisms of osmoregulation, blood-pressure modulation, and detection of toxins. In addition, the hypothalamus is a major input node for the limbic system, modulating endocrine function, behavior, emotional responses, sexual function, and autonomic control. From the reticular formation in the brainstem, the hypothalamus also receives information about skin temperature. Finally, the ascending fibers from the nucleus of the solitary tract relay a wide range of information to the hypothalamus, including gut distension and satiety, taste, and blood pressure.

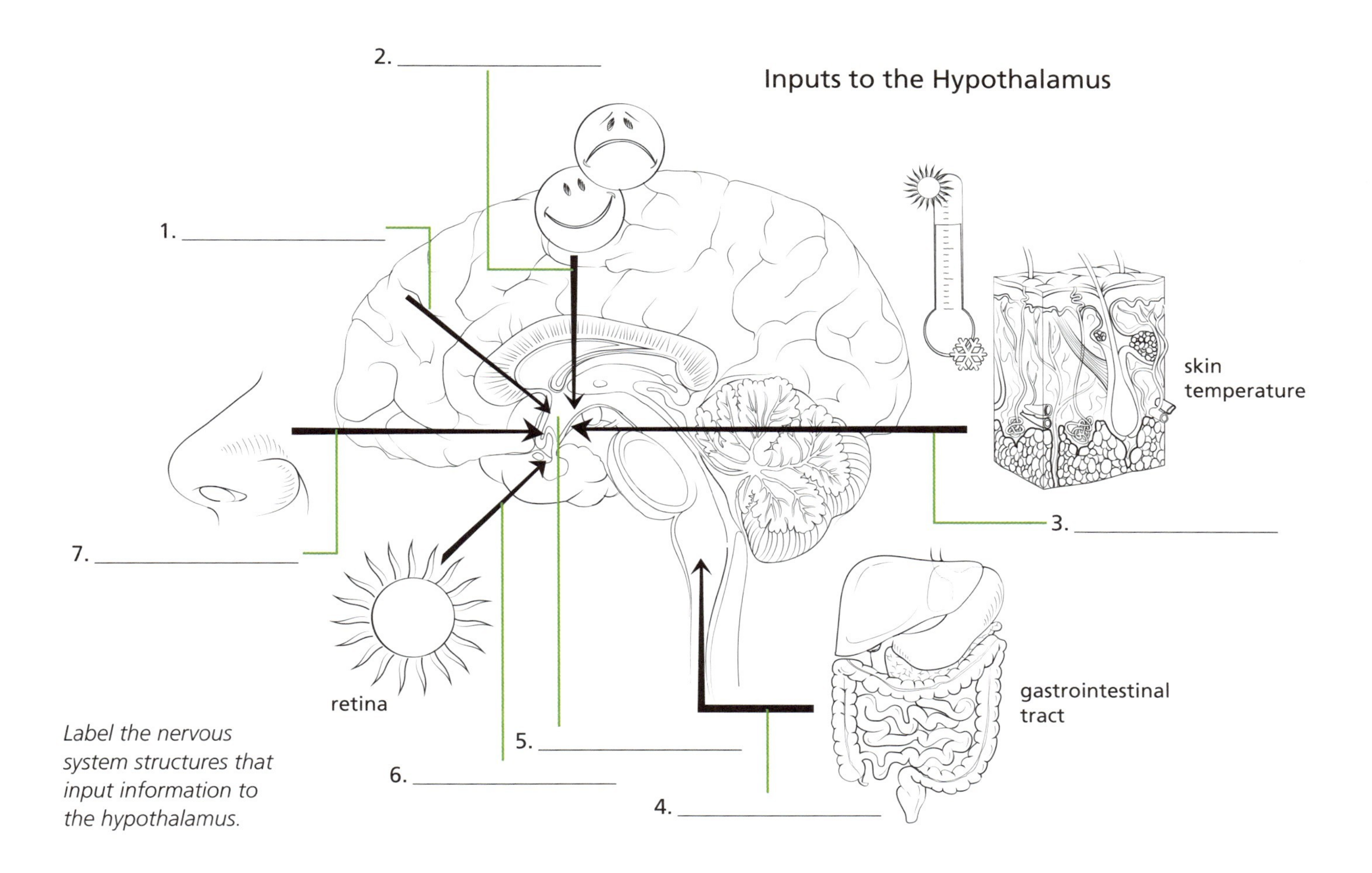

Label the nervous system structures that input information to the hypothalamus.

Answers

1. circumventricular organs, 2. limbic system, 3. reticular formation, 4. nucleus of the solitary tract, 5. hypothalamus, 6. suprachiasmatic nucleus, 7. olfactory bulb

Outputs from the Hypothalamus

Despite its many different inputs, there are only two major outputs from the hypothalamus. First, the hypothalamus is the only link between the body's nervous and endocrine systems through its connection with the pituitary gland, located just below. Releasing hormones (excitatory) or inhibiting hormones from the hypothalamus travel to the anterior or posterior lobes of the pituitary, via the hypophyseal portal system and infundibulum, respectively. These hormones bind to receptors on pituitary cells, stimulating or inhibiting the secretion of endocrine hormones. The various endocrine hormones, in turn, bind to and activate receptors on target organs throughout the body.

The second major output of the hypothalamus is the autonomic nervous system. The autonomic nervous system has two branches: sympathetic ("fight or flight") and parasympathetic ("rest and digest"). The hypothalamus connects to the sympathetic nervous system via projections to preganglionic cell bodies in the lateral horn of the thoracic and lumbar segments of the spinal cord. The connection to the parasympathetic branch, on the other hand, is via projections to preganglionic cell bodies in the sacral region of the spinal cord, as well as the motor components of cranial nerves VII, IX, and X.

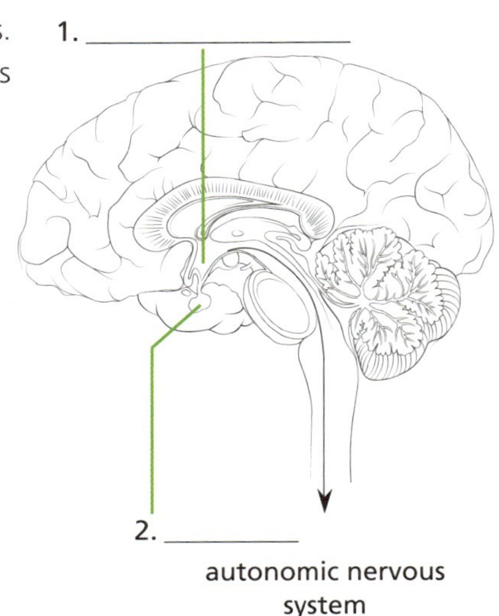

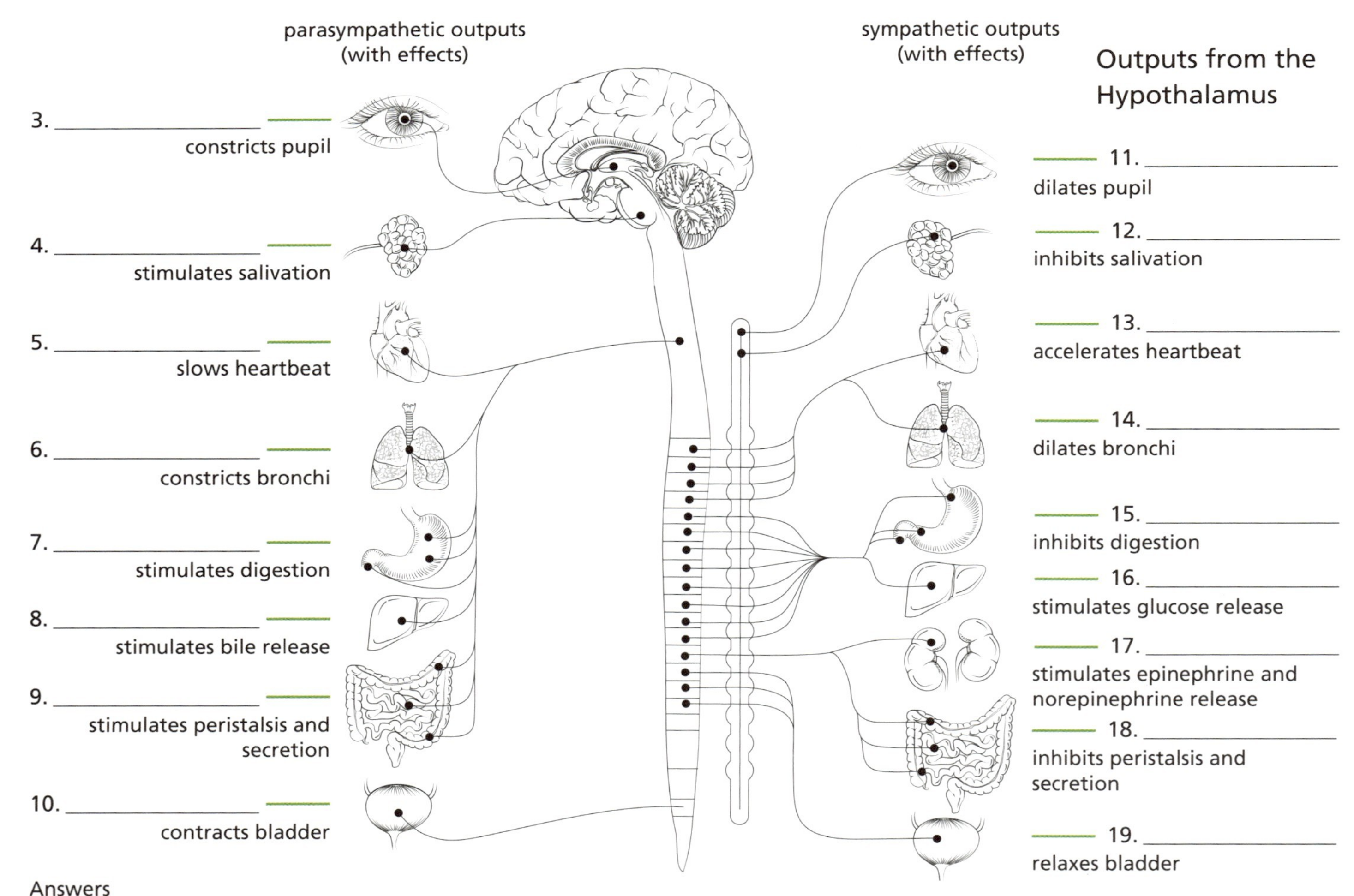

Answers

1. hypothalamus, 2. pituitary gland, 3. eyes, 4. salivary glands, 5. heart, 6. lungs, 7. stomach, 8. liver, 9. intestines, 10. bladder, 11. eyes, 12. salivary glands, 13. heart, 14. lungs, 15. stomach, 16. liver, 17. kidneys, 18. intestines, 19. bladder

Pituitary Gland

Neurosecretory cells in the hypothalamus—so named because their neurons secrete hormones—are connected to the endocrine system via the pituitary gland, a pea-sized structure just below the hypothalamus. Surrounded by a protective bony cavity called the sella turcica, the pituitary gland has two major lobes. The anterior pituitary, or adenohypophysis, receives hormonal messages from the hypothalamus via the hypophyseal portal system, a system of fenestrated (leaky) capillaries that allows for rapid communication. The posterior pituitary, or neurohypophysis, is connected to the hypothalamus by a structure called the infundibulum, where hormones are carried down axons and then released into the blood.

The anterior and posterior pituitary glands secrete different types of hormones into the bloodstream, which then act on target organs in the body. The anterior pituitary produces gonadotropins, including follicle-stimulating hormone and luteinizing hormone, which act synergistically in the gonads during pubertal development and sex hormone production; growth hormone, which promotes muscle and bone growth; prolactin, involved in milk production; adrenocorticotropic hormone (ACTH), which acts on the adrenal gland to produce a stress response; and thyroid-stimulating hormone, involved in metabolism. The posterior pituitary, on the other hand, secretes oxytocin, involved in uterine contractions and lactation, and antidiuretic hormone (ADH, or vasopressin), which stimulates water retention in the kidneys.

Anterior and Posterior Pituitary Gland

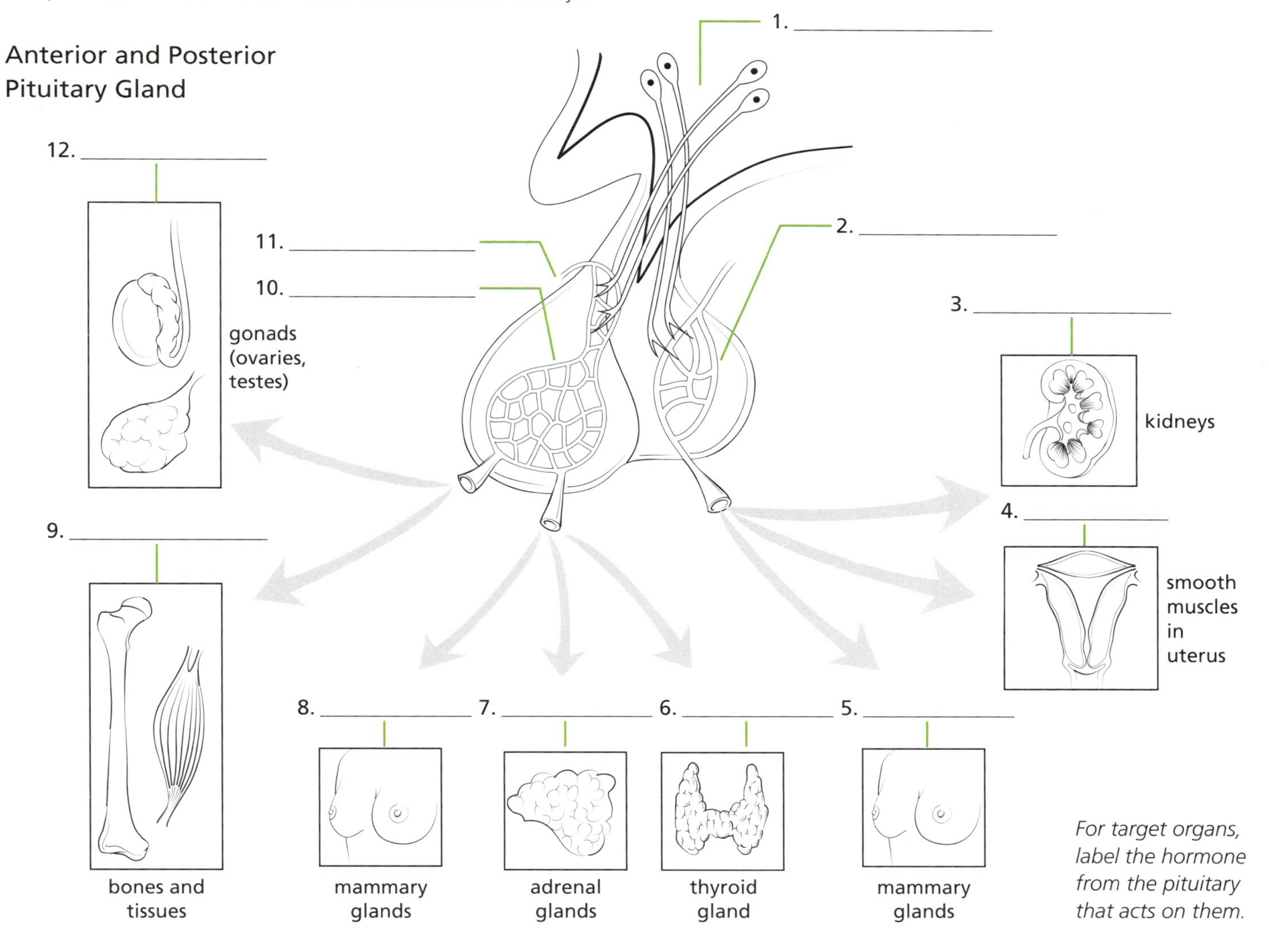

For target organs, label the hormone from the pituitary that acts on them.

Answers

1. hypothalamus, 2. posterior pituitary gland, 3. antidiuretic hormone, 4. oxytocin, 5. oxytocin, 6. thyroid-stimulating hormone, 7. adrenocorticotropic hormone, 8. prolactin, 9. growth hormone, 10. anterior pituitary gland, 11. hypophyseal portal system, 12. gonadotropins

Hypothalamic-Pituitary-Adrenal (HPA) Axis

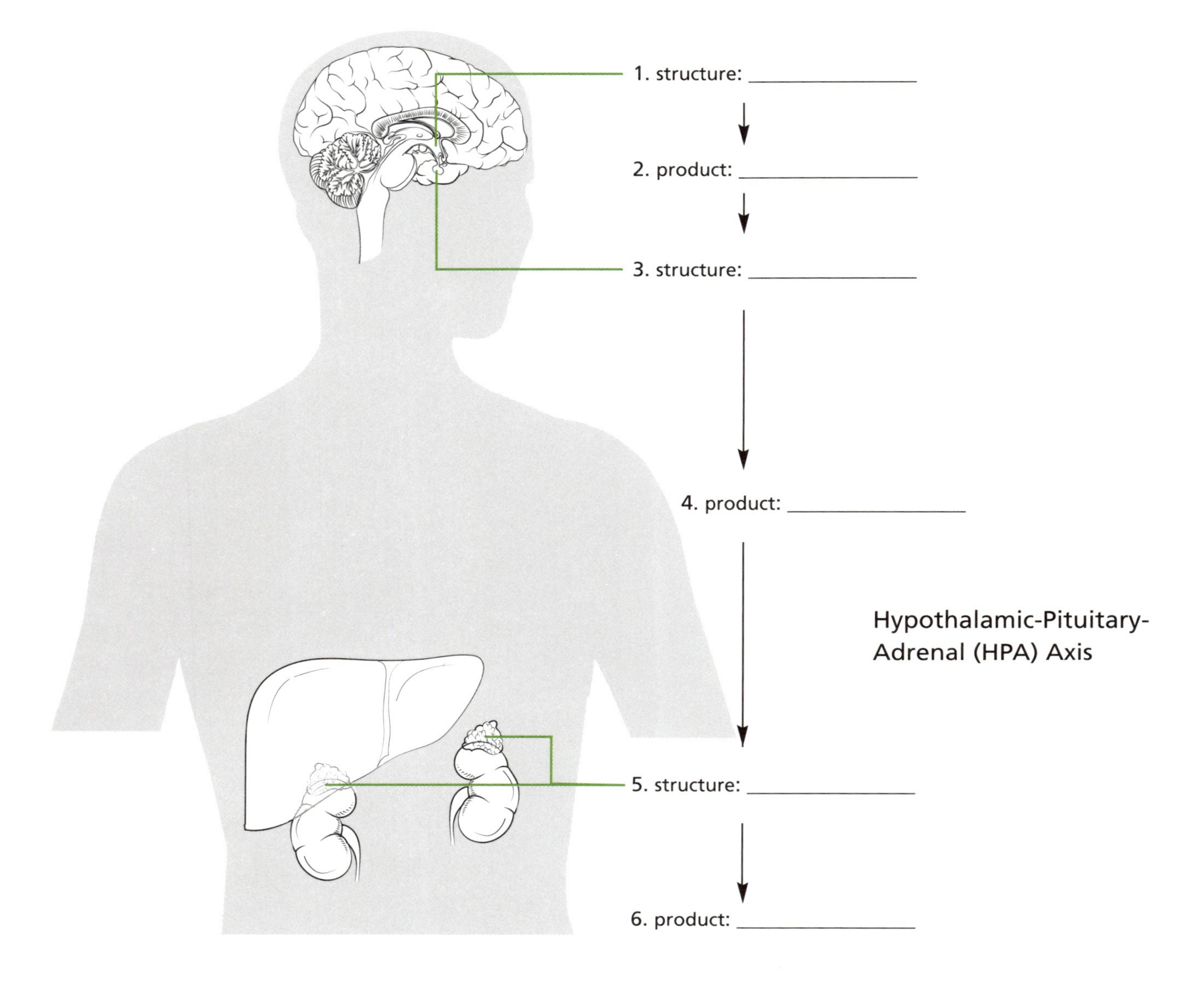

The hypothalamic-pituitary-adrenal axis, or HPA axis, is a system of feedback interactions between three endocrine glands: the hypothalamus, the pituitary gland, and the adrenal glands. Together, this system works to control the body's reaction to stress. In response to a stressor, neurons of the paraventricular nucleus of the hypothalamus synthesize corticotropin-releasing factor (CRF), which is transported to the anterior pituitary. In turn, CRF stimulates secretion of adrenocorticotropic hormone (ACTH), which travels through the bloodstream and targets the adrenal glands, located on top of the kidneys. ACTH then stimulates secretion of the steroid hormone cortisol by the adrenal cortex, the outer layer of the gland. Cortisol has many different functions, such as increasing glucose availability (via gluconeogenesis) and suppressing the immune system, which both function to raise blood sugar levels and increase energy availability, thereby facilitating the "fight or flight" response. Importantly, after its release, cortisol also acts on the hypothalamus and pituitary gland to suppress further CRF and ACTH production, creating a negative feedback cycle.

Answers

1. paraventricular nucleus of the hypothalamus, 2. corticotropin-releasing factor, 3. anterior pituitary gland, 4. adrenocorticotropic hormone, 5. adrenal cortex, 6. cortisol

Topography of the Basal Ganglia

When the basal ganglia were first described, they were grouped together based on the fact that damage in these areas resulted in movement disorders. Now we know that these areas have a much more complex function and play roles in motivation, attention, mood disorders, and other behaviors.

The basal ganglia can be divided into six major nuclei: caudate, putamen, nucleus accumbens, and globus pallidus (with internal and external segments), which are grouped together along the top and sides of the thalamus; and the subthalamic nucleus and substantia nigra, which are ventral and further back toward the brainstem, running parallel to each other.

The caudate includes a large head rostral, tapering to a curved tail caudally. The nucleus accumbens is located at the most rostral point of the basal ganglia, where the caudate and putamen join. The globus pallidus occupies the place of the accumbens in more caudal slices. The caudate and putamen are separated along their length by the internal capsule.

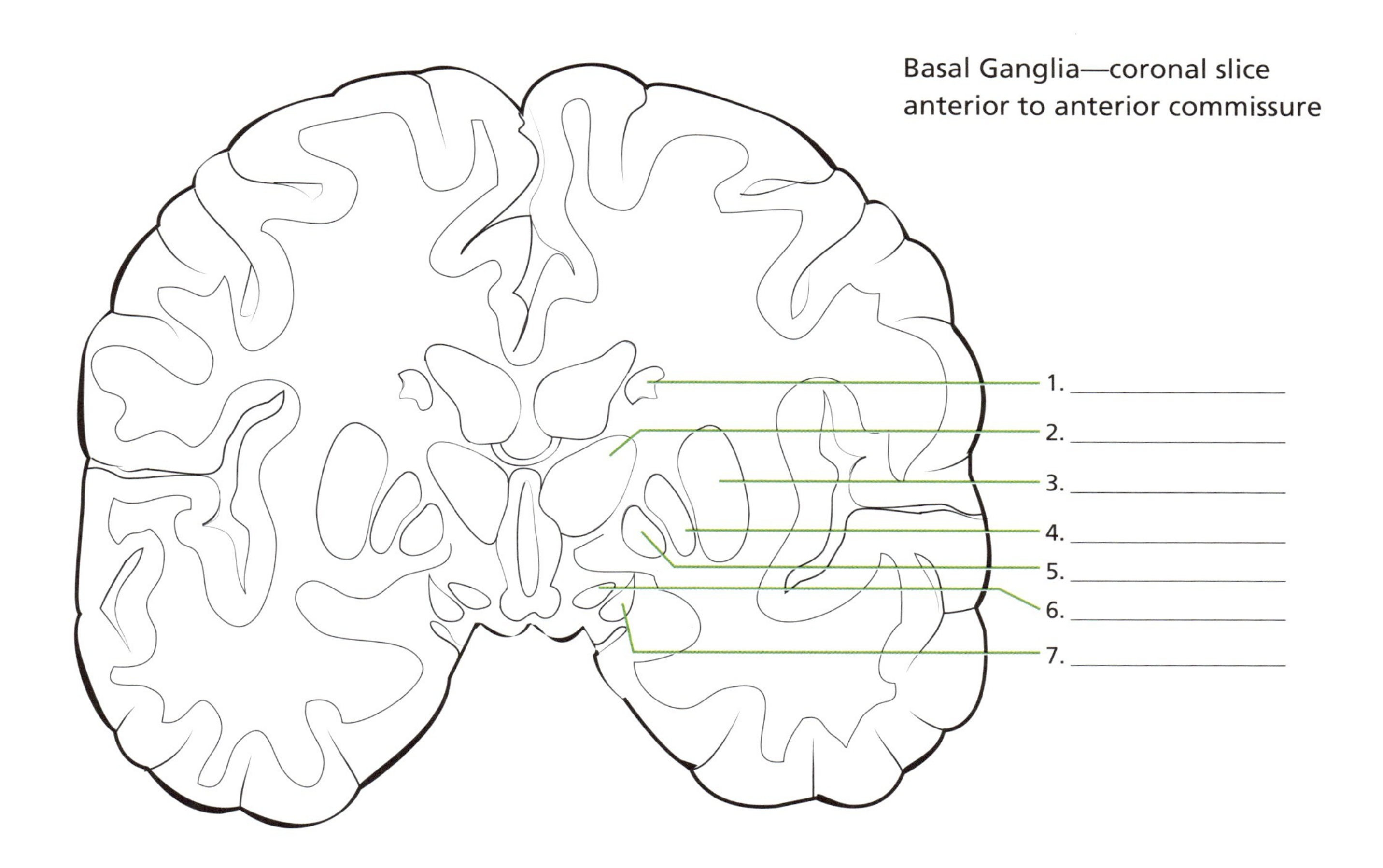

Answers

1. caudate, 2. thalamus, 3. putamen, 4. globus pallidus external, 5. globus pallidus internal, 6. subthalamic nucleus, 7. substantia nigra

Ventral Striatum: Nucleus Accumbens and Olfactory Tubercle

Ventral Striatum—coronal slices at level of anterior commissure (top) and anterior horn of lateral ventricle (bottom)

1. ____________
2. ____________
3. ____________
4. ____________
5. ____________
6. ____________
7. ____________
8. ____________

The ventral striatum is the portion of the basal ganglia that is often heavily linked with feelings of reward and motivation. It is made up of the nucleus accumbens and the olfactory tubercle. While the nucleus accumbens is easily located where the caudate and putamen meet rostrally, the olfactory tubercle is hard to spot in humans and nonhuman primates.

The olfactory tubercle receives input from limbic regions of the brain, including the amygdala, and also receives direct input from the olfactory bulb. Despite its small size, it is a processing center that plays an important role in motivated behaviors, including attention and reward.

The nucleus accumbens can be divided into core and shell regions, and it plays important roles in motivation, reinforcement, and reward. It is active both in strong senses of high (from drugs, sex, food, or other pleasurable sensations) and in negative states (such as drug withdrawal). The accumbens receives dopamine signals from the dopamine-producing neurons in the ventral tegmental area. The dopamine release from these neurons in this area can be monitored in animals and has been shown to increase during reward-prediction and decrease when expected rewards are not delivered.

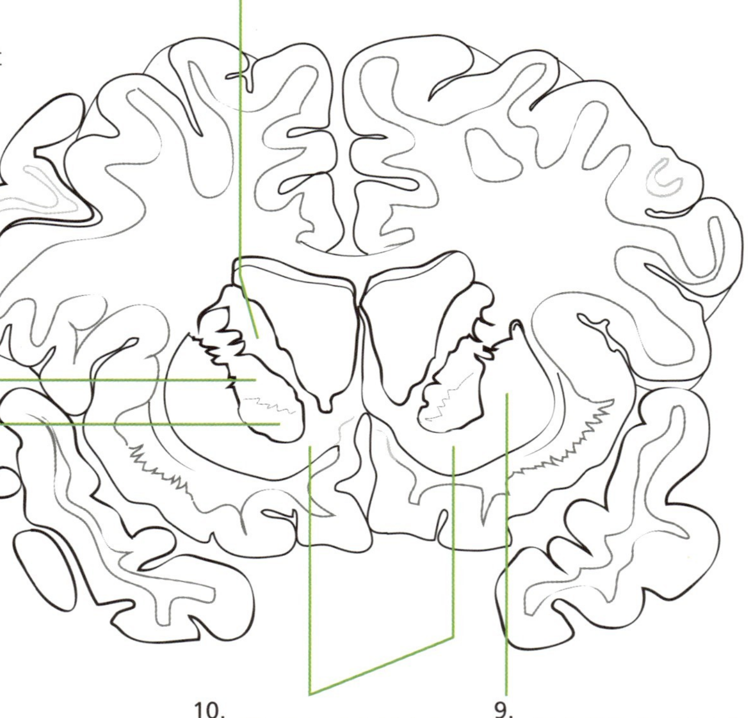

Answers

1. head of caudate, 2. internal capsule, 3. putamen, 4. globus pallidus external, 5. globus pallidus internal, 6. caudate, 7. olfactory tubercle, 8. caudate, 9. putamen, 10. nucleus accumbens, 11. globus pallidus, 12. internal capsule

Dorsal Striatum (Body of Caudate)

The dorsal striatum is usually described as including the putamen and the caudate, separated along their length by the internal capsule.

The putamen receives inputs from the motor and sensory areas of the cortex, from the dopaminergic neurons of the substantia nigra, and from the centromedian nucleus of the thalamus. It is heavily involved in motor function.

The caudate is heavily involved in decision-making. It is the main input region for the basal ganglia and receives information from the prefrontal cortex, relays to the rest of the basal ganglia structures, receives returning information, and projects back to the cortex in what is known as a corticostriatal loop.

Striatum—surface view

1. ________________
2. ________________
3. ________________
4. ________________
5. ________________
6. ________________
7. ________________
8. ________________

Striatum—parasagittal section

9. ________________
10. ________________
11. ________________
12. ________________
13. ________________
14. ________________
15. ________________

Answers

1. head of caudate, 2. body of caudate, 3. tail of caudate, 4. nucleus accumbens, 5. putamen, 6. globus pallidus, 7. internal capsule, 8. thalamus, 9. tail of caudate, 10. head of caudate, 11. putamen, 12. globus pallidus external, 13. globus pallidus internal, 14. amygdala, 15. thalamus

Connections of the Basal Ganglia

The movement-processing pathways through the basal ganglia can be subdivided into direct and indirect pathways. Both of these pathways converge at the internal segment of the globus pallidus and exit the basal ganglia via the thalamus. These pathways receive inputs from the cortex and send them out through the thalamus to the motor cortex after processing. The processing helps to determine which movement in a series should be performed first, resulting in smooth initiation and follow-through of voluntary movements.

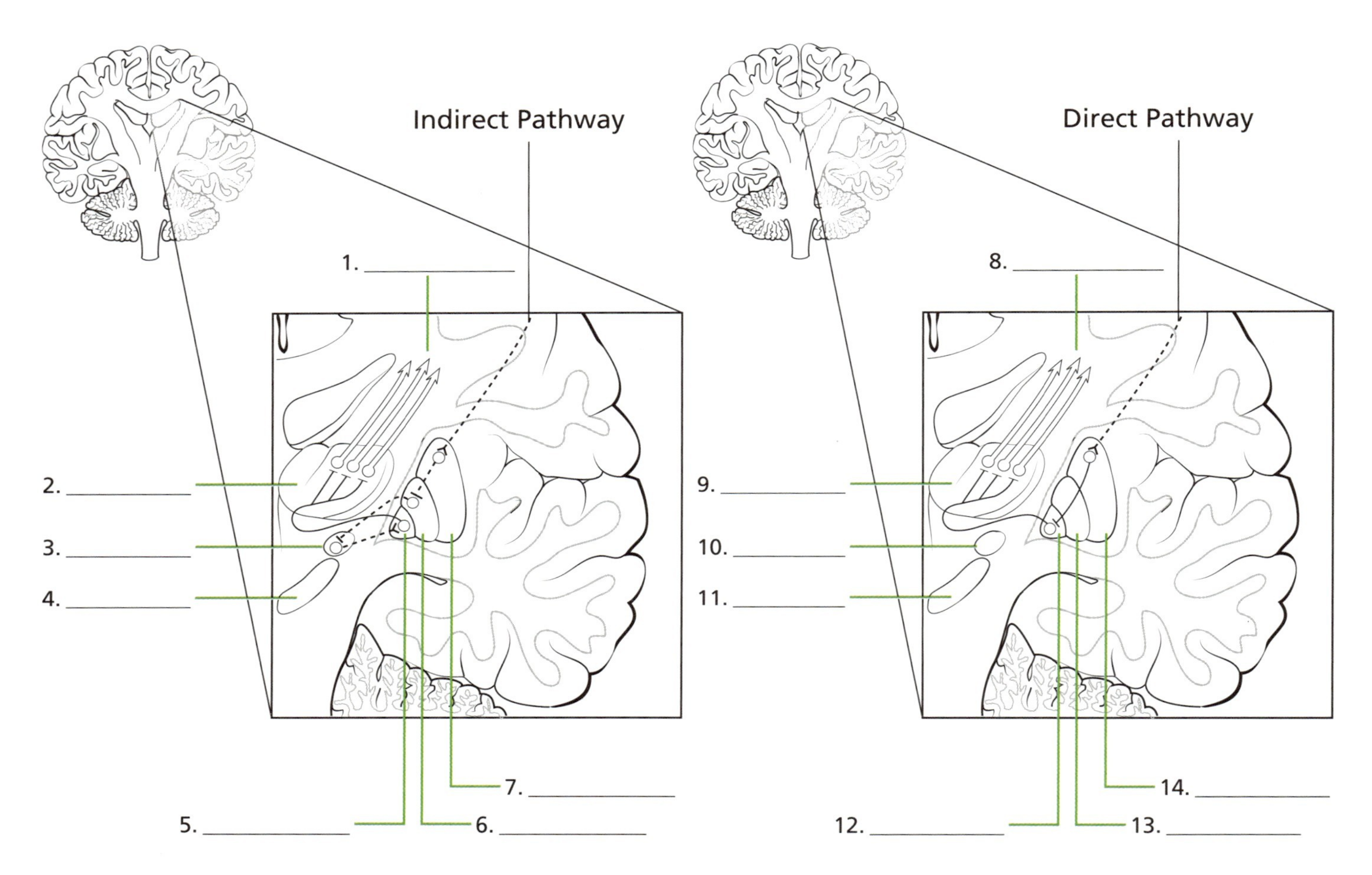

The indirect pathway synapses first in the striatum, which makes an inhibitory connection to the external segment of the globus pallidus. This makes a further inhibitory connection to the subthalamic nucleus, which then makes an excitatory connection onto the internal segment of the globus pallidus. This final connection means the globus pallidus is excited, and its inhibitory connection inhibits neurons in the thalamus, preventing excitatory signals to the cortex.

The direct pathway synapses first in the caudate and putamen on an inhibitory neuron. That neuron makes an inhibitory connection in the internal segment of the globus pallidus. This next inhibitory neuron synapses in the thalamus onto final excitatory inputs back toward the cortex. Normally, inhibitory neurons in the internal segment of the globus pallidus inhibit neurons in the thalamus. But when the direct pathway is active, the inhibitory connections from the caudate and putamen inhibit the globus pallidus, releasing the thalamus and allowing it to excite the cortex.

Answers

1. to cortex, 2. thalamus, 3. subthalamic nucleus, 4. substantia nigra, 5. globus pallidus internal, 6. globus pallidus external, 7. caudate and putamen, 8. to cortex, 9. thalamus, 10. subthalamic nucleus, 11. substantia nigra, 12. globus pallidus internal, 13. globus pallidus external, 14. caudate and putamen

Effects of Lesions of the Basal Ganglia Circuitry

Two of the most well-known disorders of the basal ganglia are Parkinson's disease and Huntington's disease. Parkinson's disease, also known as idiopathic parkinsonism, is a progressive and degenerative disorder with no known direct cause. Signs include tremor, muscle rigidity, slow gait, and difficulty initiating movements. Advanced cases may also involve depression and dementia. It arises from the death of the dopaminergic cells in the substantia nigra. Current treatments involve giving exogenous dopamine as L-dopa (which is metabolized to dopamine in the CNS), or deep brain stimulation. In brain slices with advanced Parkinson's, the disease can easily be spotted by the loss of the dark-stained band of dopaminergic neurons in the compact part (pars compacta) of the substantia nigra.

Huntington's disease is another disorder that affects muscle coordination. It is most known for producing involuntary writhing movements known as Huntington's chorea. The disease is the result of a genetic mutation that affects the huntingtin protein, causing misfolding and accumulation. It affects the whole brain, but the basal ganglia is exceptionally vulnerable, especially spiny neurons of the caudate and putamen. In affected patients, the condition can be seen by atrophy in the caudate and putamen, and subsequent increase in ventricular volume.

Huntington's Disease

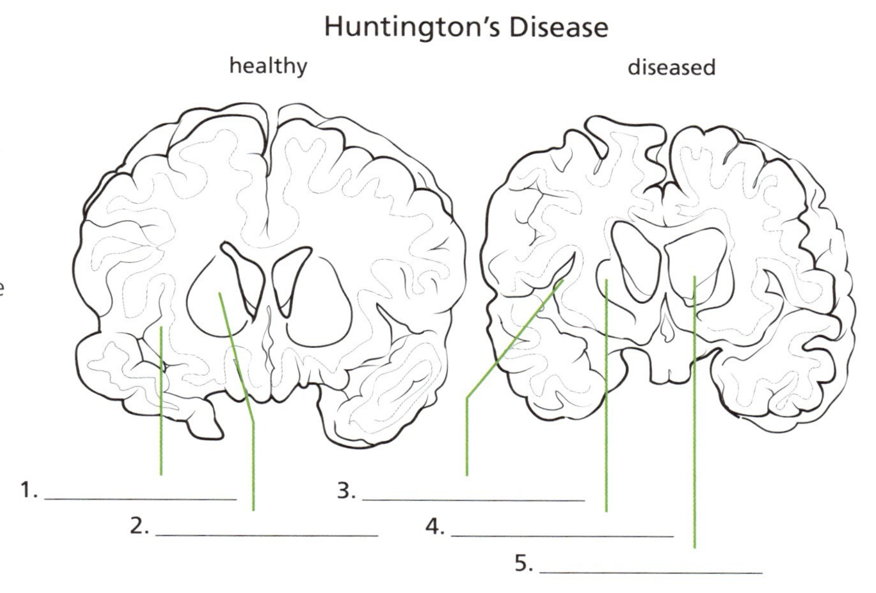

Parkinson's Disease

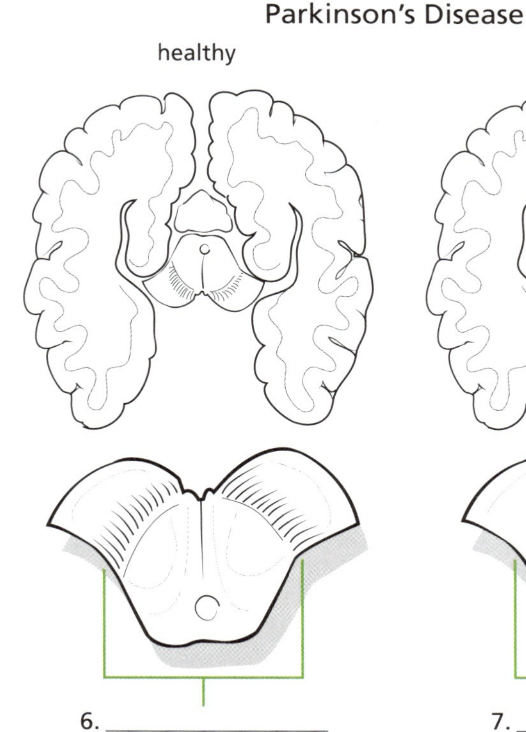

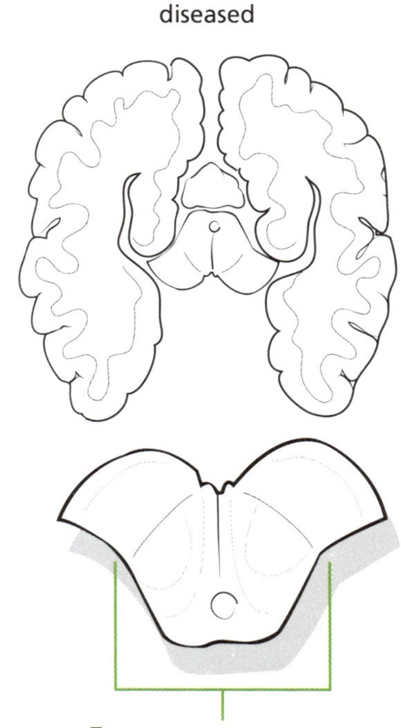

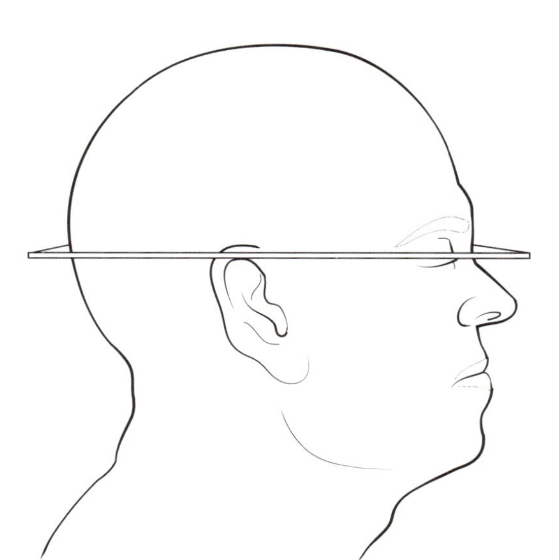

Cut Section of Midbrain Where Portion of Substantia Nigra Is Visible

Answers

1. healthy putamen, 2. healthy caudate, 3. putamen in Huntington's patient, 4. caudate in Huntington's patient, 5. ventricle in Huntington's patient, 6. healthy substantia nigra, 7. substantia nigra in Parkinson's patient

Overview of the Topography of the Cerebral Hemispheres: Lateral Surface

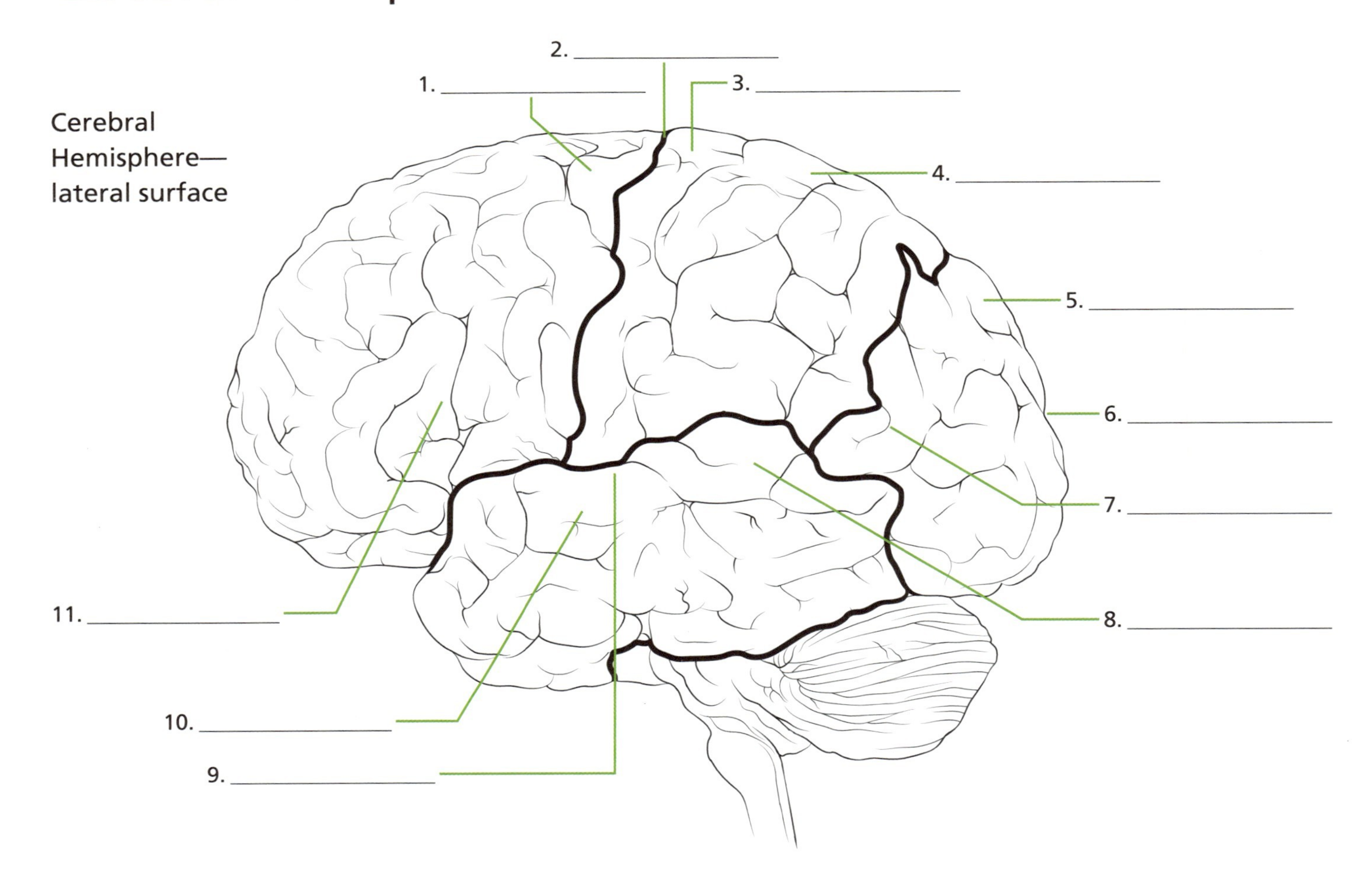

The lateral surface is the best perspective for viewing the four lobes of the cerebral hemispheres: frontal, temporal, parietal, and occipital. The frontal lobe is the most anterior, separated from the parietal lobe by the central sulcus, and contains the motor cortex, which is the origin of voluntary muscle impulses and also initiates responses to stimuli. In addition, it is responsible for intellectual processes, such as planning and decision-making. The temporal lobe is separated from the frontal and parietal lobes by the lateral sulcus and is the location of the auditory cortex, hippocampus, and other major sensory and cognitive structures. Its functions include language comprehension, memory, and recognition of stimuli. The parietal lobe is posterior to the central sulcus and houses the somatic sensory cortex. This lobe is responsible for bodily sensation and attending to stimuli, as well as expression of thoughts and emotions. The occipital lobe is the posterior-most lobe and is the location of the visual cortex, where visual information is processed. Major functions include focusing the eye, conscious perception of vision, and correlating visual images with previous experiences.

Answers

1. primary motor cortex, 2. central sulcus, 3. primary sensory cortex, 4. somatic sensory association area, 5. visual association area, 6. visual cortex, 7. reading comprehension area, 8. Wernicke's sensory speech area, 9. auditory cortex, 10. auditory association area, 11. motor speech (Broca's) area

Overview of the Topography of the Cerebral Hemispheres: Medial Surface

The principal structures on the medial surface of the brain include the corpus callosum, fornix, and parieto-occipital fissure. The corpus callosum is a large bundle of neural fibers that connects the left and right hemispheres of the brain to facilitate communication. It is also the largest white matter structure in the brain. The posterior end of the corpus callosum is known as the splenium, and the anterior end is known as the genu. A condition known as agenesis of the corpus callosum is one of the most common brain malformations observed in humans. Agenesis of the corpus callosum occurs during the first trimester and is thought to be caused by a variety of factors including chromosomal defects, toxins, or prenatal infections. Individuals with agenesis of the corpus callosum have difficulty transferring information between cerebral hemispheres.

Cutting the corpus callosum to disrupt the connection is a common treatment for refractory epilepsy. Other key areas on the medial surface play major roles in awareness, memory, emotions, and visual processing. For example, the fornix is a bundle of nerve fibers that carries signals from the hippocampus to other parts of the limbic system. Damage to the fornix has been shown to cause memory loss. The cingulate cortices are major players in emotional and cognitive processing via the limbic system.

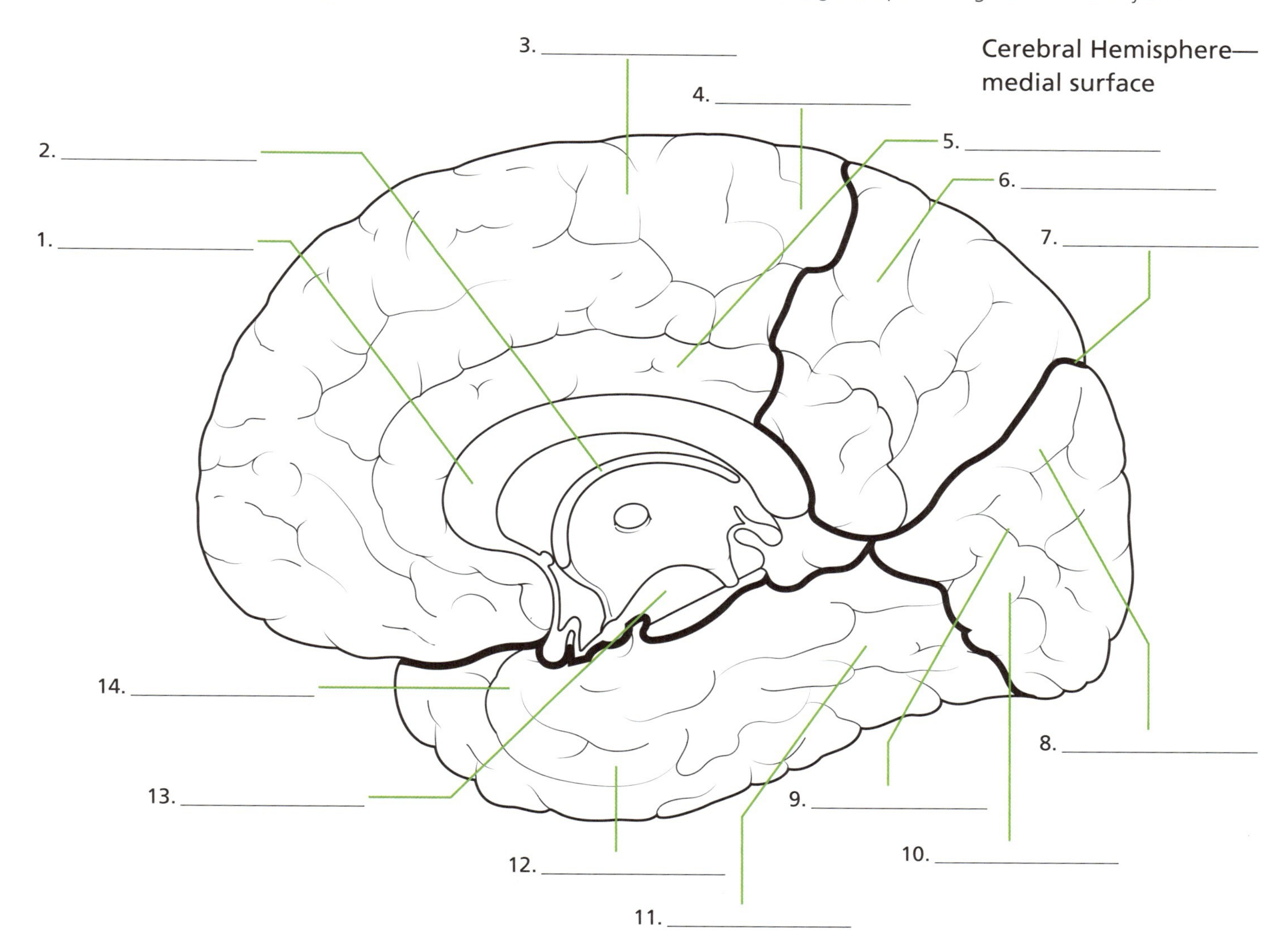

Answers

1. corpus callosum, 2. fornix, 3. superior frontal gyrus, 4. paracentral lobule, 5. cingulate gyrus, 6. precuneus, 7. parieto-occipital fissure, 8. cuneus, 9. calcarine fissure, 10. lingual gyrus, 11. parahippocampal gyrus, 12. fusiform gyrus, 13. cut diencephalon, 14. uncus

Overview of the Topography of the Cerebral Hemispheres: Inferior Surface

The inferior surface offers a view of various regions on the "underside" of the brain that are involved with the senses. An exception to this is the view of the mammillary bodies, which are a part of the limbic system and are involved in recollective memory. The sense of smell is a major player from this angle, based on the presence of the olfactory bulb, olfactory tract, and lateral and medial olfactory stria. The olfactory tract is unique in that it connects parts of the olfactory bulb to a number of target regions of the brain, including the piriform cortex, amygdala, and entorhinal cortex. The olfactory bulb functions using a neural circuit, which begins with input from the olfactory nerves in the nose and ends in the olfactory cortex in the brain. Other sensory areas in this part of the brain are the inferior olivary nucleus, involved with motor control, and the gyrus rectus, involved with higher cognitive function. The orbitofrontal cortex (OFC) is another region involved with higher cognitive processing. It is responsible for sensory integration and decision-making; damage to the OFC results in disinhibited behavior. The optic chiasm is caudal to the OFC and is the location where the optic nerves cross. Processing information from both nerves allows for binocular and stereoscopic vision.

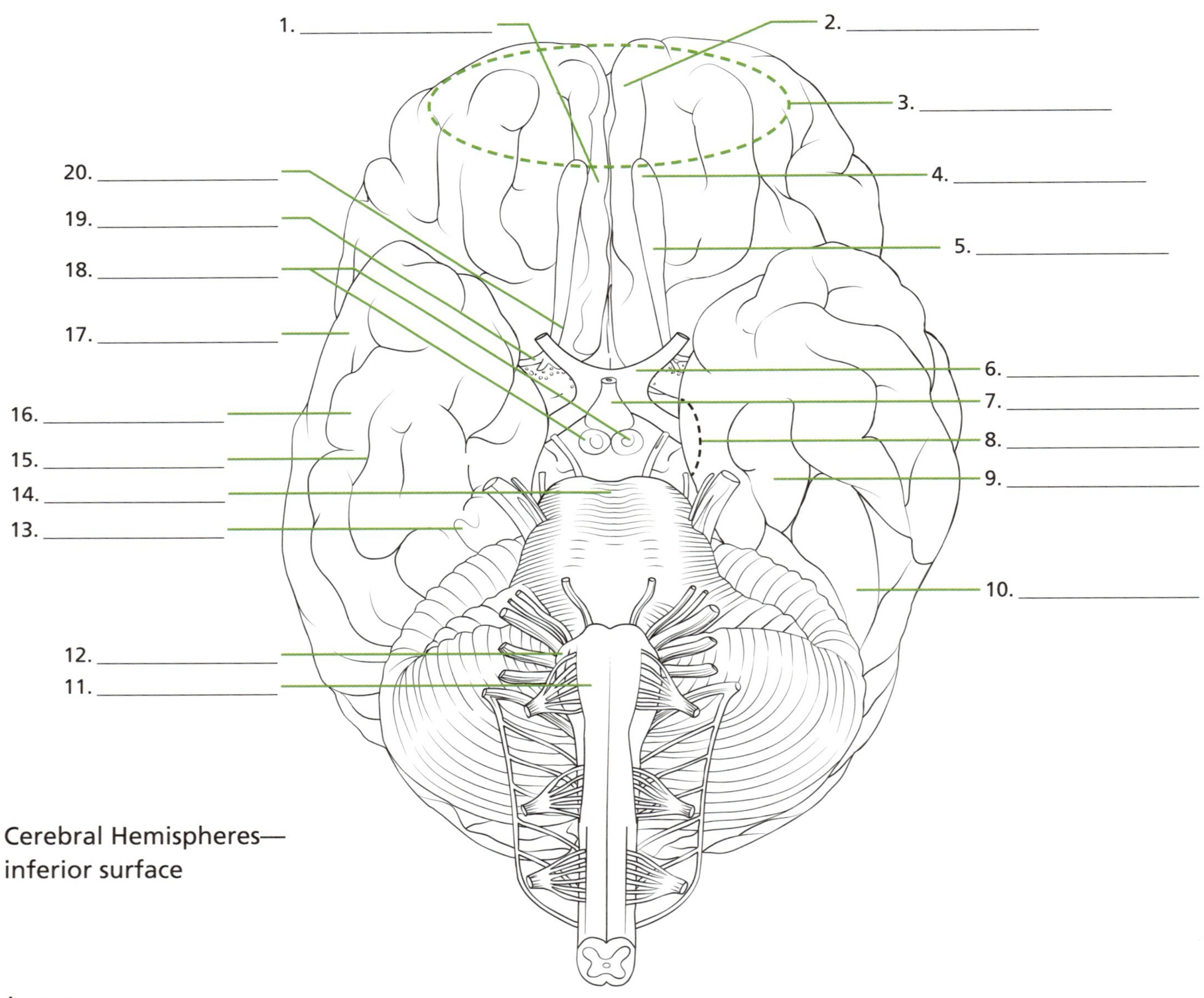

Cerebral Hemispheres—inferior surface

Answers

1. gyrus rectus, 2. prefrontal cortex, 3. orbitofrontal cortex, 4. olfactory bulb, 5. olfactory tract, 6. optic chiasm, 7. infundibular stalk, 8. uncus, 9. parahippocampal gyrus, 10. occipitotemporal gyrus, 11. pyramid, 12. inferior olivary nucleus, 13. collateral sulcus, 14. cerebral peduncle, 15. inferior temporal sulcus, 16. inferior temporal gyrus, 17. middle temporal gyrus, 18. mammillary bodies, 19. lateral olfactory stria, 20. medial olfactory stria

Overview of the Topography of the Cerebral Hemispheres: Superior Surface

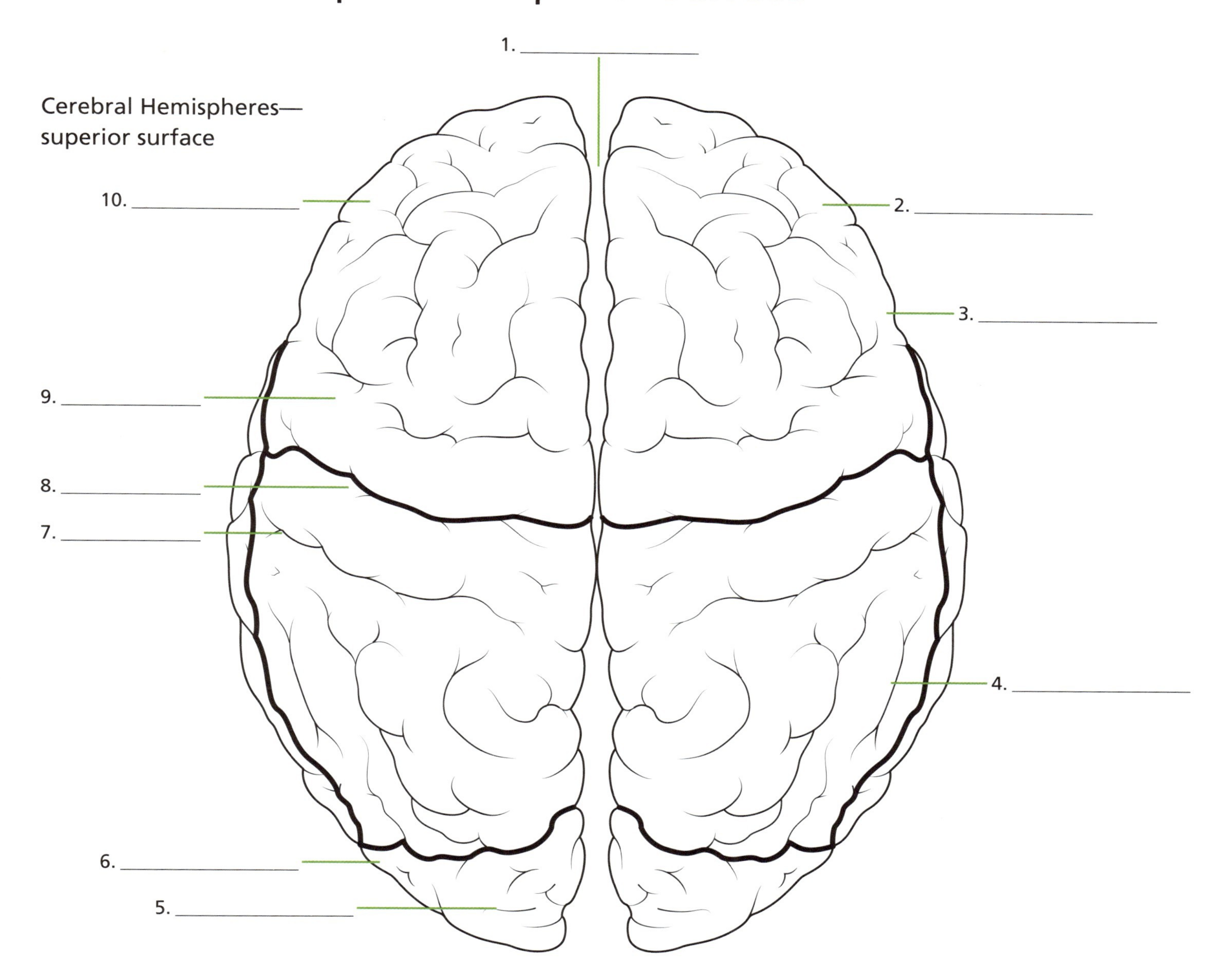

The superior surface of the brain, like the lateral surface, highlights the four lobes of the cortex. The frontal, temporal, parietal, and occipital lobes are visible, and in this view the left and right hemispheres can also be seen. The lobes are separated by the longitudinal and lateral cerebral fissures, as well as by the parieto-occipital fissure. Each hemisphere consists of an outer layer of gray matter, known as the cerebral cortex, and an inner layer of white matter. The two hemispheres are connected by a large white matter tract called the corpus callosum. Generalizations are often made about certain functions being specific to the right or left hemisphere, but most functions are actually distributed across both hemispheres. In some cases, however, the major areas involved in perception are found in one hemisphere. An example of this is language, in which the two major areas, Broca's area and Wernicke's area, are located in the left hemisphere.

Answers

1. longitudinal fissure, 2. right hemisphere, 3. frontal lobe, 4. parietal lobe, 5. occipital lobe, 6. parieto-occipital fissure, 7. postcentral gyrus, 8. central sulcus, 9. precentral gyrus, 10. left hemisphere

Levels of the Body of the Corpus Callosum and Head of the Caudate Nucleus

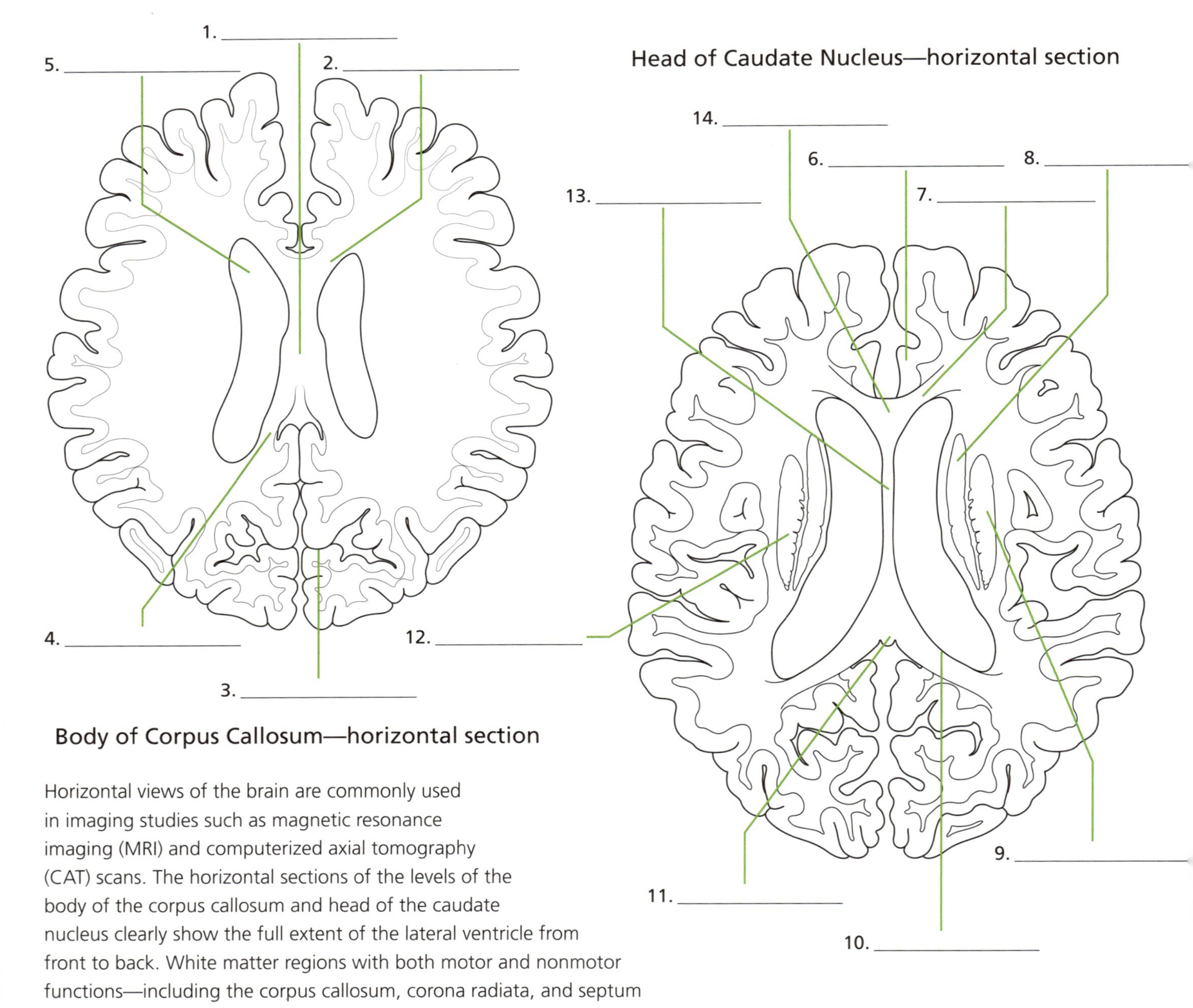

Head of Caudate Nucleus—horizontal section

Body of Corpus Callosum—horizontal section

Horizontal views of the brain are commonly used in imaging studies such as magnetic resonance imaging (MRI) and computerized axial tomography (CAT) scans. The horizontal sections of the levels of the body of the corpus callosum and head of the caudate nucleus clearly show the full extent of the lateral ventricle from front to back. White matter regions with both motor and nonmotor functions—including the corpus callosum, corona radiata, and septum pellucidum—can be observed. The corpus callosum is a large bundle of neural fibers that connects the left and right hemispheres of the brain to facilitate communication. It is also the largest white matter structure in the brain. The body of the corpus callosum is the long, narrow region between the splenium and the genu. In extreme cases of epilepsy, the corpus callosum is severed as a means of treatment. This can result in split-brain disorders, which typically involve speech irregularities, problems with object recognition, and difficulties with tasks that require interdependent movement of both hands. The caudate nucleus is one of the three structures that make up the basal ganglia, along with the putamen and globus pallidus. Disorders involving the caudate nucleus are numerous and include Parkinson's disease, Alzheimer's disease, and Huntington's disease.

Answers

1. body of corpus callosum, 2. anterior forceps, 3. parieto-occipital sulcus, 4. posterior forceps, 5. lateral ventricle, 6. cingulate gyrus, 7. anterior forceps, 8. head of caudate nucleus, 9. putamen, 10. crus of fornix, 11. splenium of corpus callosum, 12. corona radiata, 13. septum pellucidum, 14. genu of corpus callosum

Levels of the Anterior Nucleus of the Thalamus and Anterior Commissure

Horizontal sections at the levels of the anterior nucleus of the thalamus and the anterior commissure feature structures of the limbic system. The anterior nucleus of the thalamus is the part of the limbic system that plays a role in alertness and episodic memory. Lesions of the anterior nucleus of the thalamus result in amnesia involving episodic memory. The fornix is a major part of the limbic system and is composed of a body, column, and fimbria. The different parts of the fornix are based on the location along the fiber bundles, with the fimbria nearest to the hippocampus and the columns behind the anterior commissure. The anterior commissure connects the two temporal lobes and the amygdala. Studies have suggested that the anterior commissure plays a part in color perception and attention, in addition to its role in the limbic system. Other regions of the brain—including the thalamus, habenula, and dorsomedial thalamic nucleus—are part of, or are involved with, the limbic system.

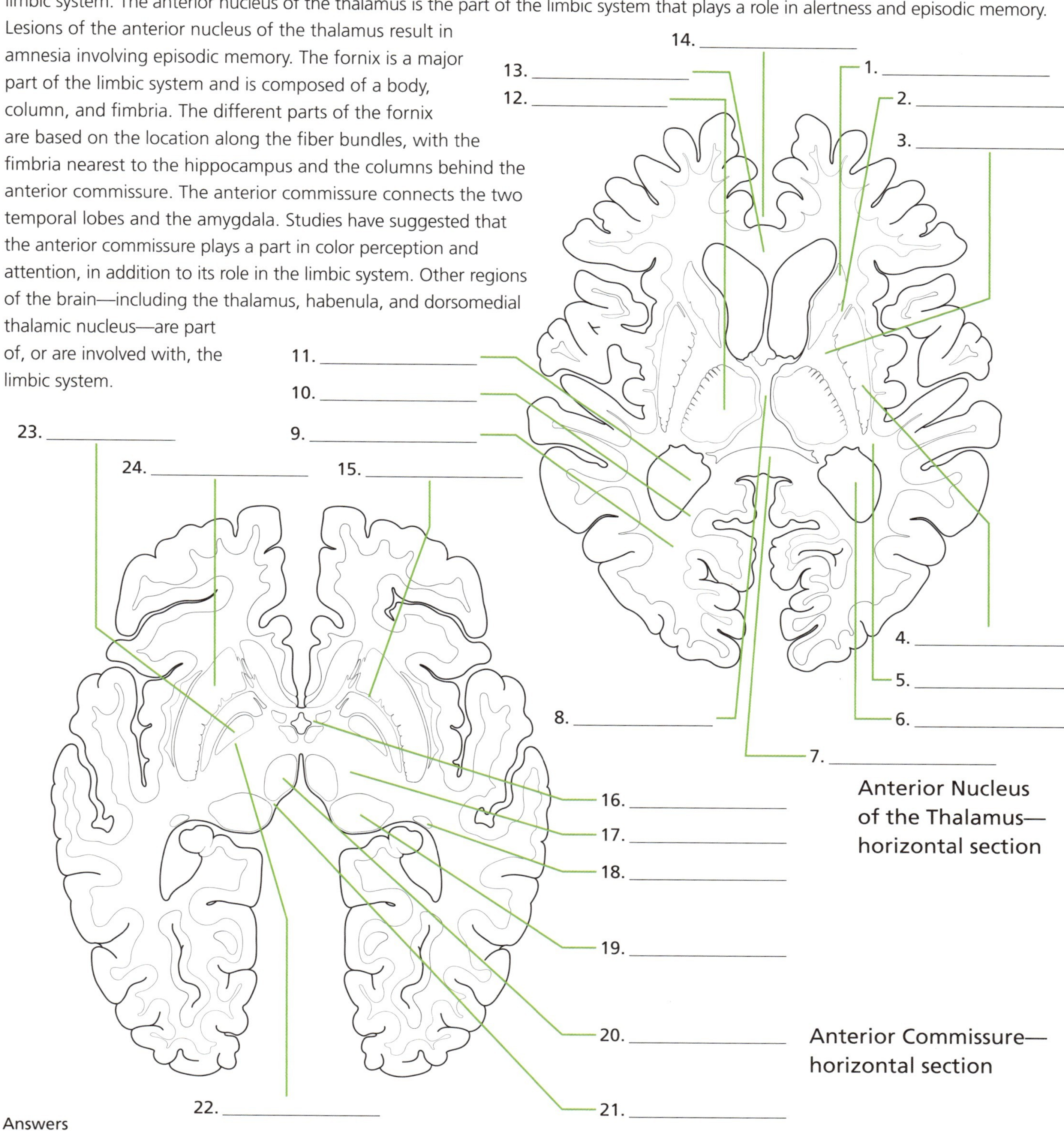

Anterior Nucleus of the Thalamus—horizontal section

Anterior Commissure—horizontal section

Answers

1. head of caudate nucleus, 2. anterior limb of internal capsule, 3. genu of internal capsule, 4. putamen, 5. posterior limb of internal capsule, 6. thalamus, 7. forceps major of corpus callosum, 8. body of fornix, 9. optic radiation, 10. posterior forceps, 11. lateral ventricle, 12. anterior nucleus of thalamus, 13. forceps minor of corpus callosum, 14. column of fornix, 15. anterior commissure, 16. columns of fornix, 17. centromedian thalamic nucleus, 18. fimbria of fornix, 19. pulvinar, 20. habenula, 21. dorsomedial thalamic nucleus, 22. internal capsule (posterior limb), 23. globus pallidus, 24. putamen

Levels of the Genu of the Corpus Callosum and Anterior Commissure

Viewing a coronal section is an excellent way to visualize the complexity of the human brain and appreciate the regions that work together for complex cognitive and emotive functions. From this perspective both hemispheres are visible, so it is easier to understand how the corpus callosum and anterior commissure work to connect the two hemispheres for communication. The anterior end of the corpus callosum is known as the genu, and it curves down in front of the septum pellucidum. The corpus callosum is the largest white matter tract in the brain and has fibers that radiate out into the cerebral cortex. The anterior commissure also connects the two hemispheres but is approximately ten times smaller than the corpus callosum. Some studies have suggested that it may be a compensatory structure for those without a corpus callosum. The levels of the genu of the corpus callosum and anterior commissure feature many structures involved with the limbic system, including the cingulate gyrus, hypothalamus, and amygdala.

Genu of the Corpus Callosum—coronal section

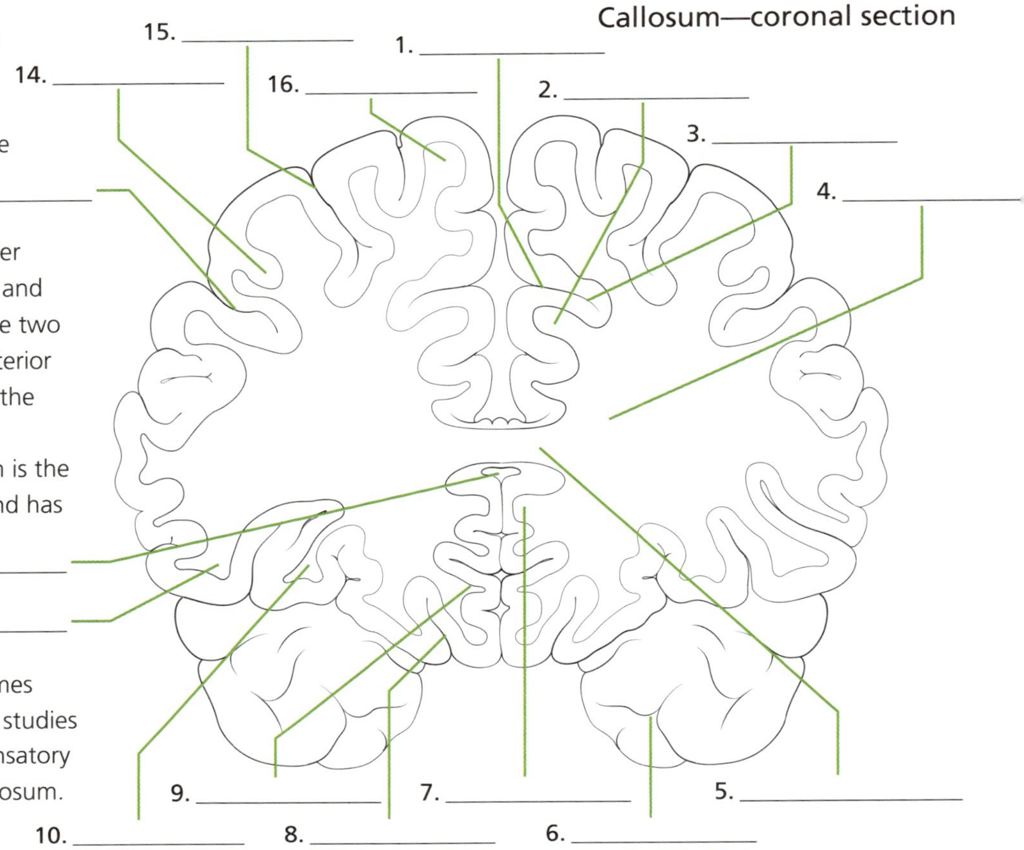

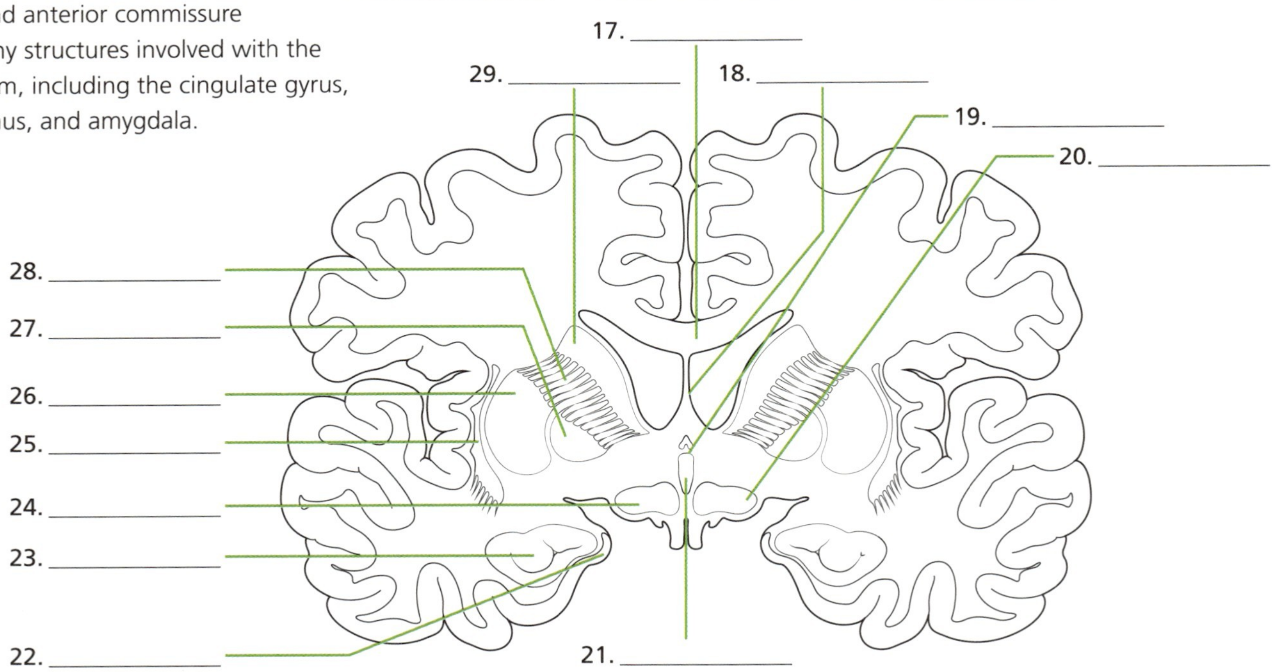

1. ________
2. ________
3. ________
4. ________
5. ________
6. ________
7. ________
8. ________
9. ________
10. ________
11. ________
12. ________
13. ________
14. ________
15. ________
16. ________

Anterior Commissure—coronal section

17. ________
18. ________
19. ________
20. ________
21. ________
22. ________
23. ________
24. ________
25. ________
26. ________
27. ________
28. ________
29. ________

Answers

1. cingulate, 2. cingulate gyri, 3. cingulate sulcus, 4. medullary center, 5. genu of corpus callosum, 6. temporal pole, 7. subcallosal gyrus, 8. olfactory sulcus, 9. gyrus rectus, 10. orbital gyrus, 11. frontal gyrus, 12. anterior cerebral artery, 13. inferior frontal sulcus, 14. medial frontal gyrus, 15. superior frontal sulcus, 16. superior frontal gyrus, 17. body of corpus callosum, 18. septum pellucidum, 19. anterior commissure, 20. optic tract, 21. third ventricle, 22. uncus, 23. amygdala, 24. hypothalamus, 25. insula, 26. putamen, 27. globus pallidus, 28. anterior limb of internal capsule, 29. caudate nucleus

Levels of the Infundibulum and Mammillary Bodies

Coronal views of the brain at the levels of the infundibulum and mammillary bodies allow for visualization of the ventricles, particularly the third and lateral ventricles. The ventricular system involves a series of interconnected fluid-filled spaces. The ventricles are particularly useful landmarks when mapping regions of the brain and act as a continuation of the central canal of the spinal cord. The central canal and ventricles contain cerebrospinal fluid, which acts as mechanical and immunological support for the CNS. The infundibulum, also known as the pituitary stalk, is the connection between the hypothalamus and the posterior pituitary. This tract is responsible for the release of the hormones vasopressin and oxytocin into the blood. Also proximal to the hypothalamus are the mammillary bodies, which are involved in recollective and spatial memory. Lesions of the mammillary bodies have been observed in amnesia-like syndromes. These lesions are associated with spatial memory impairment and anterograde amnesia, with affected individuals unable to convert new information from short-term to long-term memories.

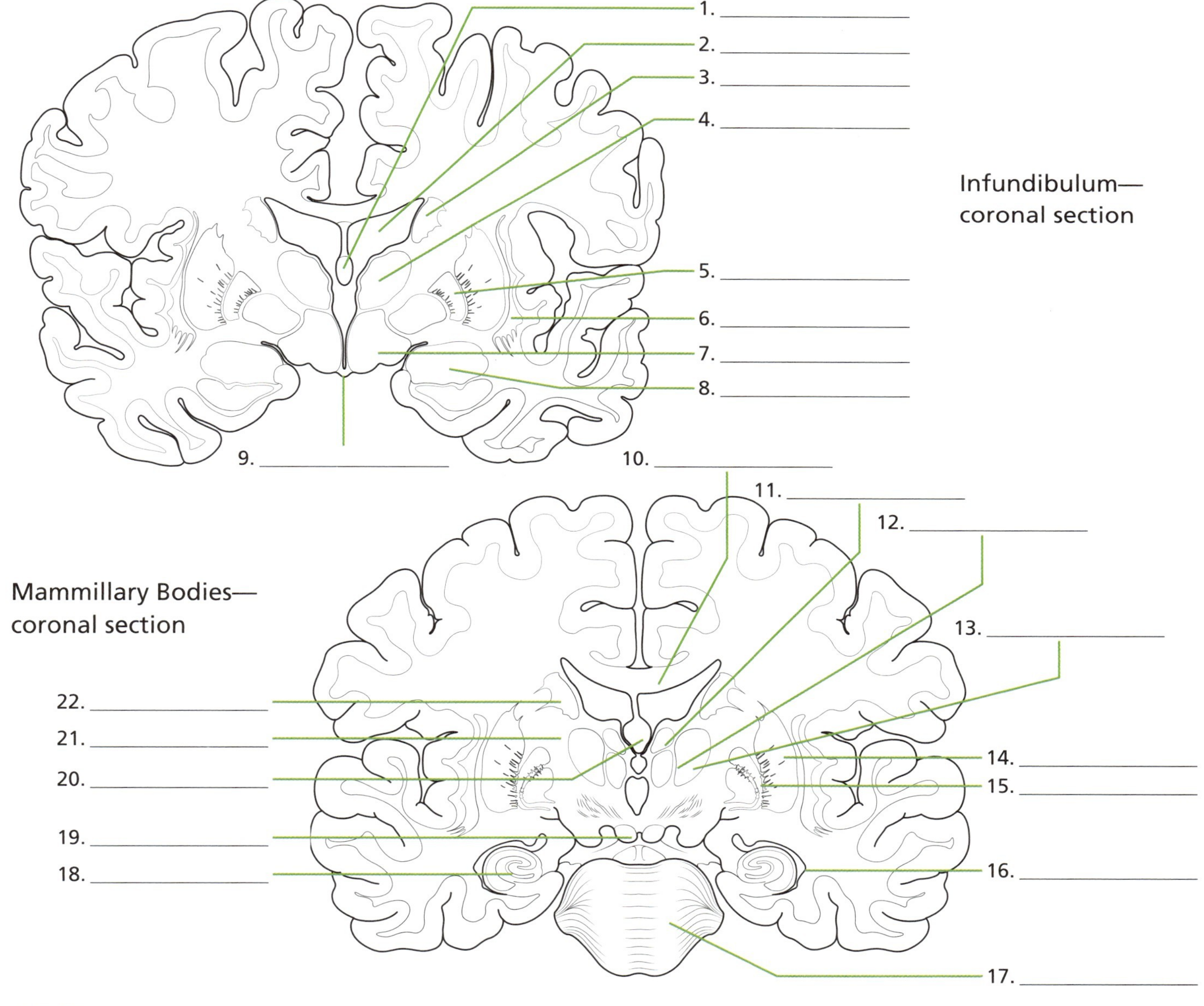

Answers

1. third ventricle, 2. lateral ventricle, 3. dorsal striatum, 4. thalamus, 5. globus pallidus, 6. claustrum, 7. hypothalamus, 8. amygdalar region, 9. infundibulum, 10. body of corpus callosum, 11. anterior nucleus of thalamus, 12. mammillothalamic tract, 13. dorsomedial thalamus, 14. putamen, 15. globus pallidus, 16. inferior horn of lateral ventricle, 17. basilar pons, 18. hippocampus, 19. mammillary body, 20. body of fornix, 21. posterior limb of internal capsule, 22. body of caudate nucleus

Levels of the Red Nucleus and the Splenium of the Corpus Callosum

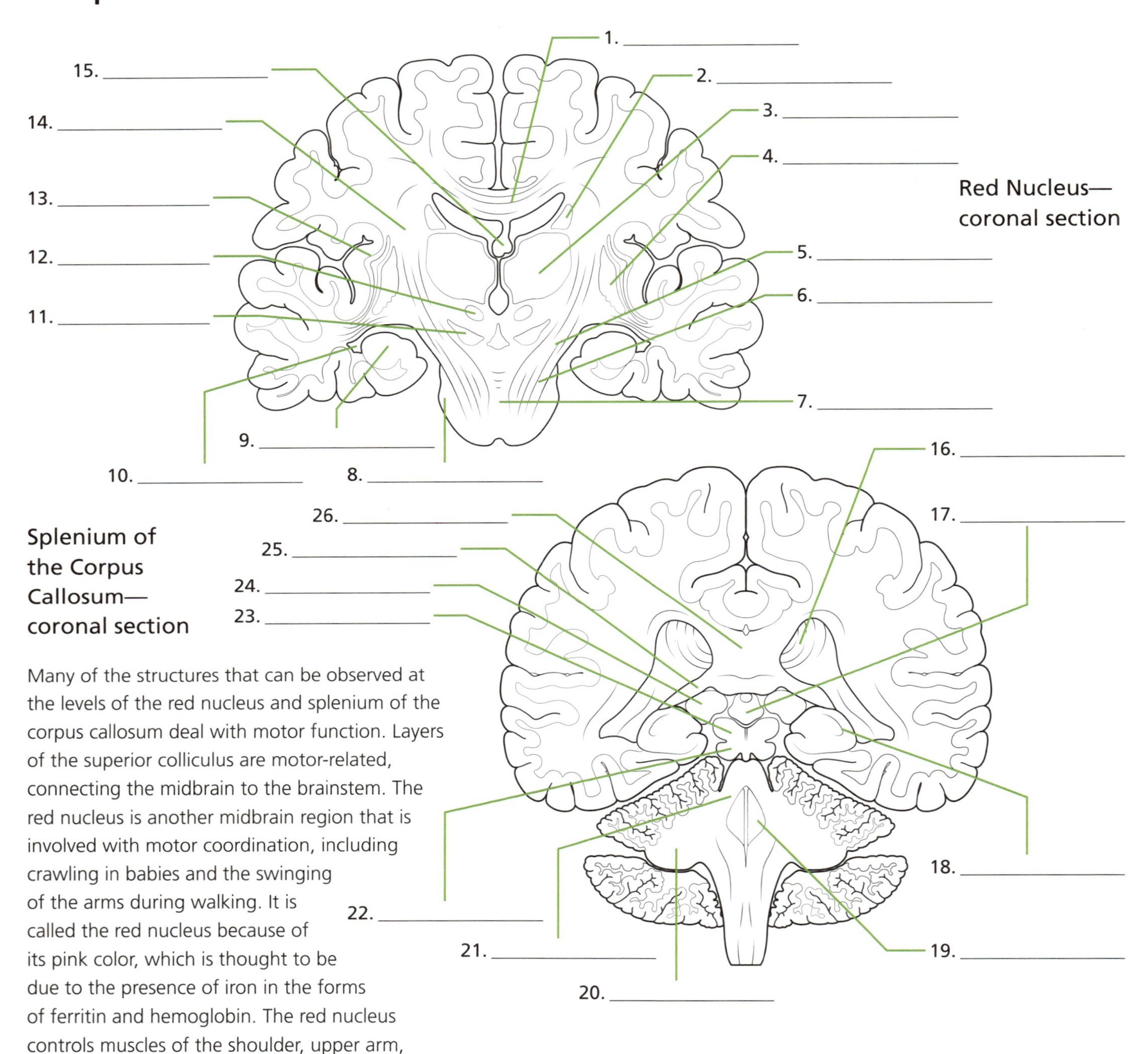

Many of the structures that can be observed at the levels of the red nucleus and splenium of the corpus callosum deal with motor function. Layers of the superior colliculus are motor-related, connecting the midbrain to the brainstem. The red nucleus is another midbrain region that is involved with motor coordination, including crawling in babies and the swinging of the arms during walking. It is called the red nucleus because of its pink color, which is thought to be due to the presence of iron in the forms of ferritin and hemoglobin. The red nucleus controls muscles of the shoulder, upper arm, and to some extent, the hand. The majority of the red nucleus axons do not project to the spinal cord, but rather relay information from the motor cortex to the cerebellum. This is achieved through the inferior olivary complex. Other key structures of the midbrain include the corticospinal fibers, substantia nigra, and crus cerebri. The substantia nigra is located in the tegmentum of the midbrain next to the red nucleus. In addition to movement, the substantia nigra plays an important role in reward and addiction. Disorders involving the substantia nigra include Parkinson's disease and schizophrenia.

Answers

1. body of corpus callosum, 2. body of caudate nucleus, 3. thalamus, 4. putamen, 5. crus cerebri, 6. corticospinal fibers, 7. basilar pons, 8. trigeminal, 9. hippocampus, 10. inferior horn of lateral ventricle, 11. substantia nigra, 12. red nucleus, 13. external capsule, 14. posterior limb of internal capsule, 15. body of fornix, 16. posterior horn of lateral ventricle, 17. pineal gland, 18. hippocampus, 19. fourth ventricle, 20. middle cerebellar peduncle, 21. superior cerebellar peduncle, 22. inferior colliculus, 23. superior colliculus, 24. pulvinar (thalamus), 25. crus of fornix, 26. splenium of corpus callosum

Fiber Bundles of the Internal Capsule

Most of the fibers that connect the cerebral cortex with the brainstem and spinal cord run through a region between the caudate nucleus and the thalamus known as the internal capsule. The internal capsule contains both ascending and descending axons, with fibers that run to and from the cerebral cortex. A large portion of the internal capsule is made up of the corticospinal tract, which conducts impulses from the brain to the spinal cord. The internal capsule is continuous with the corona radiata and crus cerebri of the midbrain. Acoustic, optic, and anterior thalamic radiations project out of this region, each of which is responsible for carrying information to the appropriate areas for processing. For example, the optic radiation carries information to the visual cortex. The corticopontine tract is made up of projections from the cerebral cortex to the pontine nuclei, which are part of the pons and are involved in motor activity. The corticonuclear tract, also known as the corticobulbar tract, connects the cerebral cortex to the brainstem.

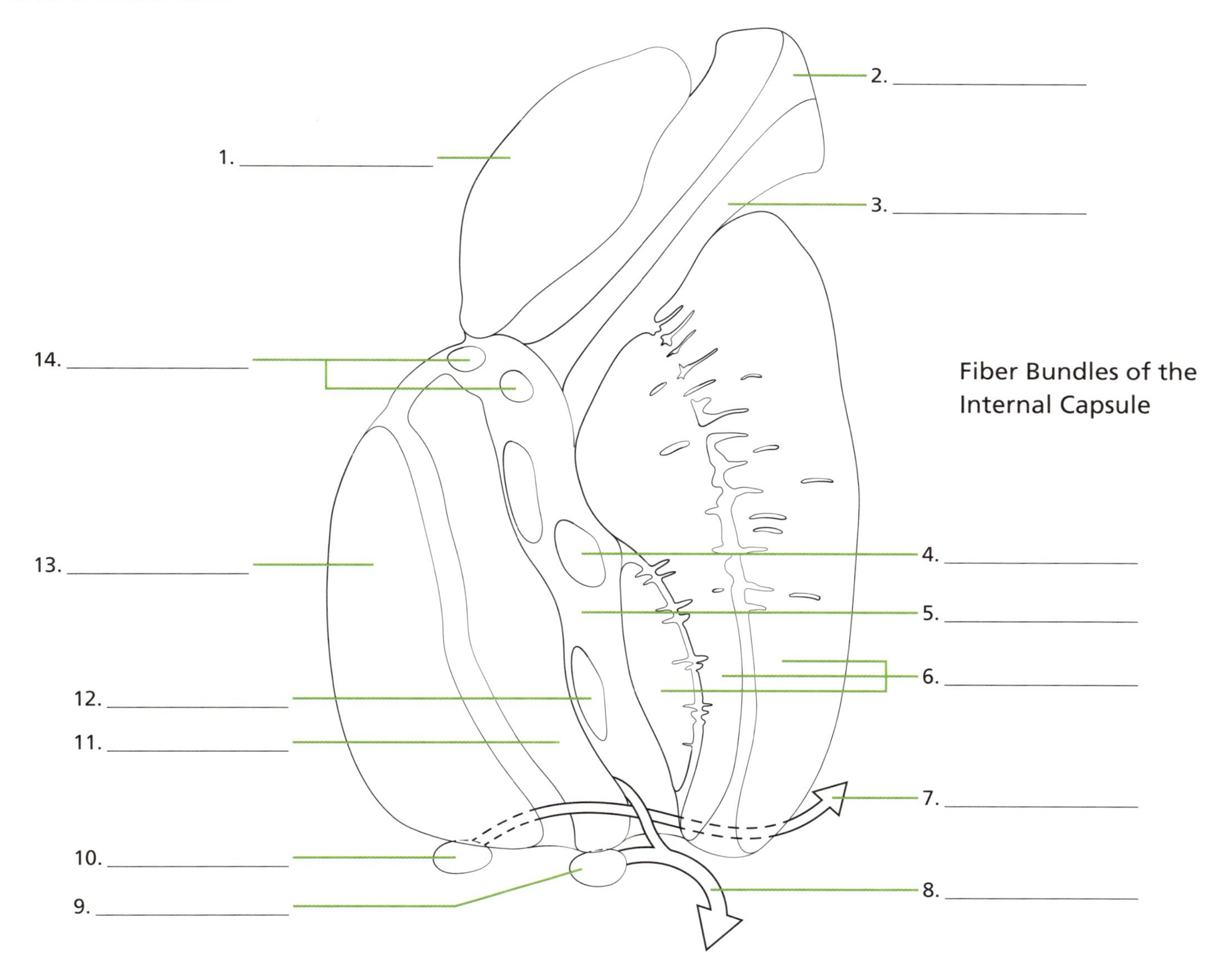

Answers

1. head of caudate nucleus, 2. anterior thalamic radiation, 3. frontopontine tract, 4. corticorubral tract, 5. corticopontine tract, 6. lentiform nucleus, 7. acoustic radiation, 8. optic radiation, 9. lateral geniculate nucleus, 10. medial geniculate nucleus, 11. central thalamic radiation, 12. corticospinal tract, 13. dorsal thalamus, 14. corticonuclear tract

Overview of the Cerebral Cortex

The cerebral cortex is the largest part of the brain, making up 77 percent of it. The cortex folds up on itself, forming crevices (called sulci) and ridges (called gyri). If you stretched all of the cerebral cortex flat (removing the sulci and gyri), it would cover an area of 350 square inches (2,300 cm^2). Prominent sulci form physical landmarks of the cortical area: The central sulcus divides the frontal lobe from the parietal lobe, and the lateral sulcus divides the frontal and parietal lobes from the temporal lobe. Some important and well-delineated areas of the cortex are the precentral gyrus (which is anterior to the central sulcus and contains the primary motor cortex) and the postcentral gyrus (which is posterior to the central sulcus and contains the primary somatosensory cortex). In addition to the topographic areas that are associated with singular functions (such as those above), many other regions have associated roles. For instance, the superior temporal gyrus contains the primary auditory cortex, while the orbital gyri are critical for decision-making and reward-learning.

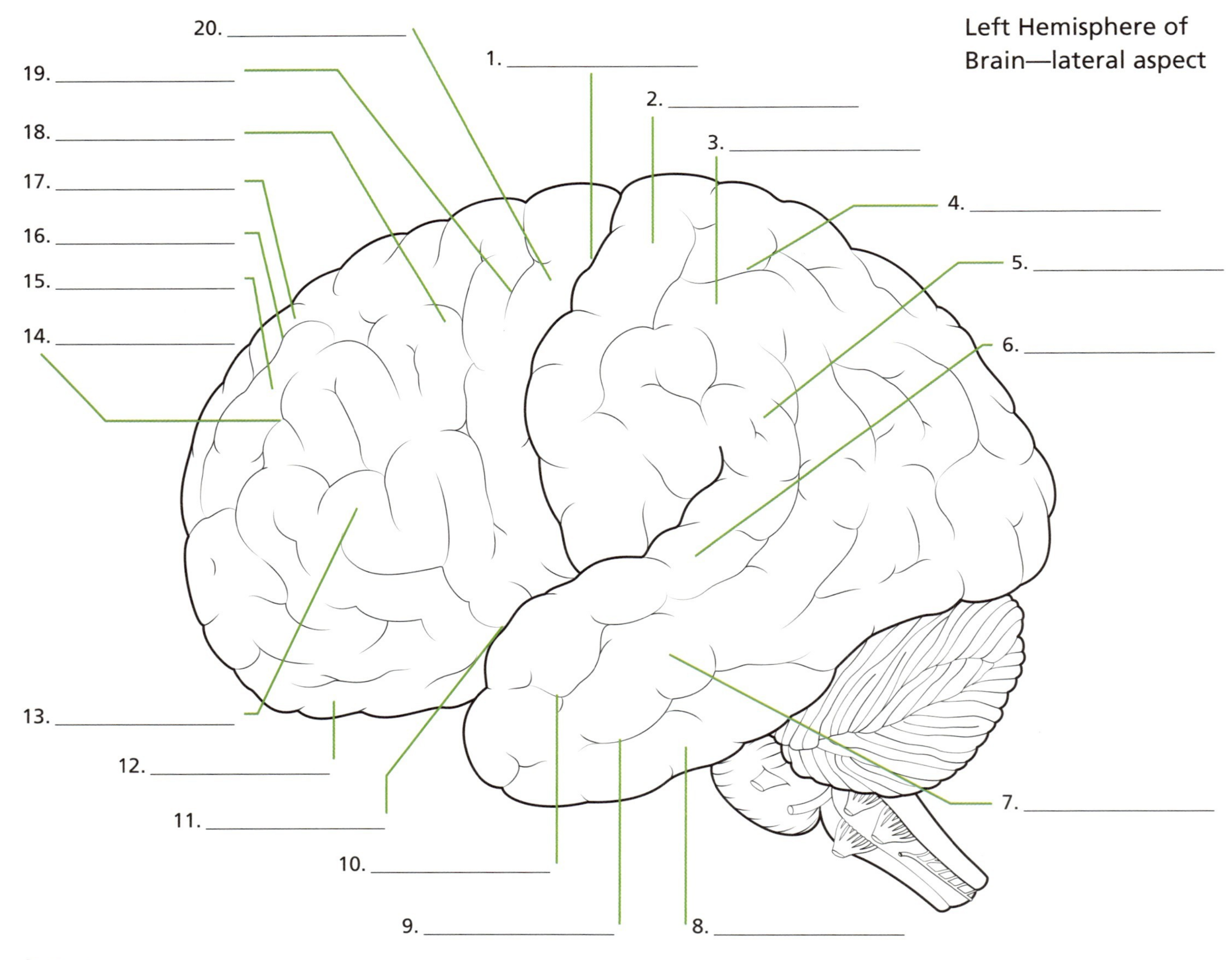

Answers

1. central sulcus, 2. postcentral gyrus, 3. supramarginal gyrus, 4. interparietal sulcus, 5. angular gyrus, 6. superior temporal gyrus, 7. middle temporal gyrus, 8. inferior temporal gyrus, 9. middle temporal sulcus, 10. superior temporal sulcus, 11. lateral sulcus/Sylvian fissure, 12. orbital gyrus, 13. inferior frontal gyrus, 14. inferior frontal sulcus, 15. middle frontal gyrus, 16. superior frontal sulcus, 17. superior frontal gyrus, 18. frontal eye field, 19. precentral sulcus, 20. precentral gyrus

The Six Layers of the Cerebral Cortex

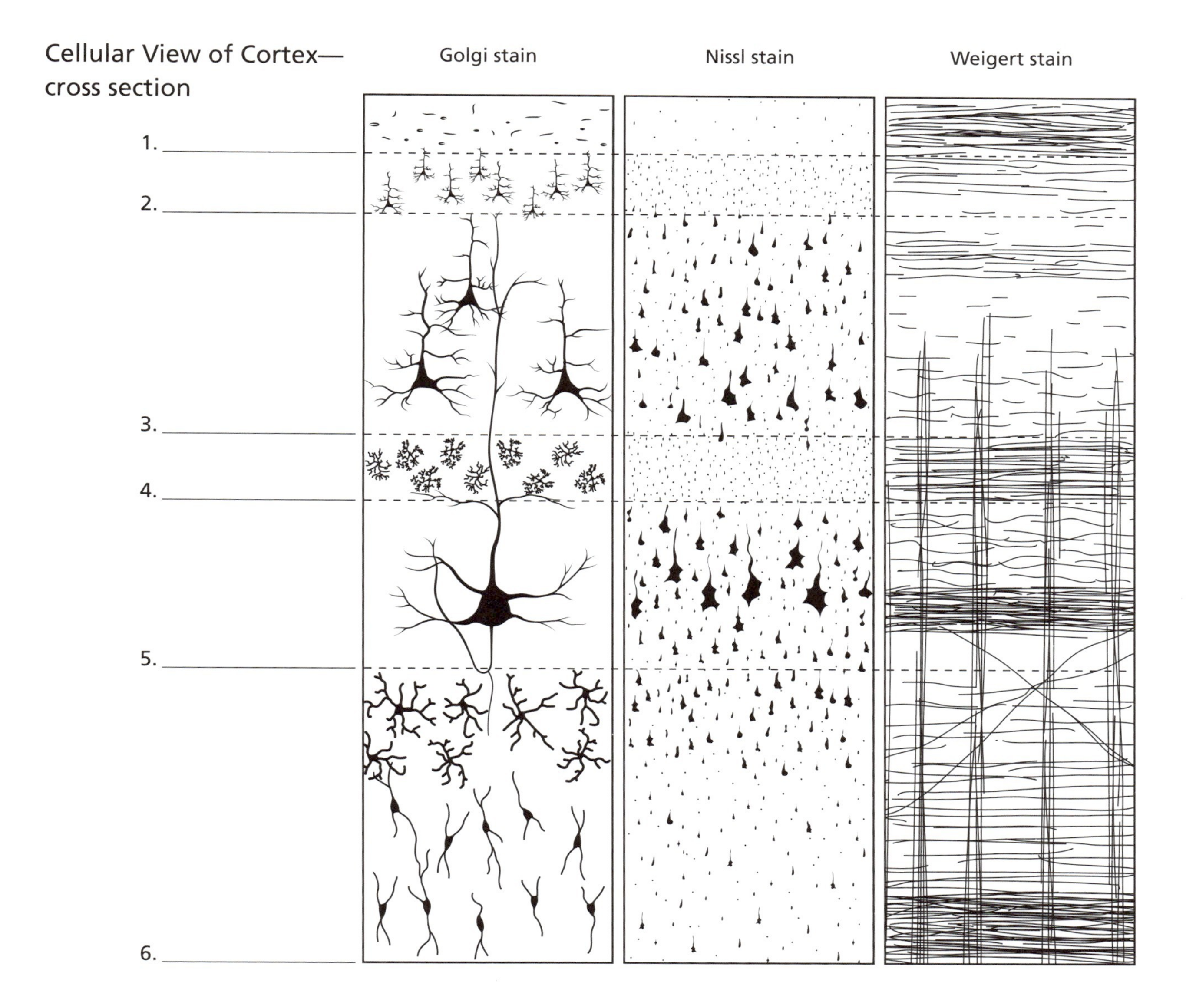

When viewed in cross section, the cerebral cortex has six layers, each containing different types of neuron or neuronal fibers. The molecular layer (layer I) is most superficial and is composed of dendrites from local pyramidal cells (whose somas are deeper), as well as axonal inputs from distal regions of the brain. The external granular layer (layer II) contains small pyramidal cells and spiny stellate interneurons. The external pyramidal layer (layer III) also contains pyramidal cells. These three layers make up the supragranular layers, which generally project to other cortical areas and allow for intercortical information transfer. The internal granular layer (layer IV) is made up of mostly stellate interneurons and some pyramidal cells. This layer is primarily involved with receiving information from the thalamus and sending signals to the supragranular layers. The internal pyramidal layer (layer V) is made up of large pyramidal cells, while the multiform layer (layer VI) is made up of smaller pyramidal and stellate interneurons. These last two deep layers make up the infragranular layers and generally send signals to the thalamus and subcortical structures.

Answers

1. layer I: molecular layer, 2. layer II: external granular layer, 3. layer III: external pyramidal layer, 4. layer IV: internal granular layer, 5. layer V: internal pyramidal layer, 6. layer VI: multiform layer

Types of Cortex Layer

Not all cortical layers are equal across different regions of the cortex. In the granular cortex (e.g., the primary sensory cortex), there is a very thick internal granular layer (layer IV), with thinner infragranular layers (layers V and VI). This is because sensory information transmitted from the thalamus is received by layer IV. The "typical" arrangement for layers of the cortex is generally found within association cortices, which comprise the majority of cortical areas. The typical cortex has a moderately thick layer IV and thicker infragranular layers compared to the granular cortex. In the agranular cortex (e.g., the primary motor cortex), there is a very small or nonexistent internal granular layer—thus the name "agranular." The largest layers within the agranular cortex are the infragranular layers, which project signals such as motor outputs to the subcortical regions. Agranular layers also have slightly thinner supragranular layers, which are mostly involved with communication between other parts of the cortex.

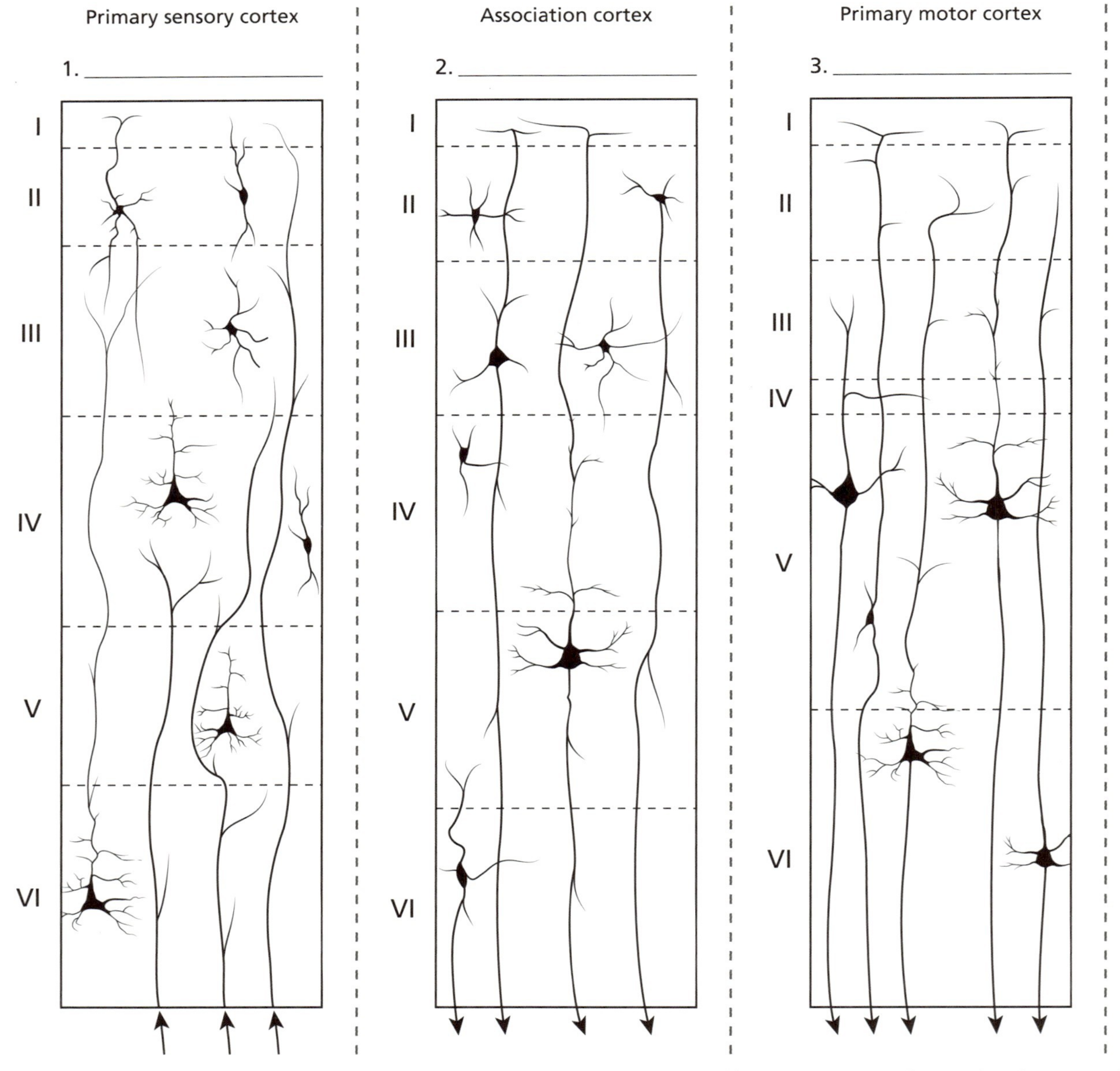

Different Types of Cerebral Cortical Layers

Answers

1. granular layer, 2. typical layer, 3. agranular layer

Gustatory and Olfactory Cortices

The areas that process taste and smell are not on the most obvious surface of the cerebral cortex but lie "under" the inferior frontal gyrus and superior temporal gyrus, respectively. The sense of taste is processed by the gustatory cortex. Much like the receptors for different types of taste on your tongue, the gustatory cortex is involved in discriminating different types of taste sensation: sweet, salty, bitter, sour, and umami, or savory. The sense of smell is processed by the olfactory cortex, a part of which overlaps with areas commonly associated with memory. Both the gustatory and olfactory cortices govern what is commonly known as the chemosensory senses, in that they involve transducing a chemical signal from the environment into an electrical neural signal.

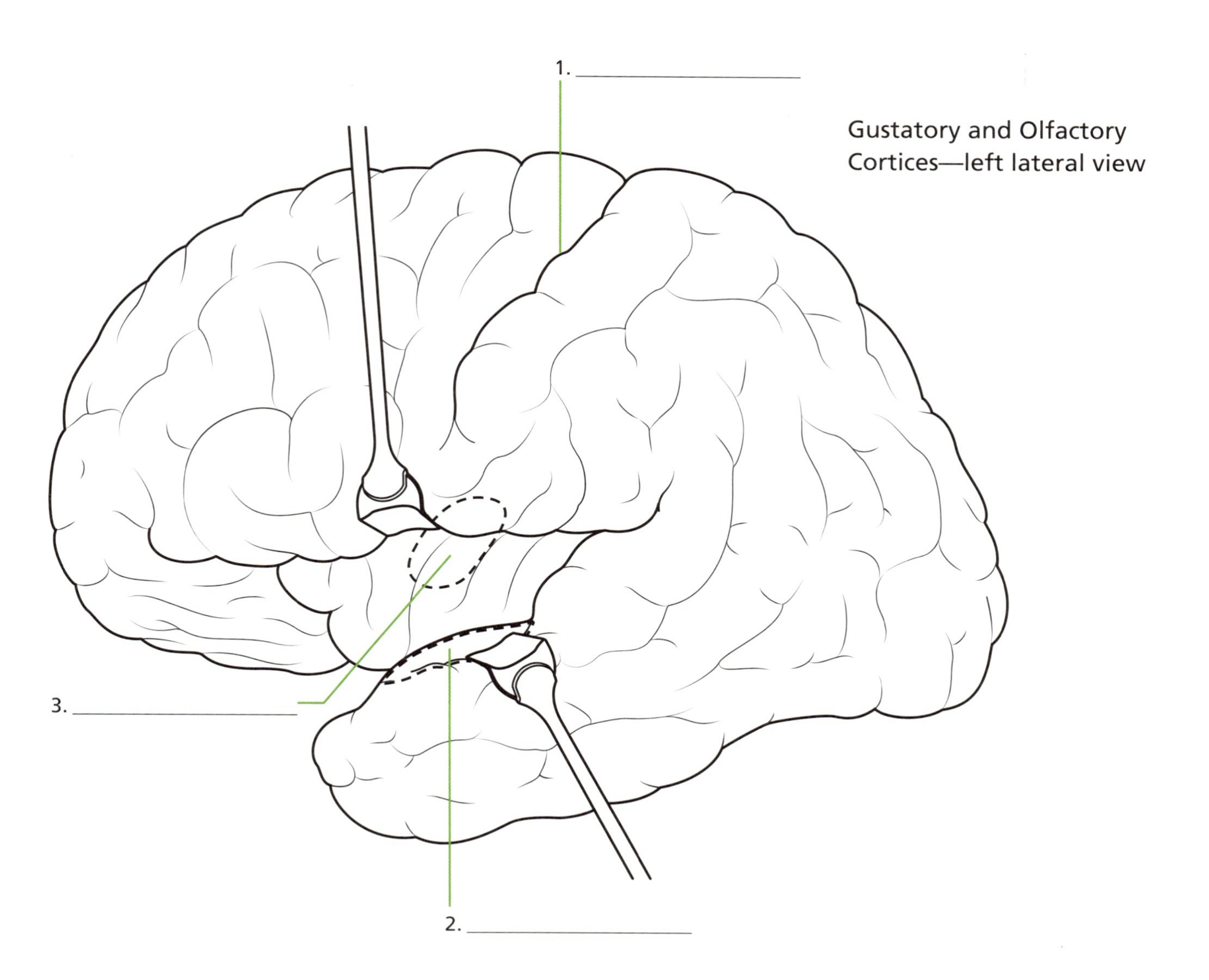

Gustatory and Olfactory Cortices—left lateral view

Answers

1. central sulcus, 2. olfactory cortex, 3. gustatory cortex

Visual Cortex

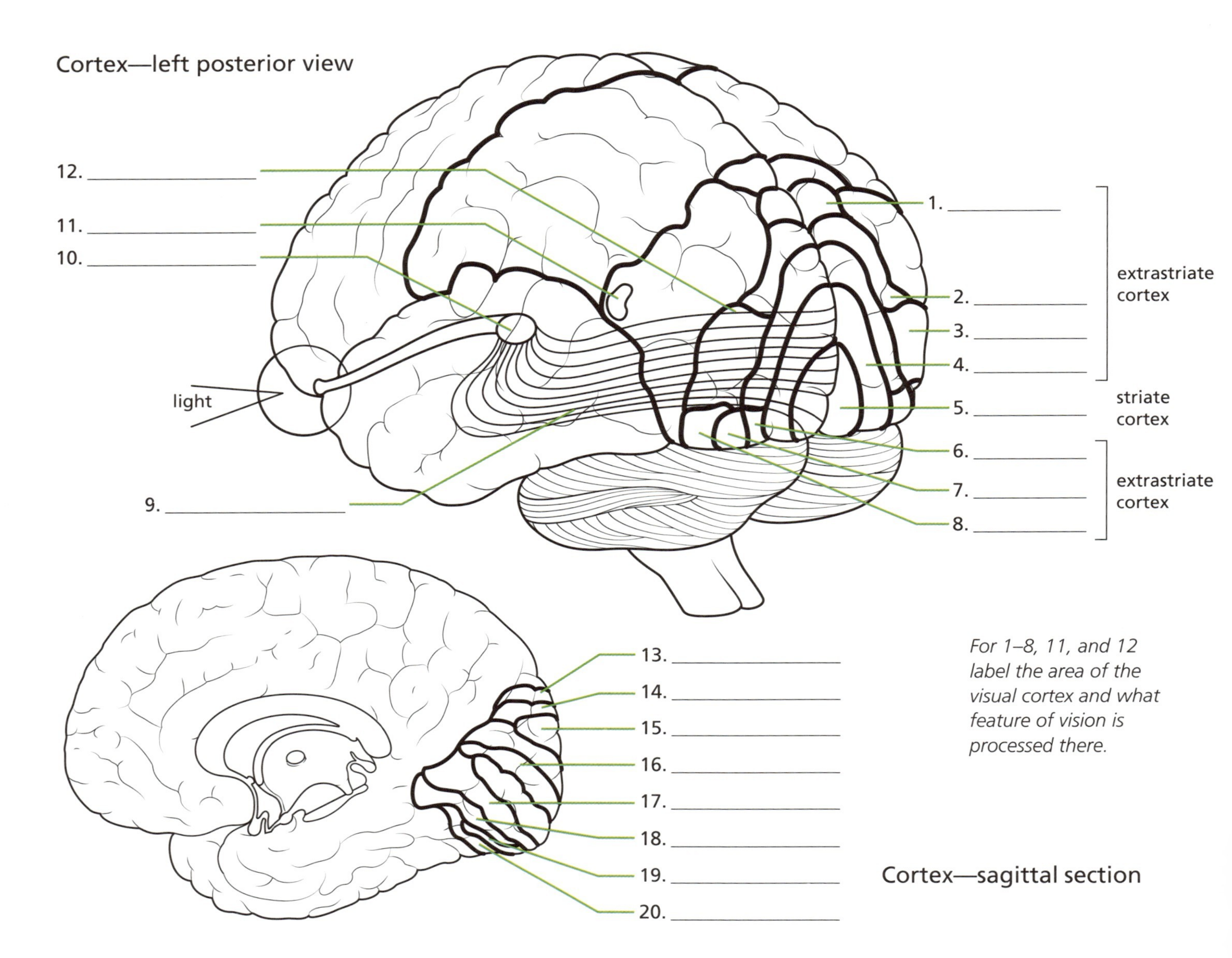

For 1–8, 11, and 12 label the area of the visual cortex and what feature of vision is processed there.

The visual cortex is located primarily in the occipital lobe. Although the occipital lobe (and subsequently the visual cortex) makes up the smallest percentage of the cerebral cortex (18 percent), just about every area in the occipital lobe functions to process vision, whereas other lobes (like the frontal lobe) serve a multitude of different processes (e.g., decision-making, emotional regulation, speech production, etc.). The areas on the visible surface of the occipital cortex seem small, but a sagittal view shows that the visual cortex extends deep between the hemispheres. The primary visual cortex (V1) is also called the striate cortex after a prominent axonal bundle that runs through layer IV of this area, forming a stripe. This area of the cortex is topographically laid out to resemble the retina, forming a retinotopic map (that is, neurons that are close to each other in the retina send information to neurons that are close to each other in V1). The neurons of V1 process small pieces of visual information, such as orientation of lines. All other areas of the visual cortex make up the extrastriate cortex and are involved in processing increasingly specific and cohesive features of vision, such as shape, color, and motion.

Answers

1. V7 (motion and space), 2. V3a (motion), 3. V3 (color and motion), 4. V2 (depth), 5. V1 (orientation), 6. VP (border representation), 7. V4 (color), 8. V8 (color), 9. visual radiation, 10. lateral geniculate nucleus, 11. V5 (motion), 12. lateral occipital (object recognition), 13. V7, 14. V3a, 15. V3, 16. V1, 17. V2, 18. VP, 19. V4, 20. V8

There is more to the visual cortex than meets the eye. Visual processing extends beyond the occipital lobe via the dorsal and ventral streams of information. The dorsal stream begins in the primary visual cortex (V1) in the occipital lobe and stretches to the parietal lobe. It is involved in spatial awareness, guidance of actions such as reaching, and the detection and analysis of movements. The ventral stream receives information from the lateral geniculate nucleus (LGN) of the thalamus and projects through visual cortex layers 2–4 (V2–V4) to the inferior temporal lobe. This stream is associated with object recognition and form representation. The dorsal and ventral streams are part of a dual-process model that integrates "where" and "what," respectively.

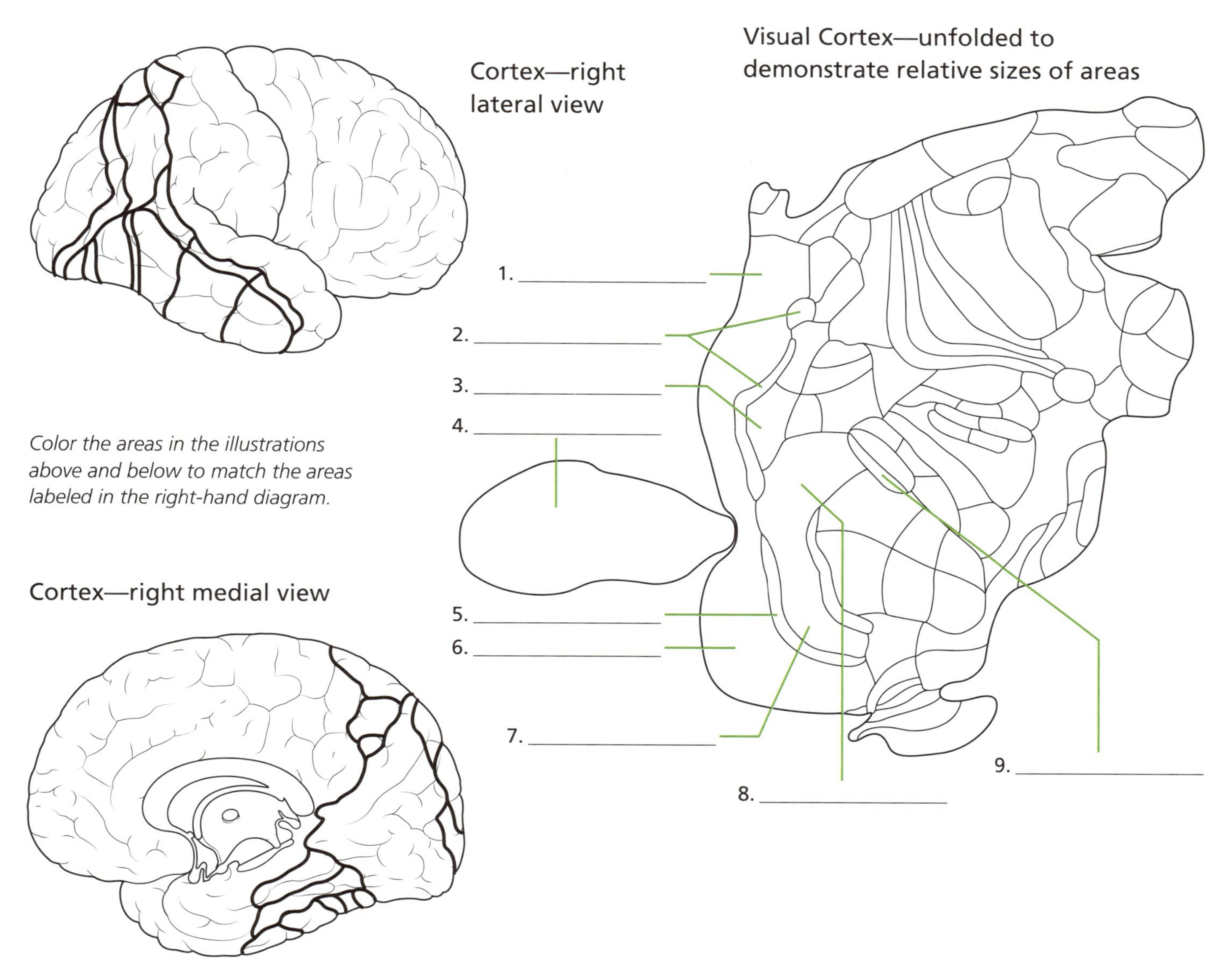

Color the areas in the illustrations above and below to match the areas labeled in the right-hand diagram.

Answers

1. V2d (depth, lower visual field), 2. V3 (color and motion), 3. V3a (motion), 4. V1 (orientation), 5. VP (border perception), 6. V2v (depth, upper visual field), 7. V4v (border perception), 8. V4d (border perception), 9. V4t (border perception), lower visual field)

Auditory Cortex

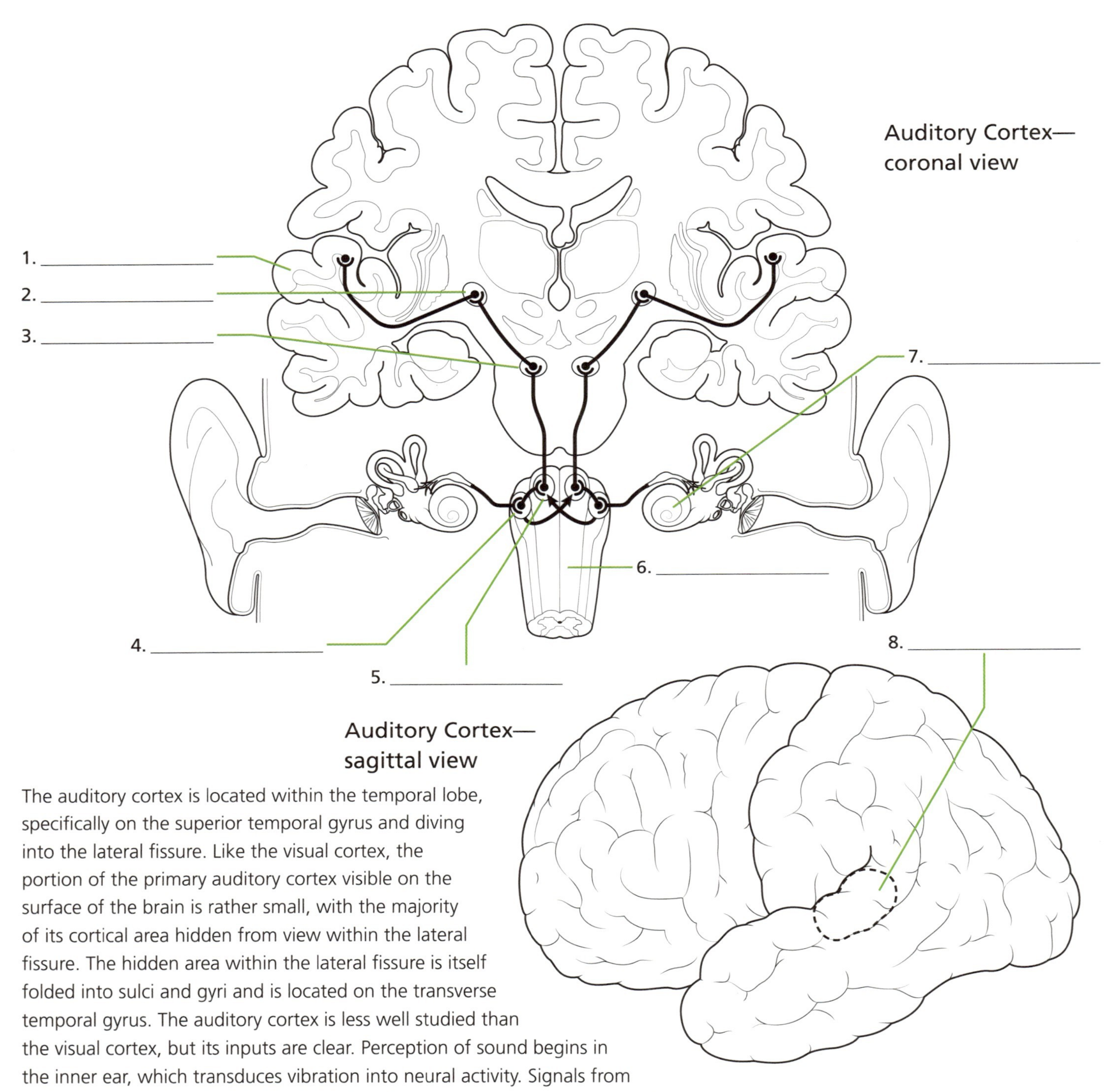

The auditory cortex is located within the temporal lobe, specifically on the superior temporal gyrus and diving into the lateral fissure. Like the visual cortex, the portion of the primary auditory cortex visible on the surface of the brain is rather small, with the majority of its cortical area hidden from view within the lateral fissure. The hidden area within the lateral fissure is itself folded into sulci and gyri and is located on the transverse temporal gyrus. The auditory cortex is less well studied than the visual cortex, but its inputs are clear. Perception of sound begins in the inner ear, which transduces vibration into neural activity. Signals from the inner ear are sent along the vestibulocochlear nerve to nuclei in the hindbrain, and from there to the inferior colliculus and then to the thalamus. Thalamocortical projections carry the auditory information to the primary visual cortex, which is required for conscious perception of auditory information (but not for reflexive responses to sounds in the environment).

Answers

1. auditory cortex, 2. medial geniculate nucleus, 3. inferior colliculus, 4. cochlear nucleus, 5. superior olivary nucleus, 6. pons, 7. cochlea, 8. primary auditory cortex

Somatosensory and Motor Cortices

The somatosensory cortex is a system made up of different types of receptors in skin and epithelial tissue. Processing occurs in the primary somatosensory area in the parietal lobe, before information is sent through spinal tracts via sensory nerves and into the brain. The motor areas of the cortex (including the premotor cortex, primary motor cortex, and supplementary motor area) are involved in planning and execution of movement-related tasks. The primary motor cortex is located on the medial surface of the anterior paracentral lobule, and the premotor cortex is located directly anterior. The supplementary motor area is located on the midline surface of the hemisphere anterior to the primary motor cortex. The motor cortex in each hemisphere governs movement on the contralateral side of the body (motor areas in the left hemisphere control muscles on the right side of the body and vice versa).

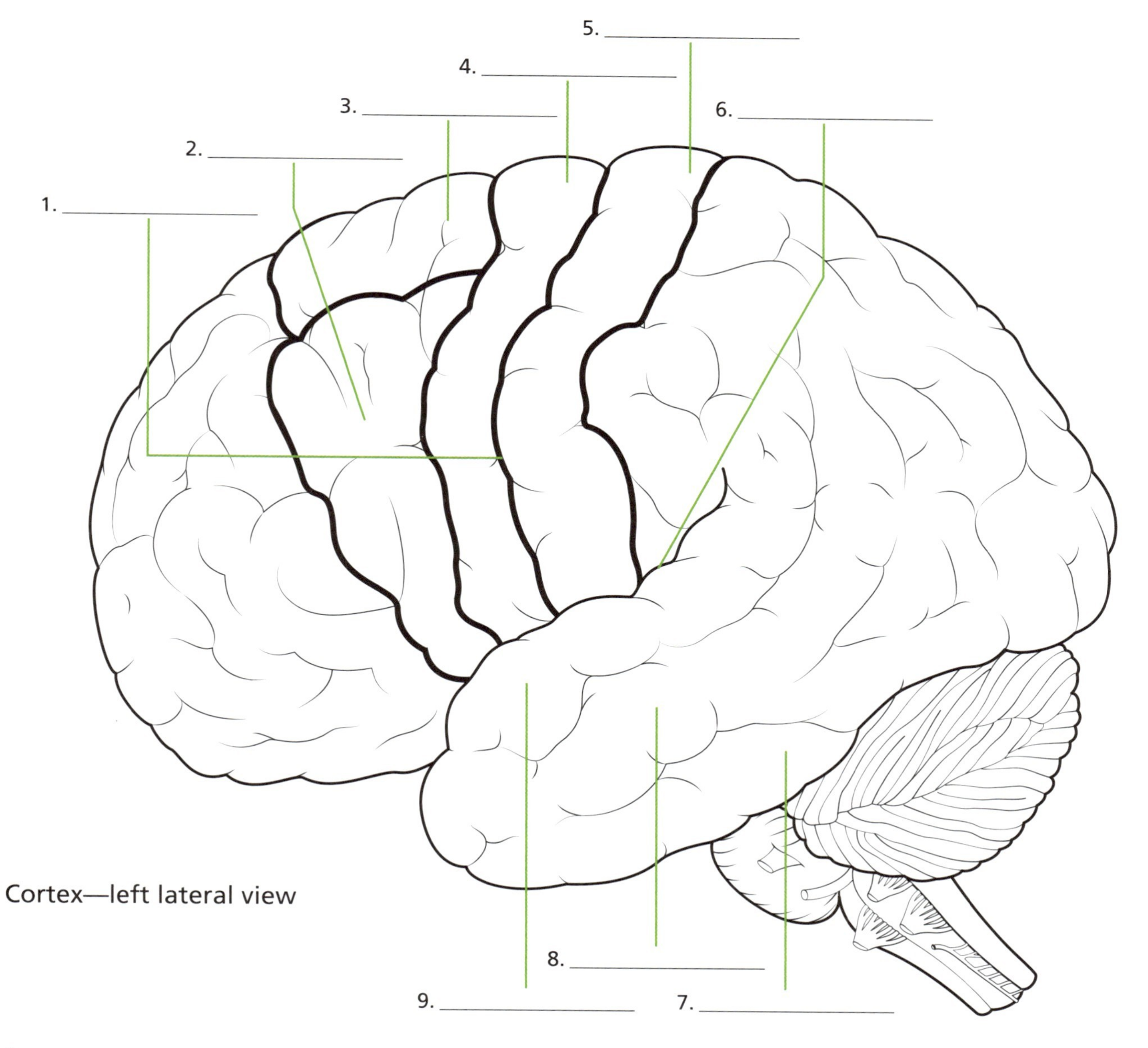

Cortex—left lateral view

Answers

1. central sulcus, 2. premotor cortex, 3. supplementary motor area, 4. primary motor cortex, 5. somatosensory cortex, 6. lateral sulcus, 7. inferior temporal gyrus, 8. middle temporal gyrus, 9. superior temporal gyrus

Frontal Cortex

In the frontal lobe, the area anterior to the motor cortex is called the prefrontal cortex. This is typically associated with executive functions (planning, attention, decision-making, emotional control, and memory), as well as personality. As such, it is much more highly developed in humans than in other species. The prefrontal cortex covers the entire visible surface of the anterior portion of the frontal lobe and also wraps around to the ventral aspect of the brain. This ventral area is called the orbitofrontal cortex. Furthermore, a large portion of the prefrontal cortex is hidden inside the longitudinal fissure and is visible only when the two hemispheres of the brain are separated. The portion of the prefrontal cortex hidden in the longitudinal fissure is referred to as the medial aspect and is separated into dorsomedial, ventromedial, and orbitofrontal areas (in order from superior to inferior). The portions of the prefrontal cortex on the visible surface of the brain are referred to as the lateral aspect and are similarly separated into dorsolateral, ventrolateral, and orbitofrontal areas (in order from superior to inferior).

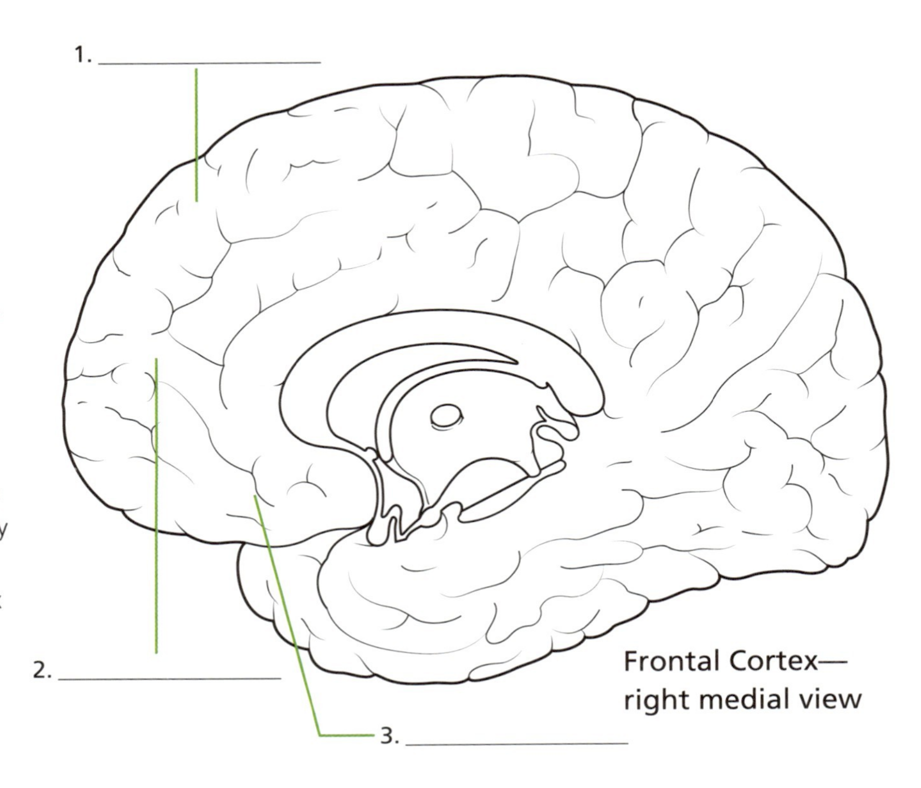

Frontal Cortex—
right medial view

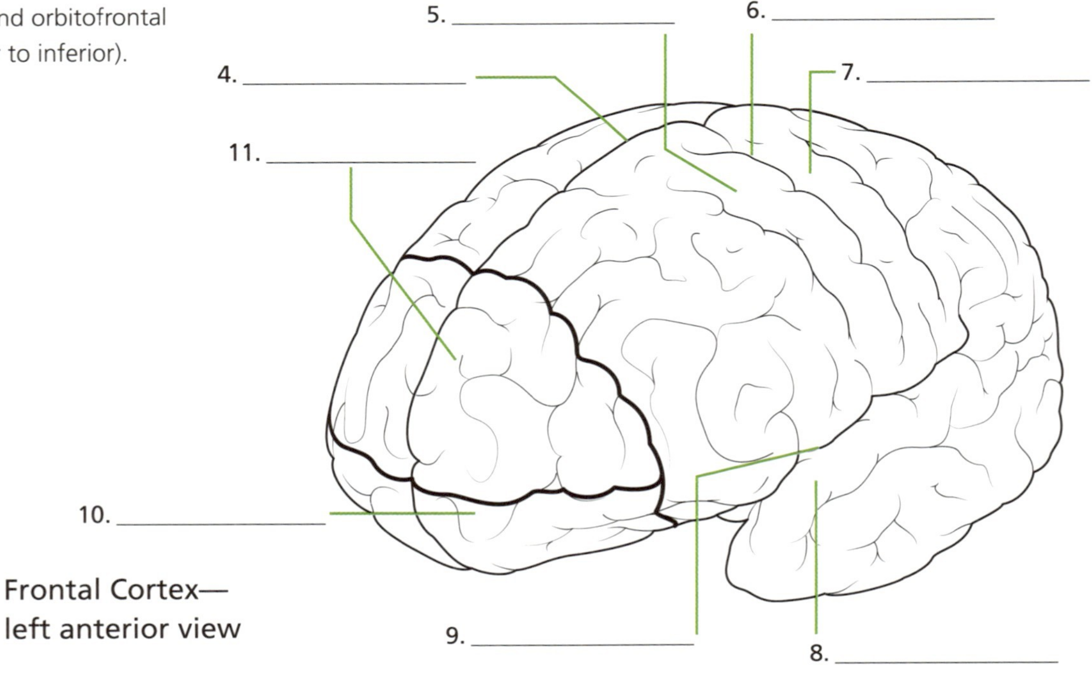

Frontal Cortex—
left anterior view

Answers

1. dorsomedial prefrontal cortex, 2. ventromedial prefrontal cortex, 3. orbitofrontal cortex, 4. longitudinal fissure, 5. primary motor cortex, 6. central sulcus, 7. somatosensory cortex, 8. superior temporal gyrus, 9. lateral sulcus, 10. orbitofrontal cortex, 11. dorsolateral prefrontal cortex

Major Connections to the Cerebral Cortex

The connections between areas of the brain are called fasciculi. The superior longitudinal fasciculus originates from the parietal lobe and terminates at the frontal lobe. Information transferred through the superior longitudinal fasciculus facilitates motor functions, attention and memory functions, language functions, and auditory perception functions. Another fasciculus that originates from the parietal lobe and terminates in the frontal lobe is the arcuate fasciculus, which is typically involved with the ability to process language (i.e., hearing or reading) and communicating language (i.e., speaking or writing). Also related to language are the extreme capsule (originating from the superior temporal gyrus and projecting to the inferior frontal gyrus) and the middle longitudinal fasciculus (originating from the parietal lobe and projecting to the superior temporal gyrus). The inferior longitudinal fasciculus originates from the occipital lobe and terminates at the temporal lobe and is associated with recognition (e.g., of words, colors, or faces). Another fasciculus that projects visual and also auditory information is the inferior fronto-occipital fasciculus. The uncinate fasciculus connects the temporal lobe to the inferior frontal lobe, networking through the hippocampus and the amygdala. The uncinate plays a role in the regulation of memory and emotion.

Cortex—left lateral view

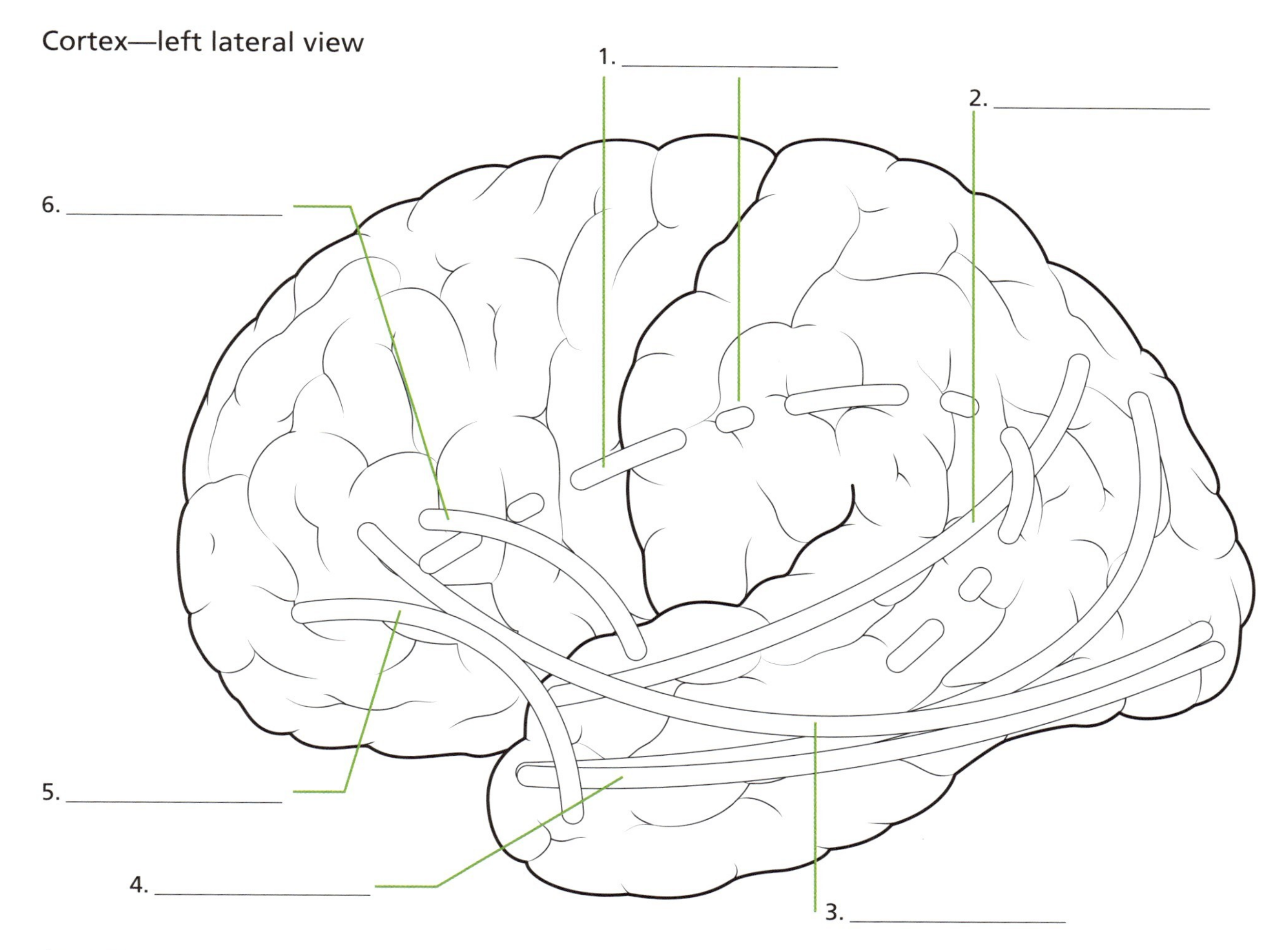

Answers

1. superior longitudinal/arcuate fasciculus, 2. middle longitudinal fasciculus, 3. inferior fronto-occipital fasciculus, 4. inferior longitudinal fasciculus, 5. uncinate fasciculus, 6. extreme capsule

Components of the Limbic System

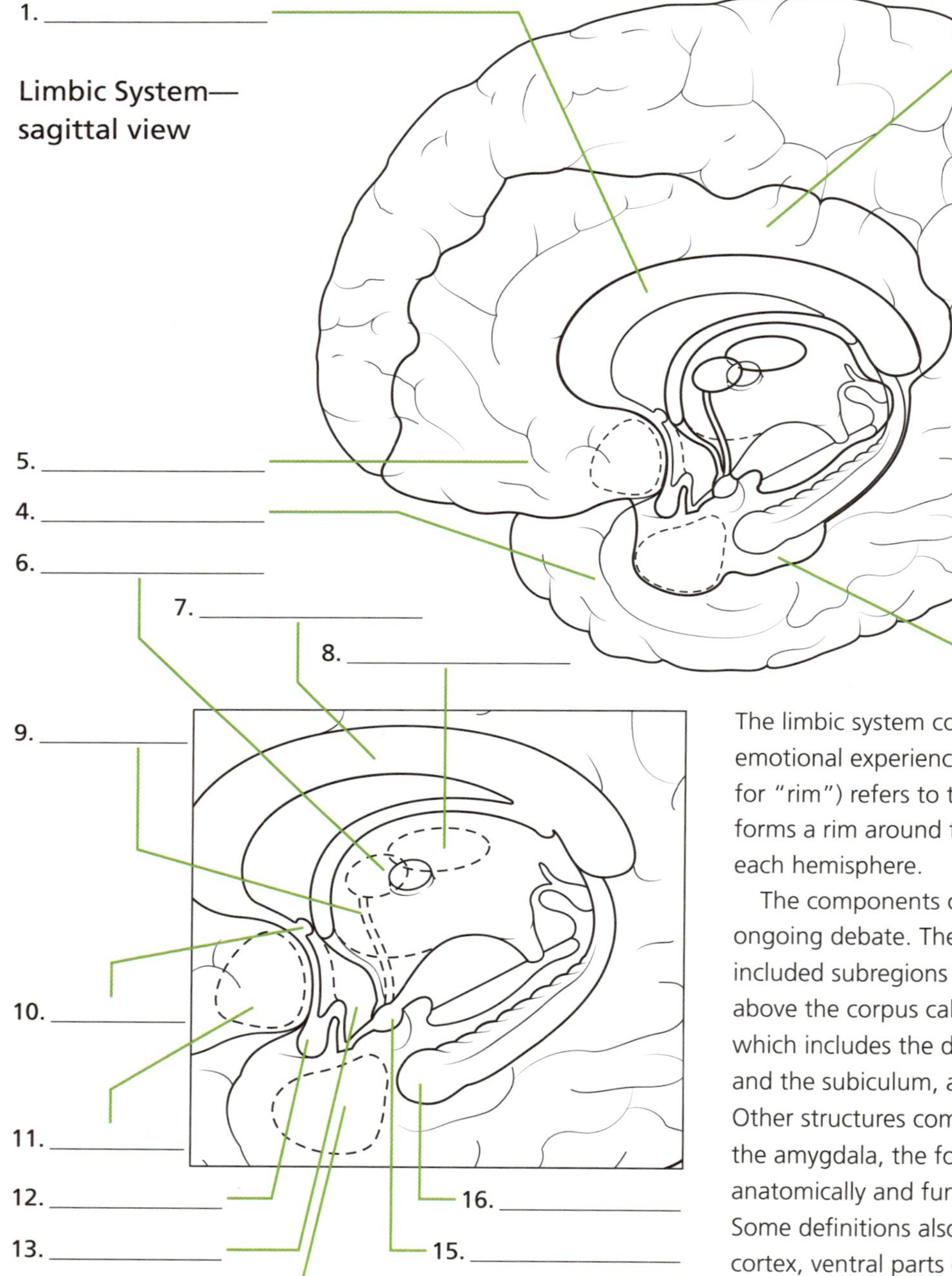

The limbic system comprises the brain circuits involved in emotional experience and expression. The term limbic (Latin for "rim") refers to the location of the limbic system, which forms a rim around the corpus callosum on the medial face of each hemisphere.

The components of the limbic system are a matter of ongoing debate. The two most prominent and commonly included subregions are the cingulate gyrus, which resides above the corpus callosum, and the hippocampal formation, which includes the dentate gyrus, the hippocampus proper, and the subiculum, all of which are in the temporal lobe. Other structures commonly included in the limbic system are the amygdala, the fornix, and cortical regions that anatomically and functionally connect with the hippocampus. Some definitions also include the orbital and medial prefrontal cortex, ventral parts of the basal ganglia, and the thalamus and hypothalamus.

The limbic system primarily contributes to behaviors related to self-preservation, such as motivation, emotional expression, autonomic regulation, and memory processing. Limbic dysfunction contributes to many psychiatric disorders such as depression and obsessive-compulsive disorder.

Answers

1. corpus callosum, 2. cingulate gyrus, 3. parahippocampal gyrus, 4. temporal lobe, 5. prefrontal cortex, 6. anterior nucleus of the thalamus, 7. fornix, 8. medial dorsal nucleus of the thalamus, 9. mammillothalamic tract, 10. anterior commissure, 11. ventral basal ganglia, 12. optic chiasm, 13. hypothalamus, 14. amygdala, 15. mammillary body, 16. hippocampus

Circuitry for Emotional Expression

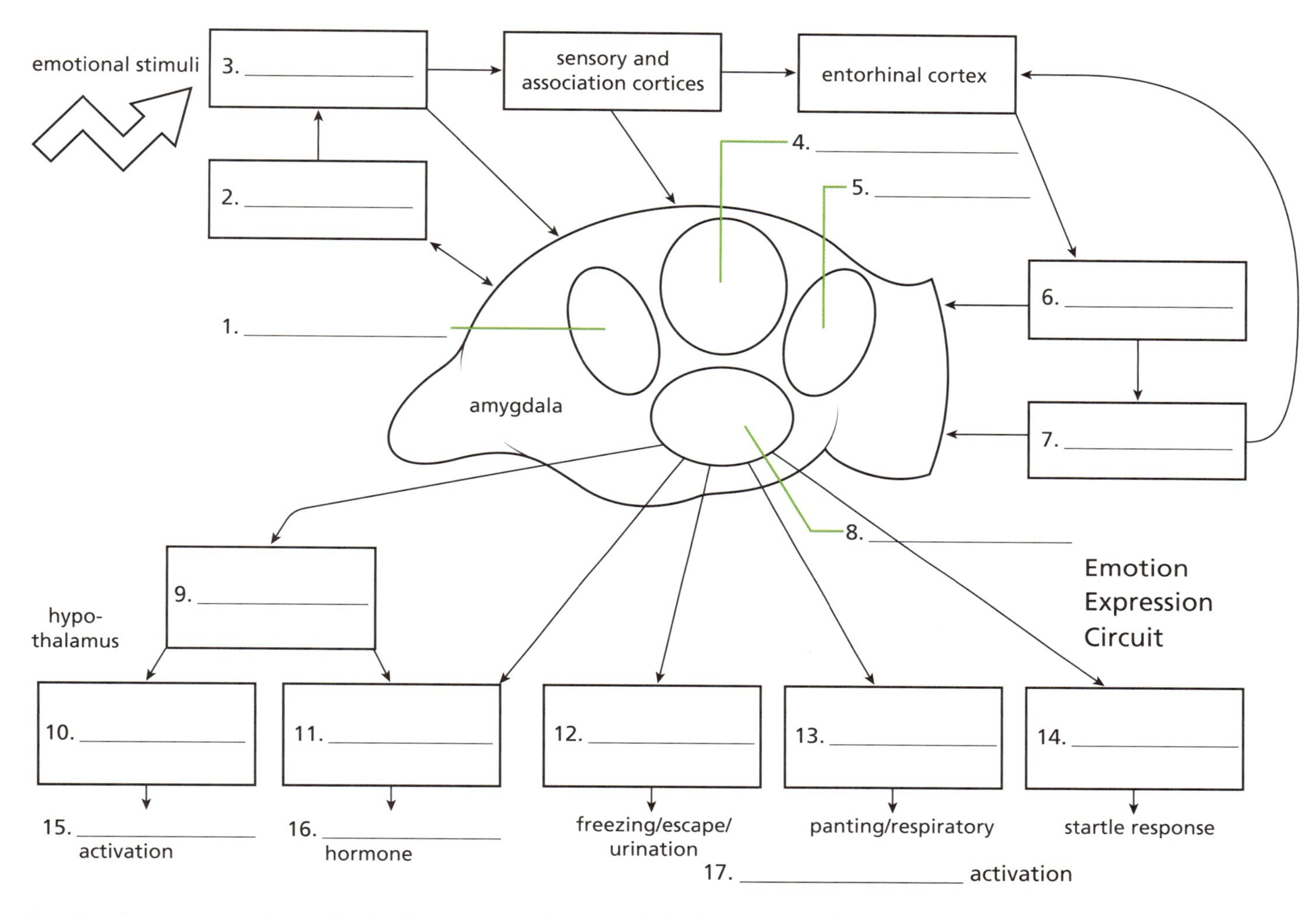

Emotional expression refers to the bodily responses that occur following emotionally salient stimuli, and it includes visceral motor changes and stereotypical motor responses, such as the movement of facial muscles. Although many regions of the limbic system contribute to emotional processing, the amygdala is often considered the central node that links together cortical and subcortical brain regions involved in the emotional brain network.

The amygdala contains around 13 nuclei that are broadly divided into four groups: lateral, basal, accessory basal, and central. The amygdala receives two main types of input: those arising in cortical and thalamic structures, such as sensory information, and those arising in the hypothalamus or brainstem. In addition, the prefrontal cortex also projects to the basal nuclei, and to a lesser extent to the other nuclei, and helps regulate emotional responses.

The main pathways involved in emotional expression arise from the amygdala's central nuclei and terminate in the hypothalamus, bed nucleus of the stria terminalis (BNST), and other areas in the midbrain and brainstem. Activation of these systems triggers the somatic, endocrine, and autonomic responses associated with emotional expression. Responses related to emotional expression may include freezing behavior, defecation, and increases in heartbeat and stress hormone levels.

Answers

1. basal nuclei, 2. prefrontal cortex, 3. thalamus, 4. lateral nuclei, 5. accessory basal nuclei, 6. hippocampus, 7. subiculum, 8. central nuclei, 9. bed nucleus of the stria terminalis, 10. lateral hypothalamus, 11. paraventricular nucleus, 12. periaqueductal gray, 13. pontine parabrachial nucleus, 14. reticularis pontis caudalis, 15. sympathetic, 16. neuroendocrine, 17. parasympathetic

Memory Formation Circuitry

There are two primary types of human long-term memory. Procedural memory is the memory for learned skills and tasks, such as how to ride a bicycle. Declarative memory includes episodic memory, such as autobiographical life events and experiences, and semantic memory, such as facts and concepts.

Episodic memory represents memory of events serially in time. This type of memory is not just facts about events but also contains the emotional charge and context surrounding the events. For example, the memory of when and where you met a loved one, what he or she was wearing, and what you felt at that time is an episodic memory.

The hippocampus plays a crucial role in creating an episodic memory. Information regarding context, time, and location is processed in a variety of brain regions and is then delivered to the perirhinal and entorhinal cortices of the parahippocampal gyrus, before being relayed to the dentate gyrus. From there, the information is transmitted to the CA3 area of the hippocampus via mossy fibers and subsequently to the CA1 area via Schaffer collaterals. CA1 sends signals to various brain areas via the subiculum, mammillary bodies, and thalamic nuclei. Following the initial learning that relies on the hippocampus, different components of the memory (such as visual, auditory, and olfactory impressions) are transferred to their respective cortices for consolidation and long-term storage. The hippocampus links these memory components into a cohesive episode, allowing all parts to be evoked when a memory is recalled.

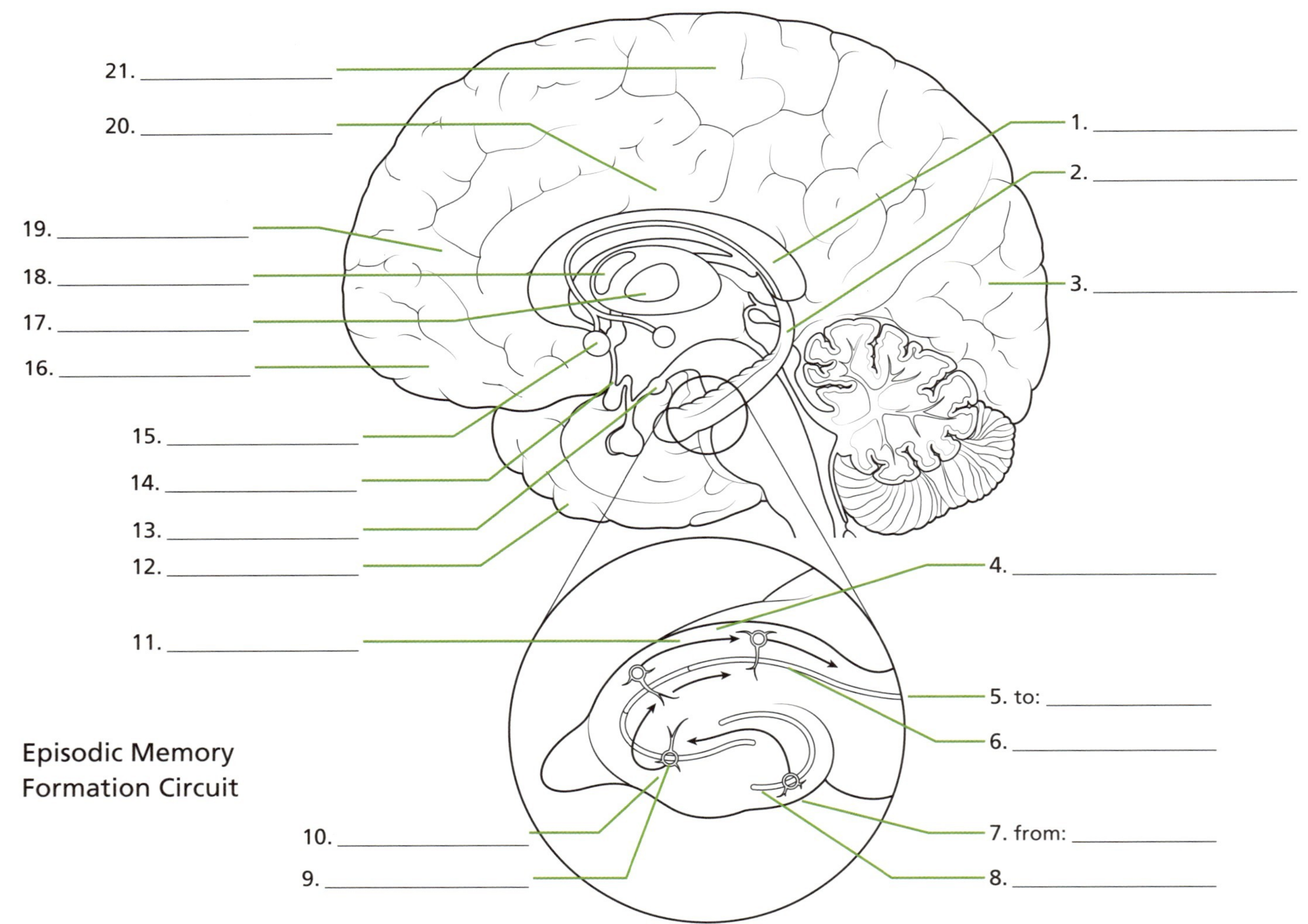

Episodic Memory Formation Circuit

Answers

1. fornix, 2. hippocampus, 3. retrosplenial cortex, 4. CA1, 5. subiculum, 6. dentate gyrus, 7. entorhinal cortex, 8. perforant path, 9. mossy fibers, 10. CA3, 11. Schaffer collaterals, 12. temporal lobe, 13. mammillary bodies, 14. amygdala, 15. septal nuclei, 16. orbitofrontal cortex, 17. medial thalamic nuclei, 18. anterior thalamic nuclei, 19. prefrontal cortex, 20. cingulum, 21. parietal lobe

Each component of the hippocampal circuit has a distinct operation. CA3 and the dentate gyrus are involved in pattern separation, whereas CA1 is crucial for spatial learning. The CA1 and CA3 contain specialized "place cells" that create a mental map of space, which underlies their role in spatial learning. This role was demonstrated by a study that showed that the size of the hippocampi of London taxi drivers increased following extensive training on how to navigate the city.

Lesions to the hippocampus and adjacent cortical structure produce severe and lasting anterograde amnesia. A famous case in neurology is that of patient H.M., who had major parts of his medial temporal lobe removed bilaterally due to intractable seizures. He subsequently developed anterograde amnesia, where he was unable to encode new life events. Interestingly, removing his hippocampi had no effect on his ability to acquire new procedural memories, his general intelligence, or his recall of episodic memories from his early childhood. This underscores the role of the hippocampus in the formation of declarative memories.

The amygdala is essential for encoding emotionally salient memories. For example, the amygdala is involved in fear conditioning, in which a neutral conditional stimulus, such as a tone, is linked to an aversive unconditional stimulus, such as an electrical shock. Repeated pairing of the tone with the shock in rodents will result in freezing and other expressions of fear when the rodent hears the tone. The circuitry for auditory fear conditioning begins with the transmission of a tone to the lateral nuclei of the amygdala via sensory cortices and the thalamus. The memory of the tone is associated with the memory of the electrical shock in the basal nuclei. This fear memory is subsequently expressed via connections between the central nuclei and efferent systems, which results in the animals freezing in place.

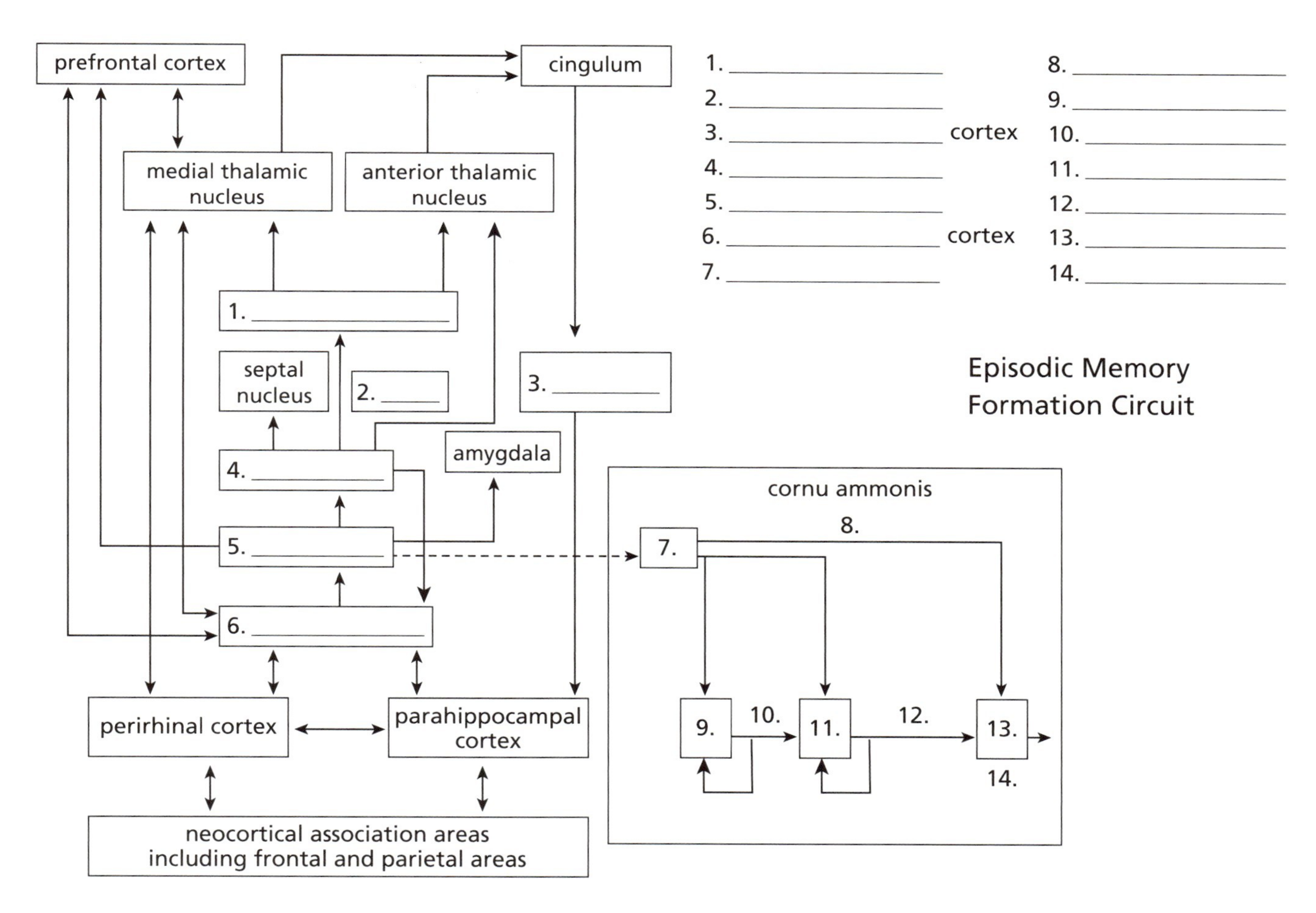

1. ____________
2. ____________
3. ____________ cortex
4. ____________
5. ____________
6. ____________ cortex
7. ____________
8. ____________
9. ____________
10. ____________
11. ____________
12. ____________
13. ____________
14. ____________

Episodic Memory Formation Circuit

Answers

1. mamillary bodies, 2. fornix, 3. retrosplenial, 4. subiculum, 5. hippocampus, 6. entorhinal, 7. from: entorhinal cortex, 8. perforant path, 9. dentate gyrus, 10. mossy fibers, 11. CA3, 12. Schaffer collaterals, 13. CA1, 14. to: subiculum

Papez Circuit

The Papez circuit was proposed in 1937 by James Papez to describe the cortical control of emotional expression. He began by studying the medial face of the hemispheres, focusing particularly on structures in the limbic system that were initially thought to be concerned with olfactory processing. According to Papez, the circuit that underlies emotional processing begins with the hippocampus, leads to the mammillary bodies by way of the postcommissural fornix, and then goes through the mammillothalamic tract to the thalamus. From there, the circuit then loops back to the hippocampus by way of the anterior thalamic nuclei, cingulate gyrus, cingulum, and parahippocampal gyrus.

Papez Circuit—components and connections

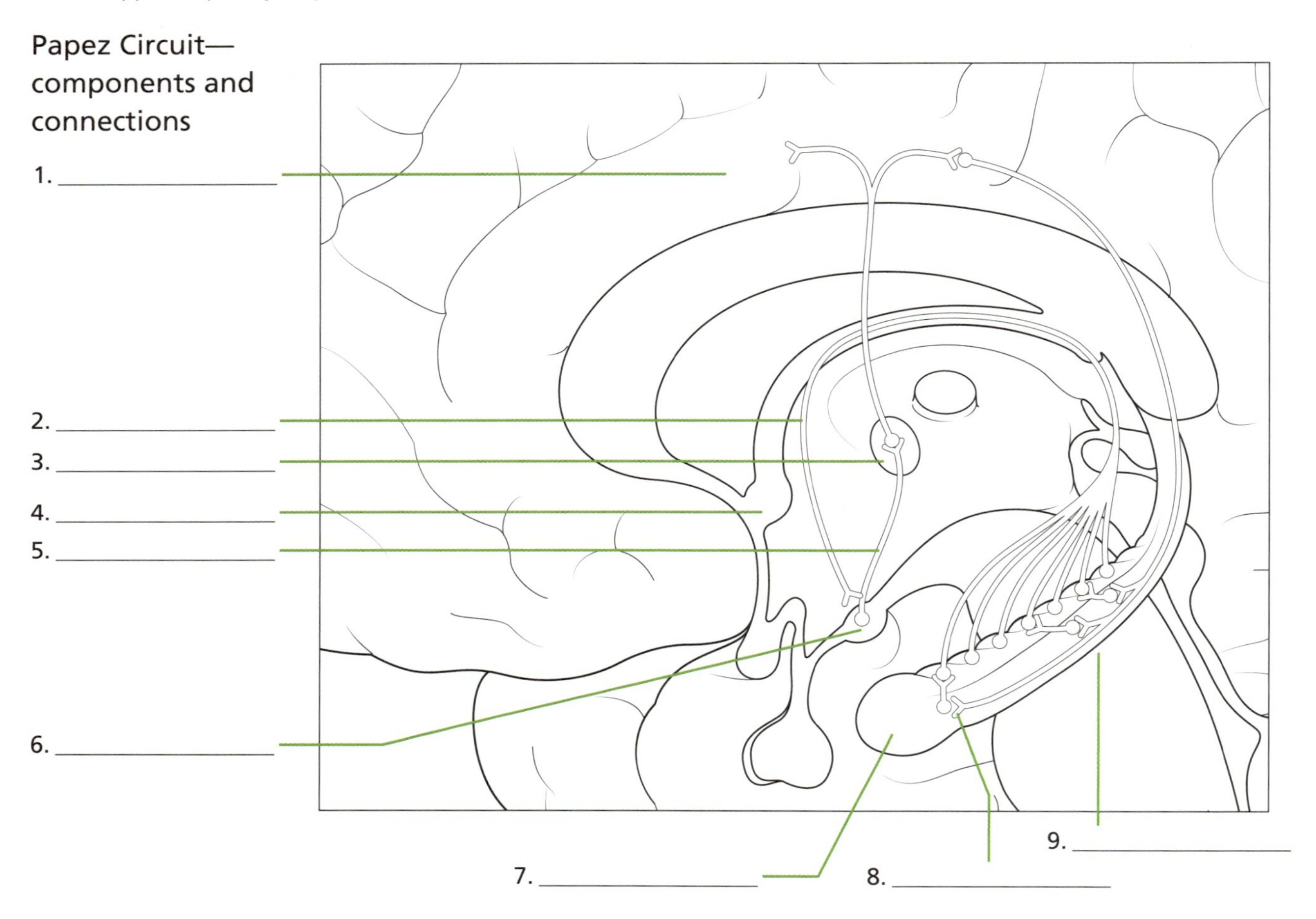

Papez believed that emotional experiences occur when the cingulate cortex integrates information from the hypothalamus, anterior thalamus, and sensory cortices, and output from the cingulate to the hippocampus and hypothalamus results in a top-down control of emotional responses. The key components of the Papez circuit form a closed loop, in which activation of the hypothalamus triggers its further activation, and the Papez circuit was believed to be the neurobiological circuit that supports sustained emotional responses, such as lingering anger.

It now appears that some of the structures in the Papez circuit have little to do with emotional behavior, whereas other structures crucially important for emotional processing, such as the amygdala, are missing from the circuit. Nevertheless, the Papez circuit laid the foundation for the modern study of the neuroanatomical underpinning of emotional perception and expression, and it represents a valiant effort in linking neuroanatomy to behavior.

Answers

1. cingulate gyrus, 2. postcommissural fornix, 3. anterior nucleus of the thalamus, 4. anterior commissure, 5. mammillothalamic tract, 6. mammillary bodies, 7. amygdala, 8. entorhinal cortex, 9. hippocampal formation

Language Areas of the Cerebral Cortex

Broca's and Wernicke's areas, along with the primary auditory cortex, make up a network called the arcuate fasciculi. This network allows people to hear, process, and execute language. Pierre Paul Broca discovered the area of the brain involved in producing coherent speech through a series of case studies, the first—and most famous—being "Tan," a man who was capable of saying only one word, which was "tan." Broca's area is also referred to as the inferior frontal gyrus, as it is located in the bottom half of the frontal cortex. Recent evidence suggests that the role of Broca's area is more dynamic than once believed; a study from Flinker and associates in 2015 indicated that speech production prior to articulation, but not single-word production, activates Broca's area.

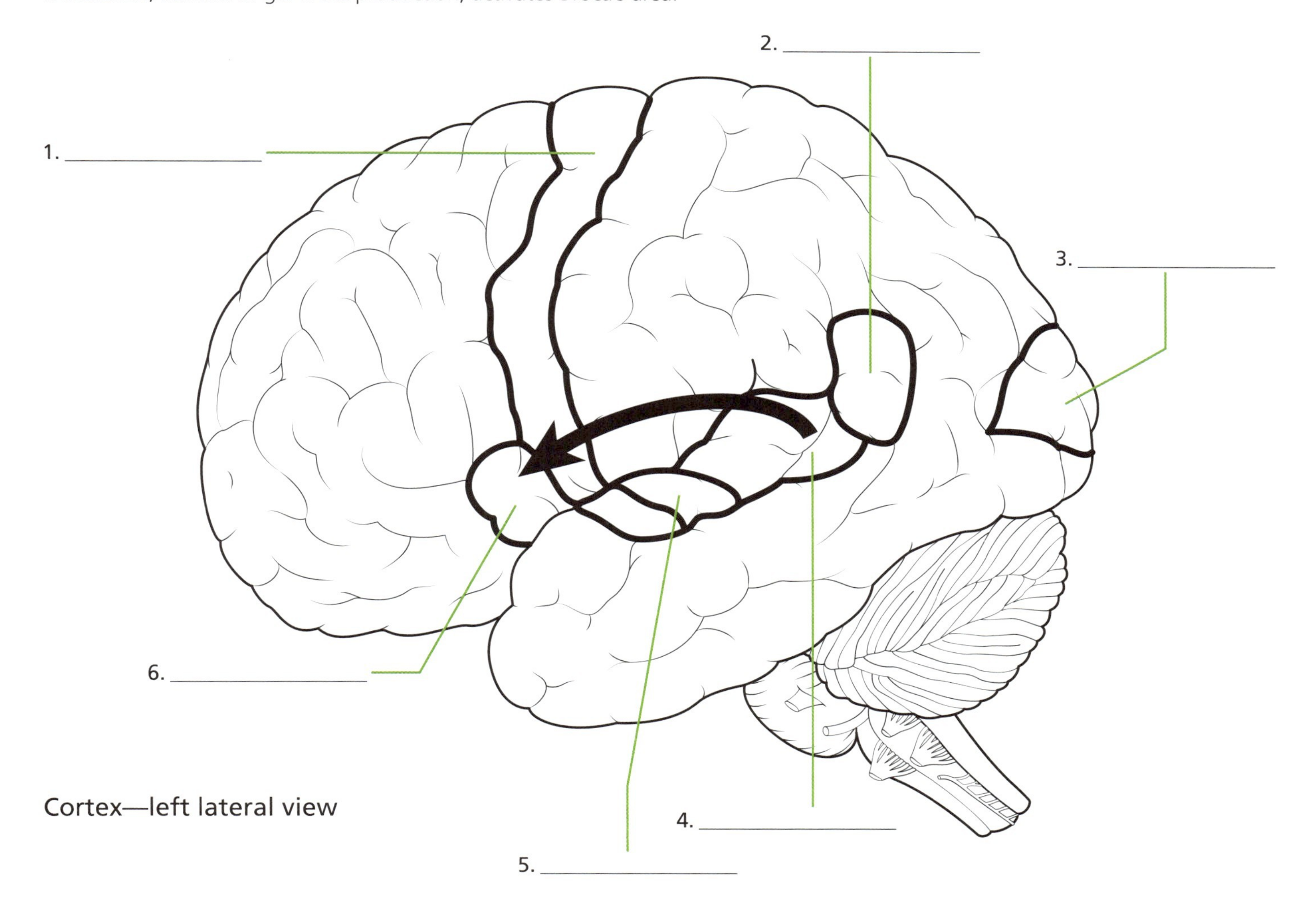

Cortex—left lateral view

Wernicke's area is located partially in the temporal lobe and partially in the parietal lobe. It is made up of two different anatomical structures: the supramarginal gyrus and the angular gyrus. Wernicke's area is related to the processing of both written and spoken speech. Individuals who have damage to Wernicke's area are usually unable to produce coherent sentences. Their speech will have proper syntax, grammar, and prosody, but the content of the sentence will have little to no meaning.

Answers

1. motor cortex, 2. angular gyrus, 3. primary visual cortex, 4. Wernicke's area, 5. auditory cortex, 6. Broca's area

Aphasia

Aphasia is a general term used to describe damage to language areas of the brain. There are multiple types of aphasia that can disrupt parts of the language network. Broca's aphasia occurs when there is damage to or around Broca's area. A person with Broca's aphasia will have trouble saying the words they want to say, but speech comprehension is still intact. Wernicke's aphasia disrupts the functions of Wernicke's area, which is related to speech processing. When people with Wernicke's aphasia produce speech, it sounds fluent but the words are nonsensical. This is commonly known as "word salad." Global aphasia occurs when there is severe damage to the language areas or to the arcuate fasciculi. This is the most severe of the aphasias, where both speech production and speech processing are compromised. People with this type of aphasia can comprehend only a few words at a time, and they may be extremely difficult to communicate with due to poor speech production. Conduction aphasia is a rare form of aphasia, in which most speech processing and production is intact. However, damage to the left parietal region, related possibly to the angular gyrus, creates great difficulty in repeating words or phrases. For example, when prompted to repeat the word "guitar," a person with conduction aphasia may say "tiguar" or struggle to say the word as if they have a stutter. In contrast, having a typical conversation without prompts shows mostly typical speech processing and production.

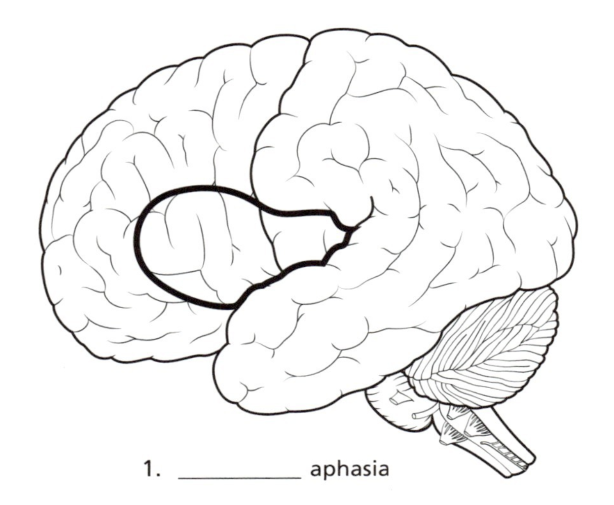

1. __________ aphasia

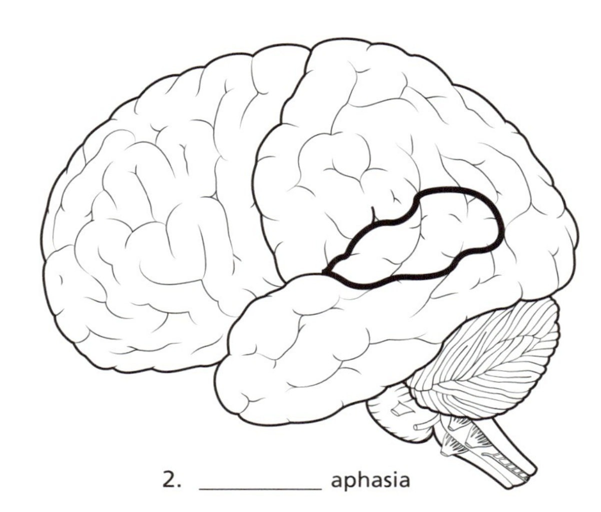

2. __________ aphasia

Cortex—left lateral views

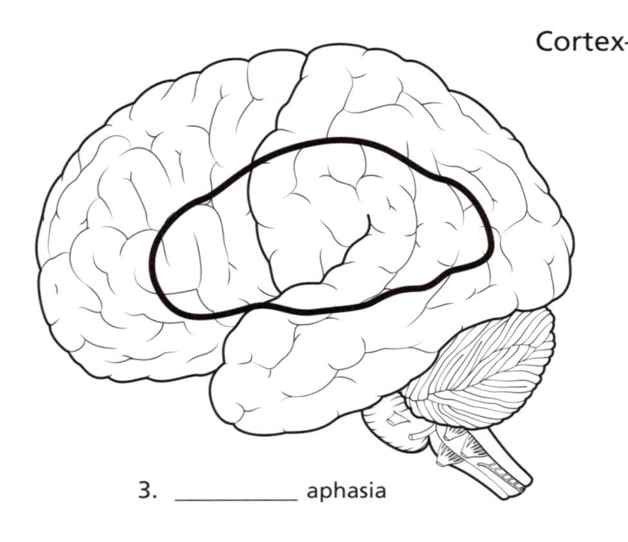

3. __________ aphasia

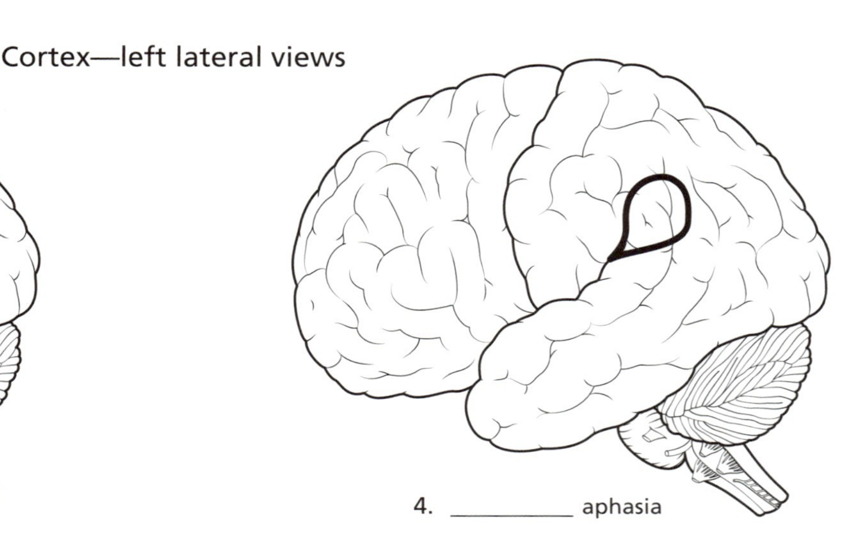

4. __________ aphasia

Answers

1. Broca's, 2. Wernicke's, 3. global, 4. conduction

Sleep Stages

EEG Patterns during the Stages of Sleep

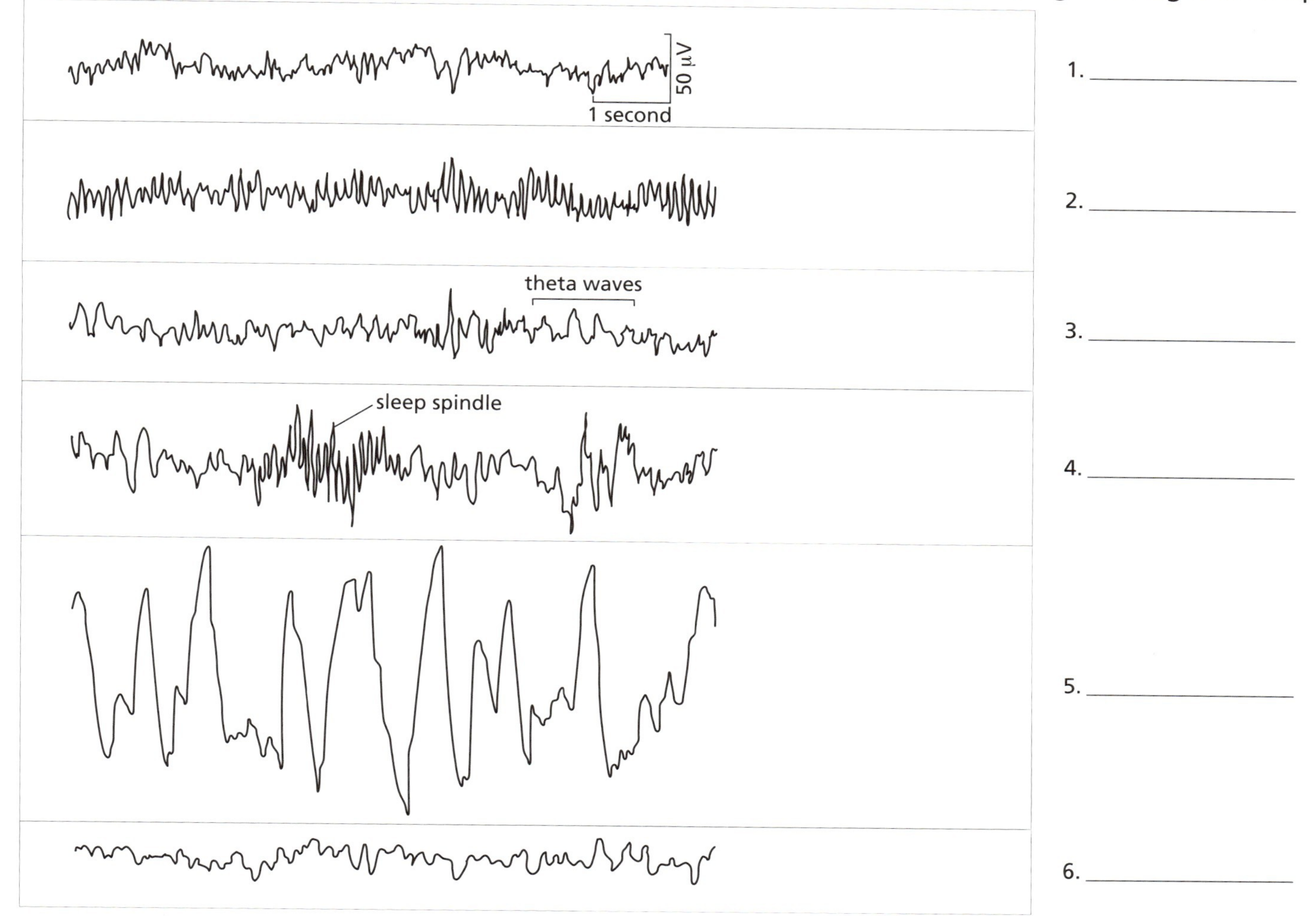

Sleep architecture is the amount of time spent in both nonrapid eye movement (NREM) sleep and rapid eye movement (REM) sleep. There are four distinct NREM sleep stages, which are characterized by the frequency and amplitude of neuronal activity. These stages are captured using electroencephalography (EEG), a noninvasive recording technique that measures the timing and synchronicity of brain electrical activity at the level of the scalp. Stage 1 sleep is the beginning of sleep and is characterized by the transition from alpha to theta waves (4–8 cycles per second [Hertz, Hz]). Stage 2 exhibits greater theta wave activity, with dispersed high-frequency sleep spindles. Behaviorally, these sleep spindles correlate with muscle twitching before falling into deeper sleep. Researchers theorize sleep spindles serve as a mechanism to transition into a sleep state. The presence of delta waves (1–4 Hz) signals the transition into stage 3 sleep. Stage 4, or REM sleep, shows lower amplitude but higher brain waves, ranging between alpha (8–12 Hz) to gamma (>24 Hz). This stage is usually associated with dreaming and is commonly connected to being able to learn cognitively demanding tasks. More recently, stage 3 and stage 4 sleep have been combined and are referred to as slow-wave sleep.

Answers

1. awake, 2. drowsy, 3. stage 1 sleep, 4. stage 2 sleep, 5. slow-wave sleep (stage 3 and stage 4 sleep), 6. rapid eye movement sleep

Brain Regions Involved in Sleep

It is easy to think of sleep as a simple process, with the brain taking a break from processing hectic input from the outside world; however, sleep is very complex. A number of areas distributed across the brain network to orchestrate sleep. The hypothalamus contributes to the body's diurnal clock, regulating both arousal and drowsiness. The neurons within the tuberomammillary nucleus of the hypothalamus utilize histamines as a neurotransmitter to increase arousal. This is the reason why many antihistamine drugs make the user drowsy—they block this arousal signal. During sleep, the thalamus aids the transition into slow-wave sleep by inhibiting stimuli from reaching the cortex. It is hypothesized that sleep spindles, which originate from thalamic activity, act as interceptors for incoming stimuli. The reticular formation aids sleep by inhibiting activation from areas contributing to wakefulness (like the frontal cortex) and engaging areas that upregulate sleep (like the thalamus). During REM sleep a different brain network comes online, including the pons, amygdala, and hippocampus. The purpose of, and the brain network involved in, REM sleep are poorly understood. One hypothesis suggests that REM facilitates the processing of important experiences from the day into long-term memory and the clearing of unimportant information from short-term memory.

Central Brain Areas Involved in Sleep—sagittal section through longitudinal fissure

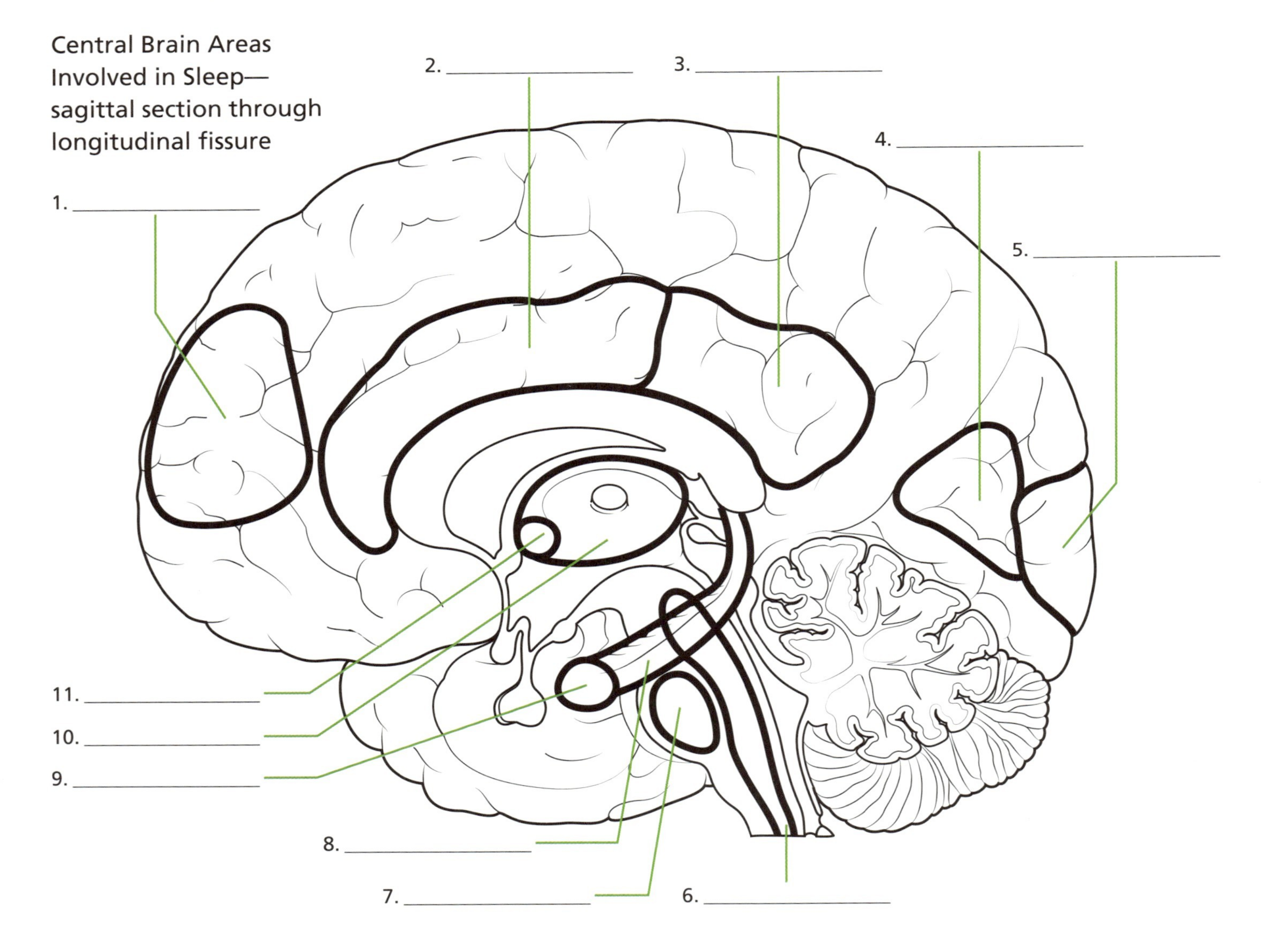

Answers

1. frontal cortex, 2. anterior cingulate gyrus, 3. posterior cingulate gyrus, 4. extrastriate visual areas, 5. primary visual cortex, 6. reticular formation, 7. pons, 8. hippocampus, 9. amygdala, 10. thalamus, 11. hypothalamus

Sleep Circuitry

Our first insight into which brain regions are involved in sleep came from the Spanish flu pandemic in 1918, which would leave victims in a sedated coma-like state or unable to sleep for days prior to death. There is a cycle of activity between areas of the brain that are active during sleep and other areas that show a reduction in activation. Oscillating activity between the reticular formation, the thalamus, and the cortex are integral to initiating and sustaining sleep. The thalamus regulates the shifting activation across the brain and how senses or cortical integration affect the onset or regulation of sleep. To promote sleep, the activity in areas like the primary visual cortex and the frontal cortex begins to reduce in order to limit interference from visual or cognitive stimulation. The anterior cingulate cortex and posterior cingulate cortex are regulated differently during sleep. The anterior cingulate cortex, which is related to emotion and reward, shows increased activity—particularly during REM sleep. This is in line with other areas related to emotion, like the amygdala, and may serve a function of producing dreamlike states. The posterior cingulate cortex is generally involved in wakefulness and awareness and is reduced in activation to reflect a reduction in wakefulness.

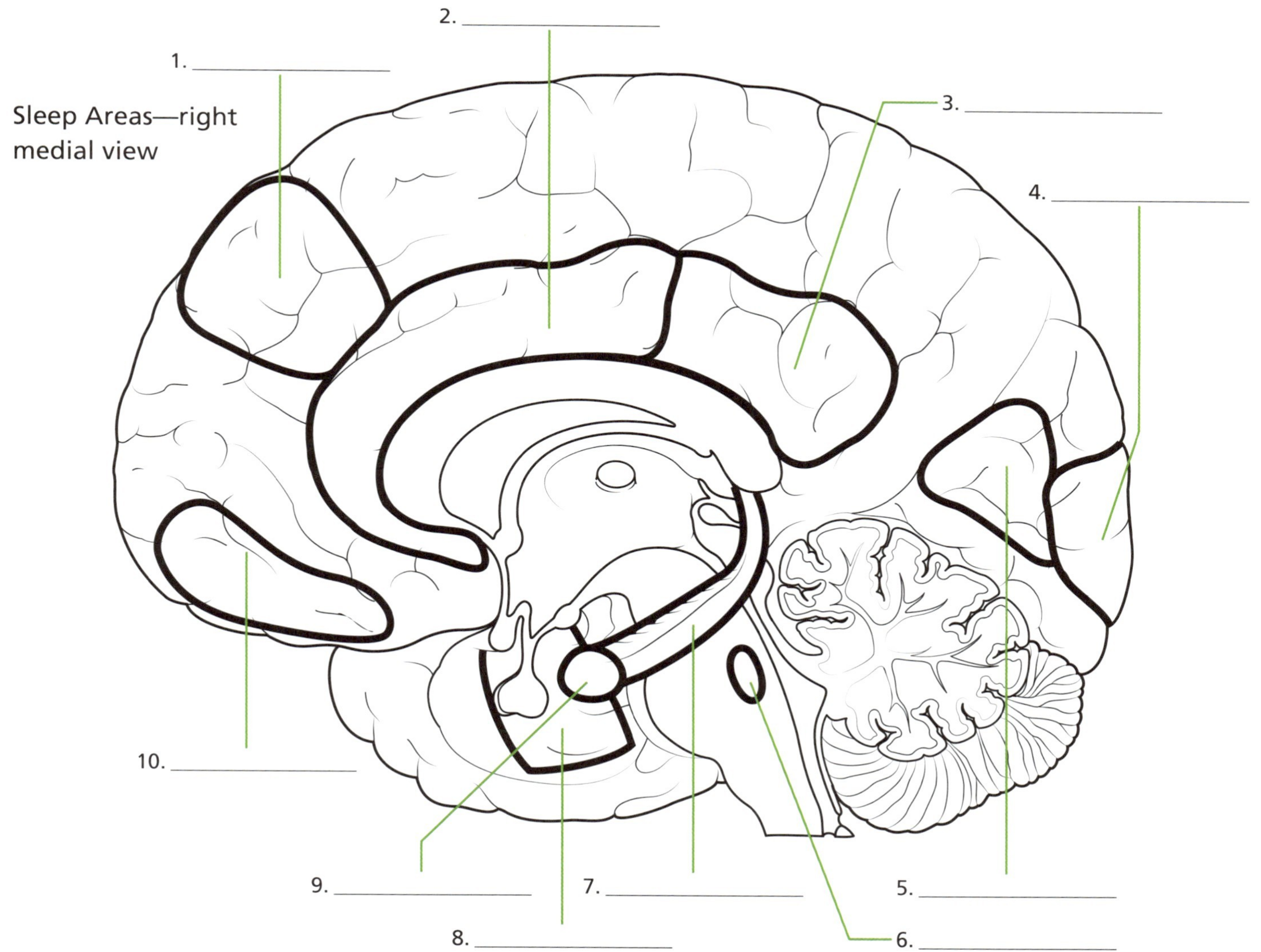

Answers

1. dorsolateral prefrontal cortex, 2. anterior cingulate cortex, 3. posterior cingulate cortex, 4. primary visual cortex, 5. extrastriate visual areas, 6. pontine tegmentum, 7. hippocampus, 8. parahippocampal cortex, 9. amygdala, 10. deep frontal white matter

Anatomical Changes During Sleep

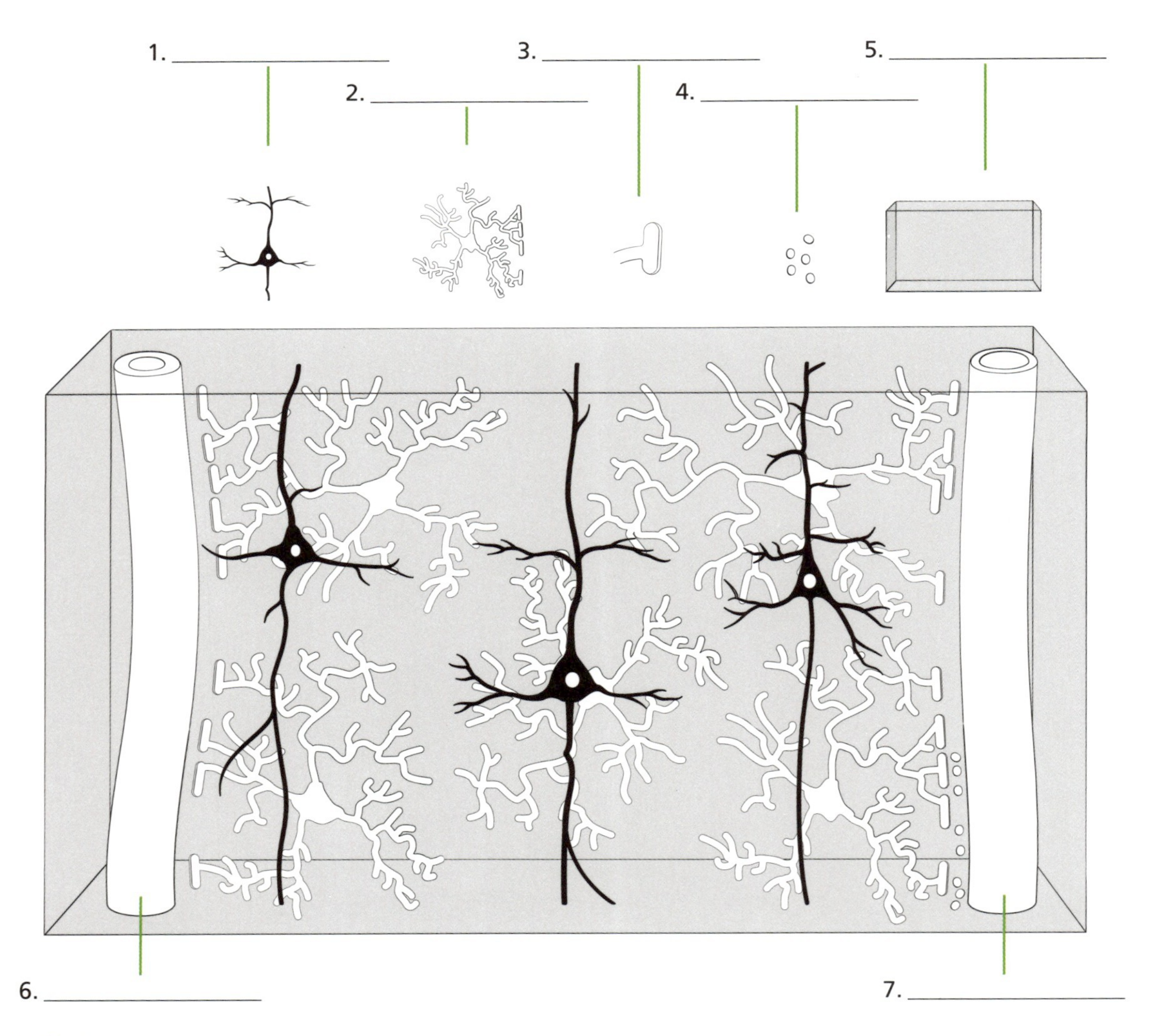

Glymphatic System

Many know about the anatomical changes in a person's body during sleep—like growing slightly when sleeping—but not many know of the anatomical changes that occur within your brain during sleep. The glymphatic system, newly described in the CNS, initiates a process that occurs in your brain where waste in the form of plaques is eliminated by the help of astrocytes and cerebrospinal fluid. The system is similar to the lymphatic system but relies on the function of glial cells. It is engaged predominantly during sleep and helps eliminate waste from the CNS, including potentially harmful substances like amyloid beta, which can aggregate into plaques, a primary pathology seen in Alzheimer's disease. The system works when the brain falls into sleep and the astrocytes begin to shrink, allowing a greater volume of cerebrospinal fluid to flow through interstitial spaces. This larger volume of cerebrospinal fluid essentially funnels waste into veins, where it is subsequently processed and excreted from the body. The system is dependent on the sleep-wake cycle and locus coeruleus-derived norepinephric signaling.

Answers

1. neuron, 2. astrocyte, 3. water channels, 4. amyloid beta, 5. cerebrospinal fluid, 6. artery, 7. vein

Spatial Perception

Have you ever wondered how certain landmarks in an environment can trigger a memory, such as when a street intersection triggers a memory of how to get home? Spatial perception allows us to recall information and orient ourselves in order to navigate a specific environment. The hippocampus, parahippocampal cortex, and entorhinal cortex are important areas for spatial perception, spatial memory, and recall. The discovery of "place cells" within the hippocampus allowed us to understand how the hippocampus and surrounding cortex locate and ground our memories within space and location. Neurons within the medial entorhinal cortex have strong connections with the place cells in the hippocampus, allowing for processing of spatial information, while neurons within the lateral entorhinal cortex have connections to the hippocampus and rhinal cortex, which facilitate identification of objects within a context. The rhinal cortex receives the contextual information from the perirhinal cortex. The medial prefrontal cortex also facilitates the recognition of objects within a place, recalling specific memories or patterns in the environment. The posterior parietal cortex is involved in processing movement of objects in space. These parallel processing streams allow us to determine spatial information and ground it within a relational learning network.

Spatial Perception Areas—right medial view

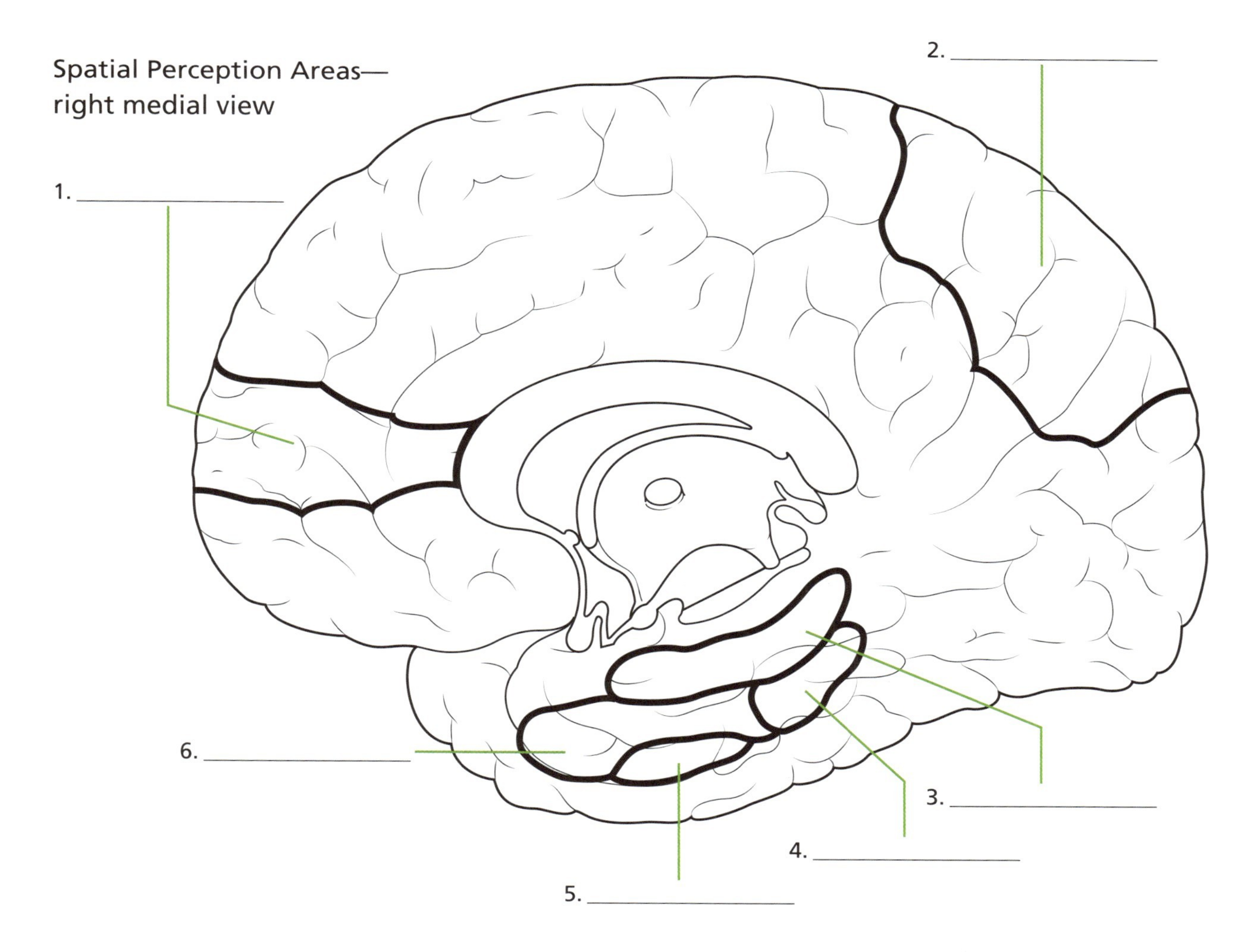

Answers

1. medial prefrontal cortex, 2. posterior parietal cortex, 3. hippocampus, 4. parahippocampal cortex, 5. entorhinal cortex, 6. perirhinal cortex

Neural Plasticity: Gross Anatomical Level

Neural plasticity is the changing of neural networks or areas due to almost any change that can occur to a person, be it physical, environmental, social, or other. Some of the most interesting studies on neural plasticity stem from amputees. For example, what happens to your brain when you lose an arm? The areas specific to hand and arm sensation are no longer receiving input or sending viable outputs. Rather than these areas of the brain simply becoming empty, unused voids, however, they begin to reorganize and become subsumed by adjacent areas—like the face or the torso. In individuals who have had limbs amputated, the reorganization of somatosensory areas can cause a phenomenon known as a phantom limb, whereby activation of ascending nerves to the reorganized cortex can cause the sensation of pain or intense itching in a limb that no longer exists. This reorganization is not specific to any area of the brain—neural plasticity can occur anywhere. In people who lose their entire sense of vision, for example, areas from the parietal lobe and temporal lobe may begin to reorganize to increase somatosensory or auditory processes, respectively.

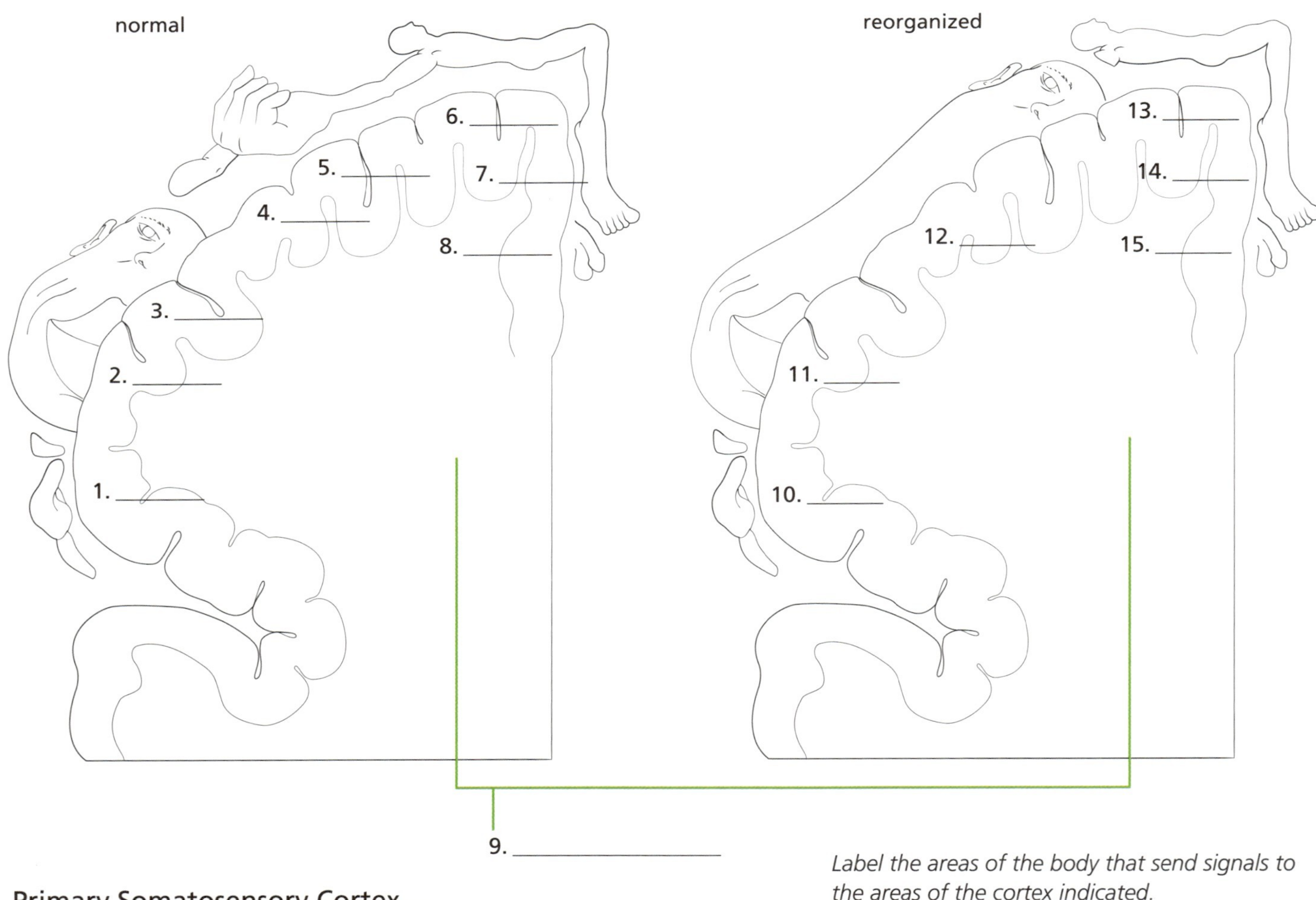

Label the areas of the body that send signals to the areas of the cortex indicated.

Primary Somatosensory Cortex—anatomical maps

Answers

1. pharynx, 2. lips, 3. face, 4. thumb, 5. hand, 6. trunk, 7. foot, 8. genitals, 9. somatosensory cortex, 10. pharynx, 11. lips, 12. face, 13. trunk, 14. foot, 15. genitals

Neural Plasticity: Cellular Level

Most people have heard the phrase "Use it or lose it" when it comes to brain power. But how does that concept translate to the level of the neuron and synapse? Neural plasticity is the ability of a neuron to strengthen or weaken connections within neural networks, and to do so it utilizes two mechanisms: long-term depression (LTD) and long-term potentiation (LTP). LTD is a systematic way to prune receptor sites on a postsynaptic terminal. This is achieved through increasing calcium levels within the postsynaptic neuron, which disrupts receptor binding in the synaptic cleft. This decreases the amount of neurotransmitters that can bind to the postsynaptic receptors, subsequently decreasing the activity of postsynaptic neurons. LTP works essentially in the opposite way as LTD, increasing receptor sites in the postsynaptic neuron. As a presynaptic neuron increasingly activates, more neurotransmitters are released into the synaptic cleft but cannot be received by receptors due to their unavailability. This demand increases the amount of receptors at the postsynaptic terminal, resulting in a greater volume of neurotransmitters being received. LTP has been associated with the learning of new behaviors in animals, from slugs right through to primates.

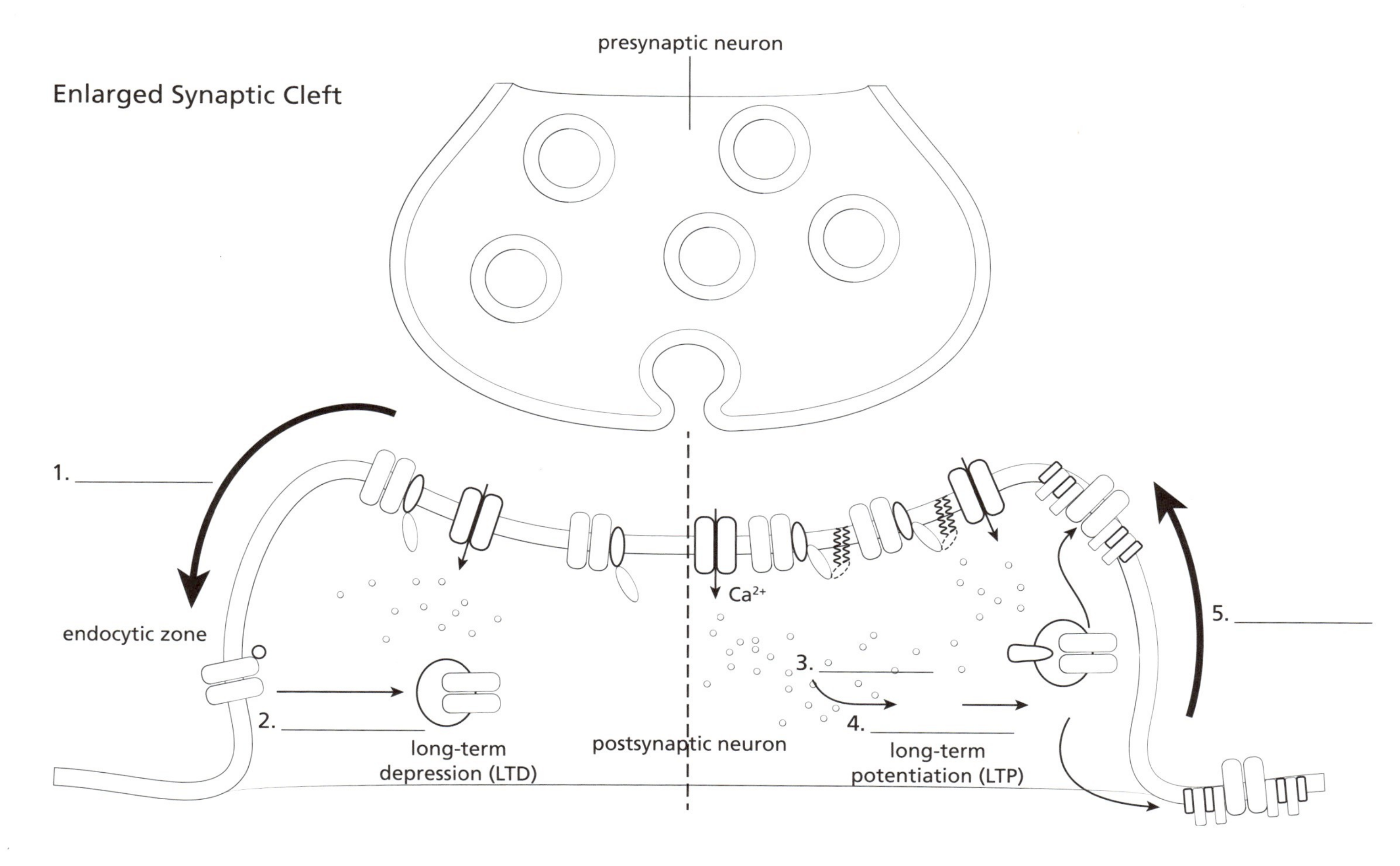

Answers

1. lateral diffusion, 2. endocytosis, 3. Ca^{2+}, 4. exocytosis, 5. lateral diffusion

Aging: Gross Anatomical Level

As the brain ages, it undergoes changes from the molecular level through large structural degradation. One of the most noticeable changes is the decreasing mass of the brain, specifically seen as less gyral surface area and greater fissure gaps within sulci. As we age, changes also occur at the level of the neuron, with cells being more likely to upregulate N-methyl-d-aspartic acid (NMDA) receptors than develop new spines. Functionally, this results in a less robust learning behavior. Across aging, white matter degeneration can cause problems in the functional connectivity between brain areas. This type of brain degeneration can cause difficulties integrating brain networks. For example, degeneration of the arcuate fasciculi can behaviorally seem as if a person is not responding, but neurally they are unable to produce speech quickly or fluently due to poor connectivity. Despite these normal changes, there is a stark difference between healthy brain aging and the development of dementia. Healthy aging is associated with decreased executive function and processing speed due to the loss of neurons within cortical areas. Dementia, through aggressive neurodegeneration, causes a loss of cognitive abilities across two or more domains and loss of the ability to function in everyday life.

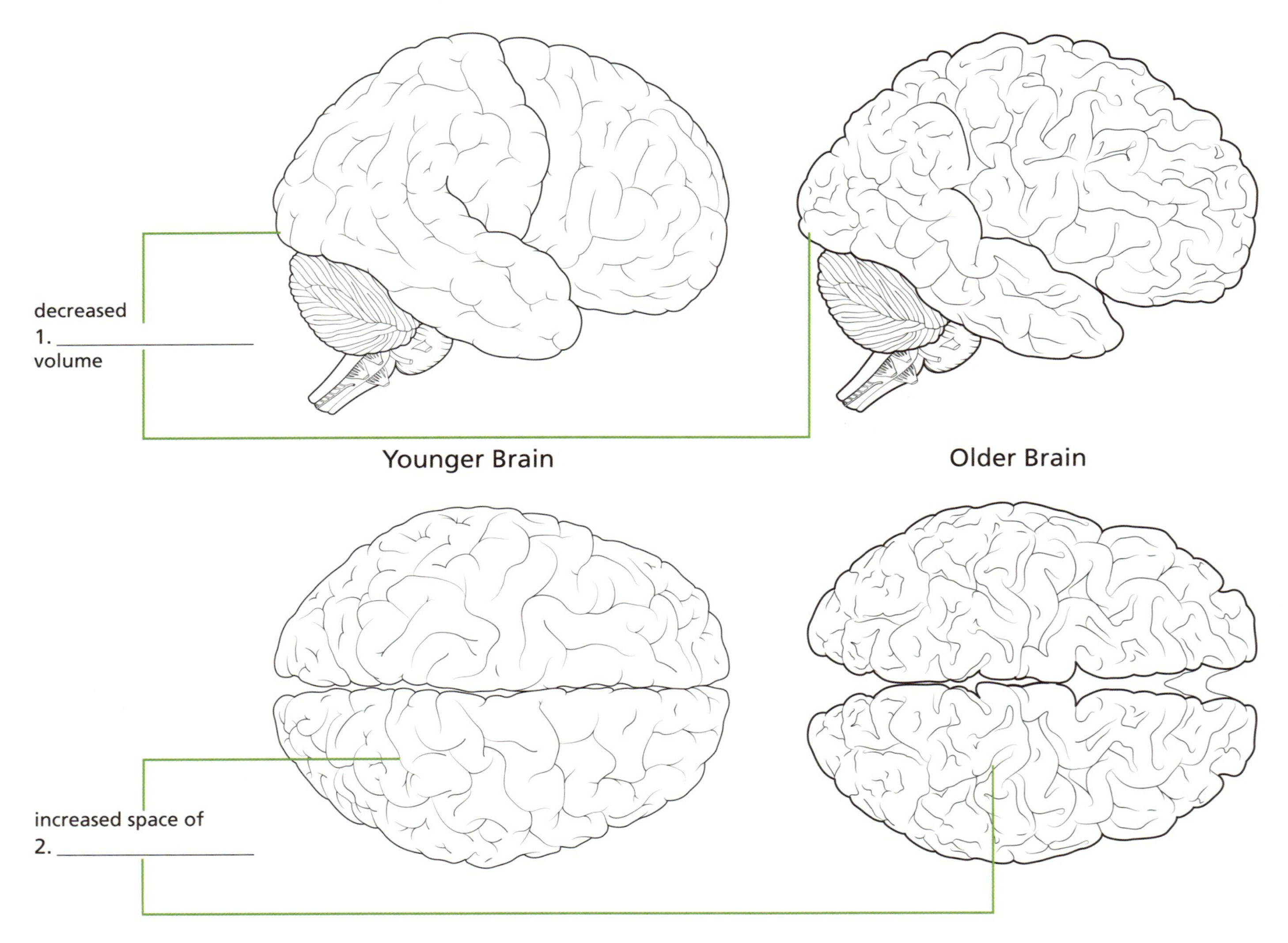

Answers

1. gyri; 2. sulci

Aging: Cellular Level

For years there was a dogma that you are born with all of the neurons you will ever have, but the discovery of adult neurogenesis has since proved this completely wrong. Aging does not begin as soon as you blow out the candles on your 40th birthday—rather, it is a process that begins at birth. You are born with an overabundance of neurons, and many are not completely developed. Before the age of five, the brain prunes a large number of neurons. It is hypothesized that one of the reasons we do not remember much of our early childhood is due to the loss of these neurons and formation of new neural networks. In childhood, the remaining neurons develop neural networks through axonal and dendritic extension, and subsequent synapse formation. Through adulthood, active neural connections are maintained and strengthened; however, connections that lie dormant are often pruned. This may explain why you lose the ability to remember information or skills you use only infrequently. Finally, in older adulthood and old age, some of these networks lose strength through neurodegeneration. Neurons can degenerate due to many different reasons, including traumatic brain injury, stroke, or pathological neurodegeneration (like dementia).

Neuron Development

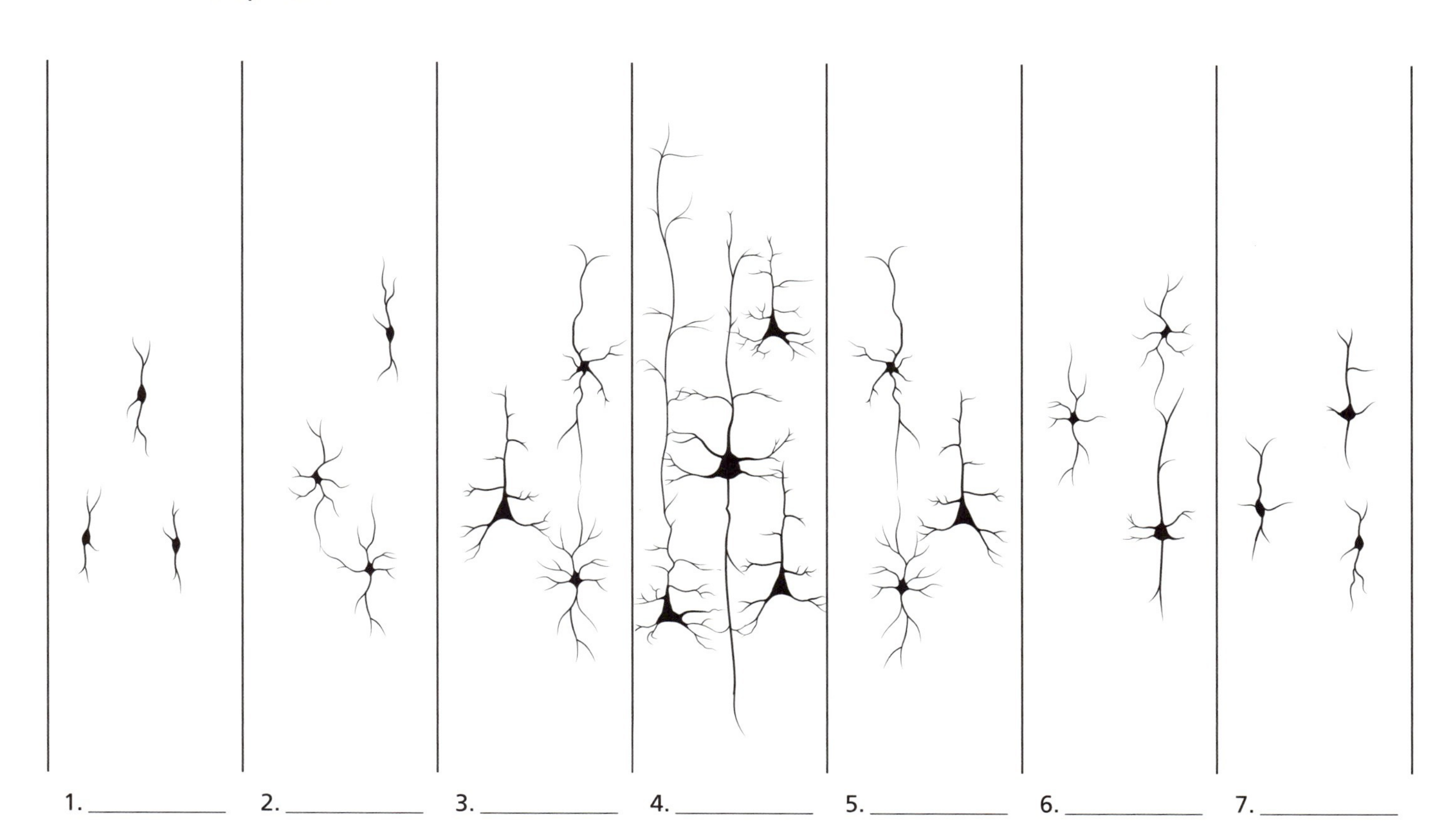

1. ________ 2. ________ 3. ________ 4. ________ 5. ________ 6. ________ 7. ________

Label the age of neuron development in each case.

Answers

1. age 1+ (birth–infant), 2. age 7+ (child), 3. age 13+ (teenager), 4. age 30–60 (adult—peak), 5. age 65+ (decay starts), 6. age 70+ (dementia), 7. age 80+ (Alzheimer's—no neural communication)

Motor Pathways—Motor Cortex to Spinal Cord Level

The motor cortex is made up of three regions: the primary motor cortex (M1), the premotor cortex, and the supplementary motor area (SMA). M1 is directly responsible for the contraction of muscles. The premotor cortex and SMA both project directly into the spinal cord, with the premotor cortex controlling trunk muscles and the SMA controlling postural stabilization. Low levels of brief stimulation elicit simple movements of individual body parts, whereas higher levels of currents elicit more complex movements. In the corticospinal/pyramidal tract, axons pass through the internal capsule, diencephalon, midbrain, pons, and, finally, the medulla, where medullary pyramids are formed. Axons passing from the medulla into the spinal cord decussate, or cross to the opposite side, and descend in the white matter of the contralateral spinal cord, where they control flexor muscles. Nondecussating axons descend in the ipsilateral cord to form ventral corticospinal tract white matter until reaching the level of the body they represent, then they decussate. These motor neurons control extensor muscles. Most descending axons terminate on interneurons rather than alpha motor neurons.

Motor Cortex to Spinal Cord Pathway

1. ____________
2. ____________
3. ____________
4. ____________
5. ____________
6. ____________
7. ____________
8. ____________
9. ____________
10. ____________
11. ____________
12. ____________
13. ____________
14. ____________
15. ____________
16. ____________
17. ____________

Answers

1. upper motor neurons, 2. primary motor area of cerebral cortex, 3. internal capsule, 4. cerebrum, 5. midbrain, 6. cerebellum, 7. pons, 8. medulla oblongata, 9. anterior corticospinal tract, 10. cervical spinal cord, 11. lumbar spinal cord, 12. lower motor neurons, 13. skeletal muscle, 14. lateral corticospinal tract, 15. decussation of pyramid, 16. pyramid, 17. cerebral peduncle

Motor Pathways—Spinal Cord to Muscle Level (Voluntary Muscle Contractions)

Motor signals travel from the spinal cord to skeletal muscles to mediate voluntary muscle contractions, including reflexes. Motor neurons have cell bodies in the spinal cord and axons within the nerves that stimulate muscle contraction. Motor neurons are activated as a result of commands originating from interneurons in the brain and refined by sensory feedback. An initial stimulus such as pain activates mechanosensitive receptors in the skin, which causes afferent impulses to travel to the spinal cord via sensory neurons. In the case of a simple reflex, sensory neurons synapse with interneurons, which in turn synapse with somatic motor neurons. Somatic motor neurons conduct impulses out of the spinal cord and to the muscle via neurotransmitters released following an action potential. Interneurons conduct impulses up the spinal cord from sensory neurons to the brain and down the spinal cord from the brain to motor neurons, which stimulate voluntary muscle contraction.

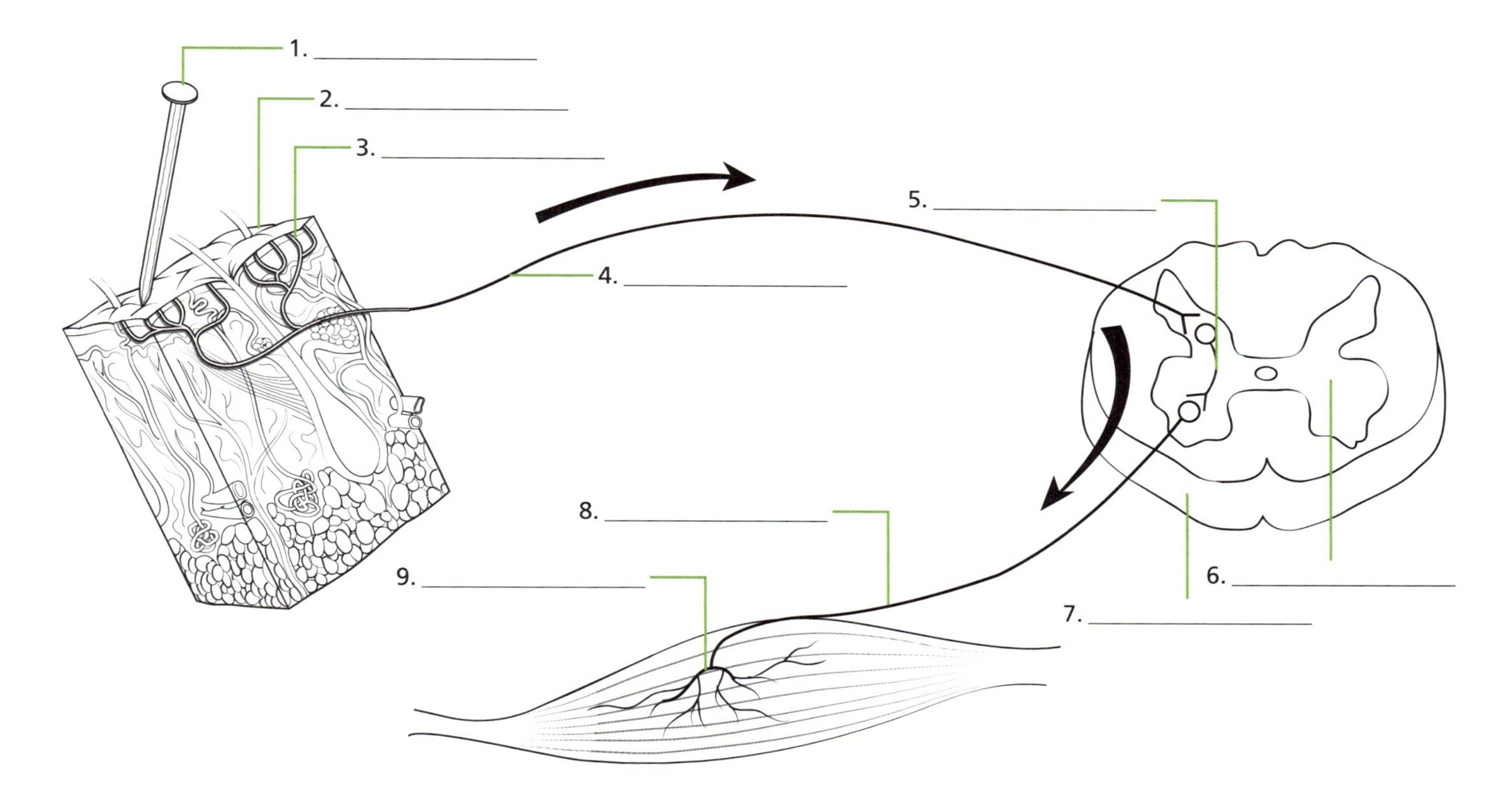

Voluntary Muscle Contraction Pathway

Answers

1. stimulus, 2. skin, 3. receptor, 4. sensory neuron, 5. interneuron, 6. integration center, 7. spinal cord, 8. motor neuron, 9. effector

Motor Pathways—Spinal Cord to Muscle Level (Involuntary Muscle Contractions)

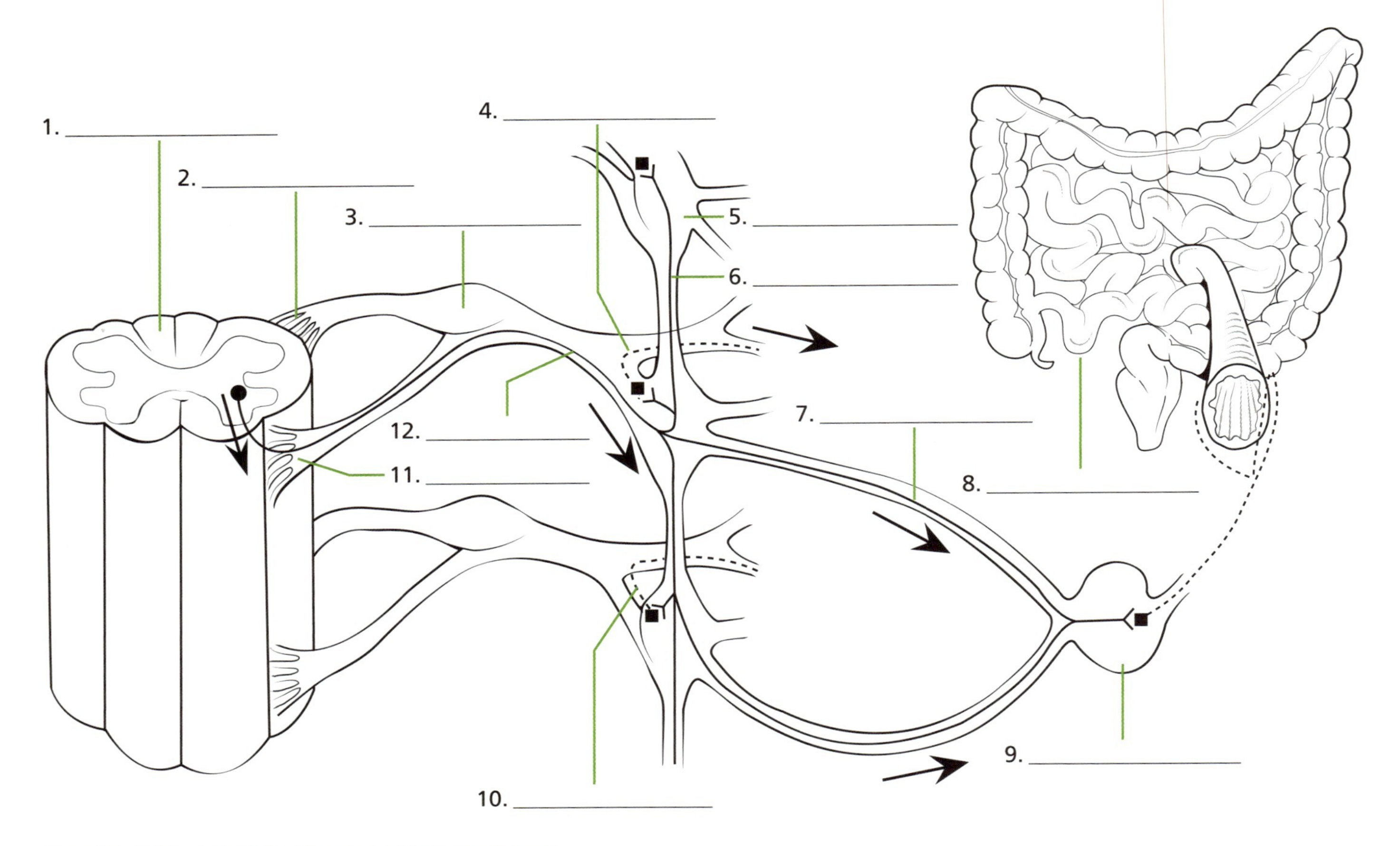

Involuntary Muscle Contraction Pathway

Cardiac and smooth muscle tissue are involuntary effectors, which are regulated by autonomic motor neurons. Autonomic motor nerves innervate organs with functions that are not under voluntary control, such as smooth muscle in the digestive tract. Preganglionic neurons have cell bodies in gray matter and enter the sympathetic chain ganglia on the white ramus. The white ramus carries sympathetic fibers from spinal nerves to sympathetic ganglia. Postganglionic neurons leave the gray ramus and rejoin the spinal nerve, while the axon extends to the effector organ. The gray ramus carries postganglionic sympathetic fibers to the spinal nerves. Through these connections, autonomic nerve fibers release neurotransmitter molecules onto the effector muscles, where they bind to receptor proteins on the smooth muscle cells to initiate contractions.

Answers

1. spinal cord, 2. dorsal root, 3. dorsal root ganglion, 4. spinal nerve, 5. sympathetic chain ganglion, 6. sympathetic chain, 7. splanchnic nerve, 8. visceral effector, 9. collateral ganglion, 10. gray ramus, 11. ventral root, 12. white ramus

Somatosensory Pathways—Pain

The pain pathway carries information about damage that may be occurring to the body from the affected region through the spinal cord and up to the brain. Nociceptors, the sensory receptors for pain, are free nerve endings that form dense networks with multiple branches. The pain pathway contains multiple types of nociceptors and nociceptive pathways. In particular, it has Aδ and C fibers, which make synaptic connections at the marginal nucleus and posterior gray column of the spinal cord, respectively. Aδ fibers are myelinated and have a faster conduction speed, which allows for their quick mechanism of action for sharp, acute pain. C fibers are nonmyelinated and have a slower conduction speed, which results in a dull, diffuse pain that persists longer. Neurons originating in the dorsal root ganglia connect to neurons in the dorsal horn of the spinal cord on the same side as the origin of the nerve impulse. Neurons in the dorsal root ganglia have axons that cross the midline to the contralateral side of the spinal cord and ascend to the brain to form the ventral spinothalamic tract. Axons of other neurons in the dorsal root ganglia make connections with neurons in the ventral posterolateral nucleus of the thalamus.

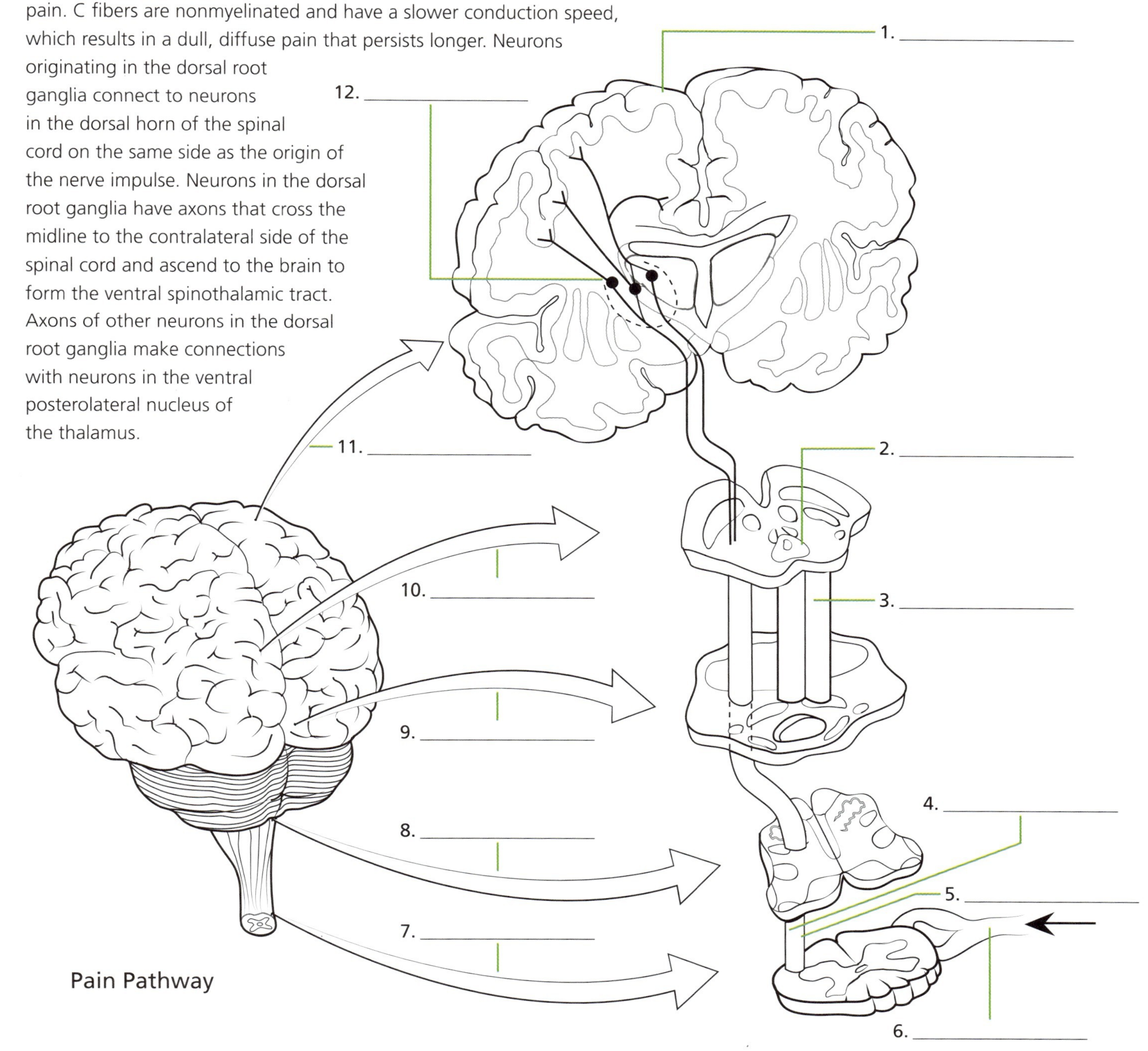

Pain Pathway

Answers

1. somatosensory cortex, 2. periaqueductal gray, 3. reticular formation, 4. neospinothalamic tract, 5. paleospinothalamic tract, 6. Aδ and C fibers, 7. spinal cord, 8. medulla, 9. pons, 10. midbrain, 11. forebrain, 12. thalamic nuclei

Somatosensory Pathways—Touch

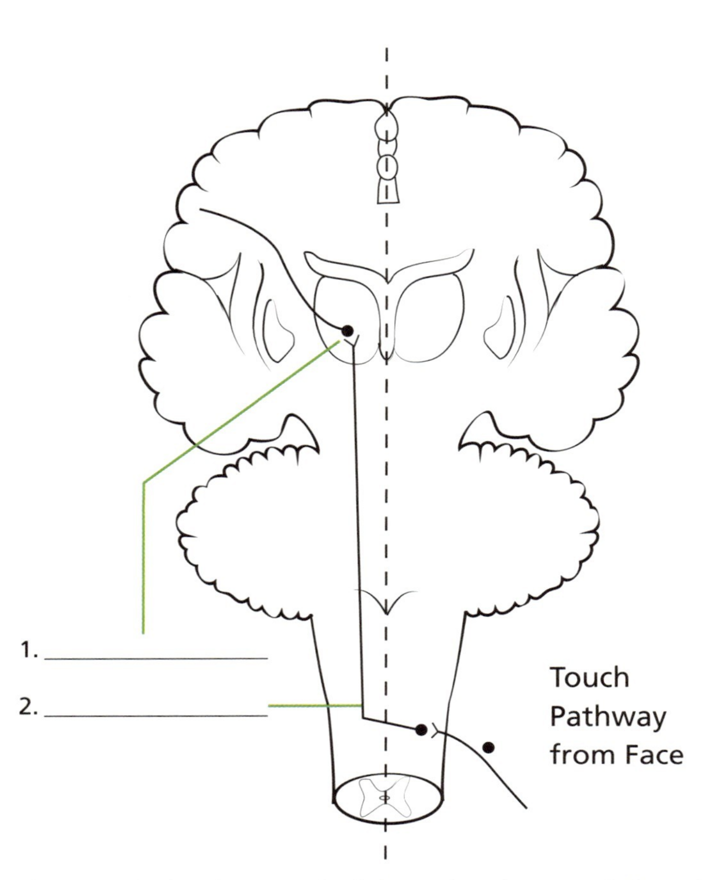

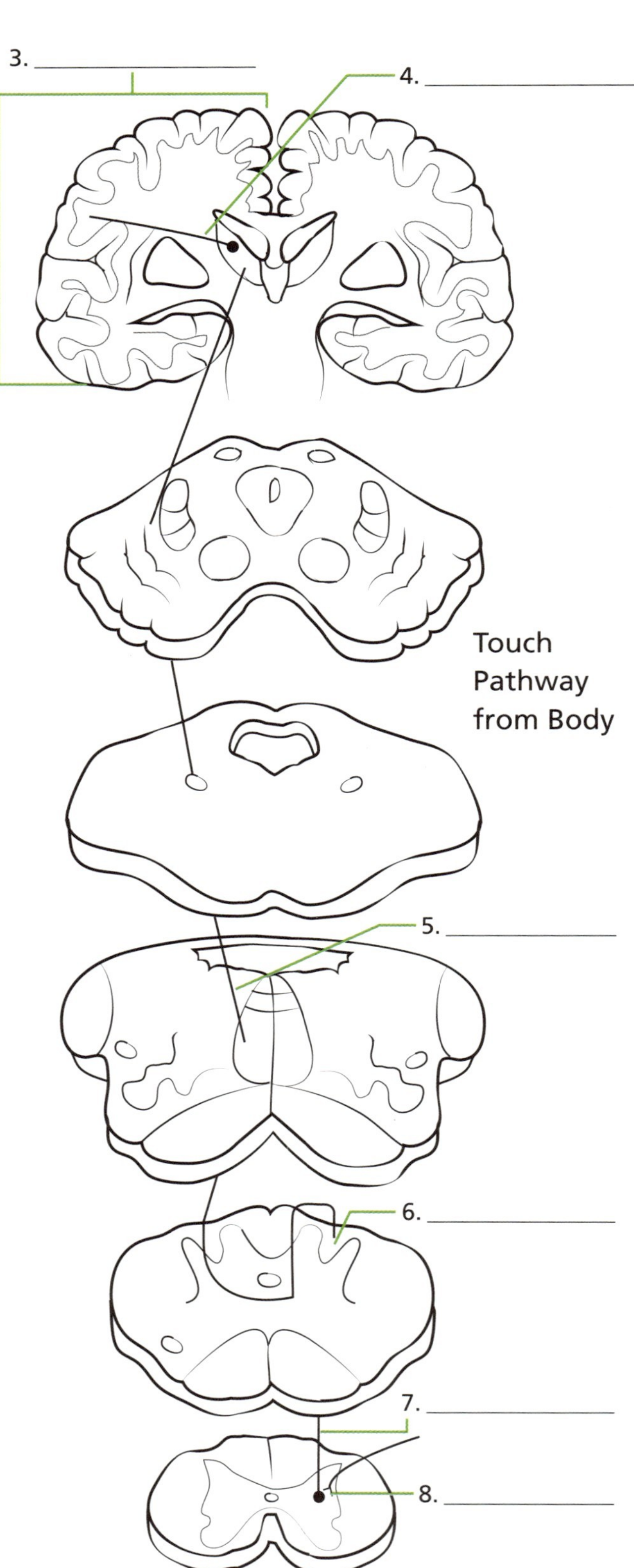

The trigeminal pathway carries information about touch from the face. The sensory fibers enter the brainstem and synapse in the medulla. The pathway decussates immediately and ascends to the ventral posteromedial nucleus of the thalamus. The spinal somatosensory pathway carries information about touch from the rest of the body. It involves Aβ sensory fibers, which are myelinated and have a very fast conduction speed. Axons run along the dorsal root of the spinal nerve and ascend in the dorsal column of the spinal cord. These axons stay on the ipsilateral side of the spinal cord until synapsing with neurons in the medulla. Axons from these medullary interneurons cross the midline and travel through the medial lemniscus to the ventral posteromedial nucleus of the thalamus, where they synapse with local thalamic neurons. These neurons project to the first somatosensory cortex, which is the main sensory receptive area for the sense of touch and is located in the postcentral gyrus of the parietal lobe. It is associated with discriminative touch and physical object recognition according to shape, texture, and size. The second somatosensory area is located in the ceiling of the lateral sulcus and is involved in functions such as tactile learning.

Answers

1. ventral posteromedial nucleus of the thalamus, 2. trigeminal pathway, 3. somatosensory cortex, 4. thalamocortical fibers, 5. medial lemniscal pathway, 6. principal sensory nucleus, 7. spinal tract, 8. spinal nucleus

Somatosensory Pathways—Proprioception

The proprioception pathway is concerned with the sensation and awareness of body position. The major fibers of this pathway are myelinated, have a large diameter, and have the fastest conduction speed among fibers involved in somatosensory pathways. The anterior and posterior spinocerebellar tracts run parallel to one another in the spinal cord and carry proprioceptive information from special receptors called Golgi tendon organs and muscle spindles. The anterior spinocerebellar tract decussates in the spinal cord near the point of entry and then again in the pons in order to reach its target in the cerebellum. The posterior spinocerebellar tract does not decussate at all, coursing through the ipsilateral spinal cord to the medulla, where it exits on its way to the cerebellum.

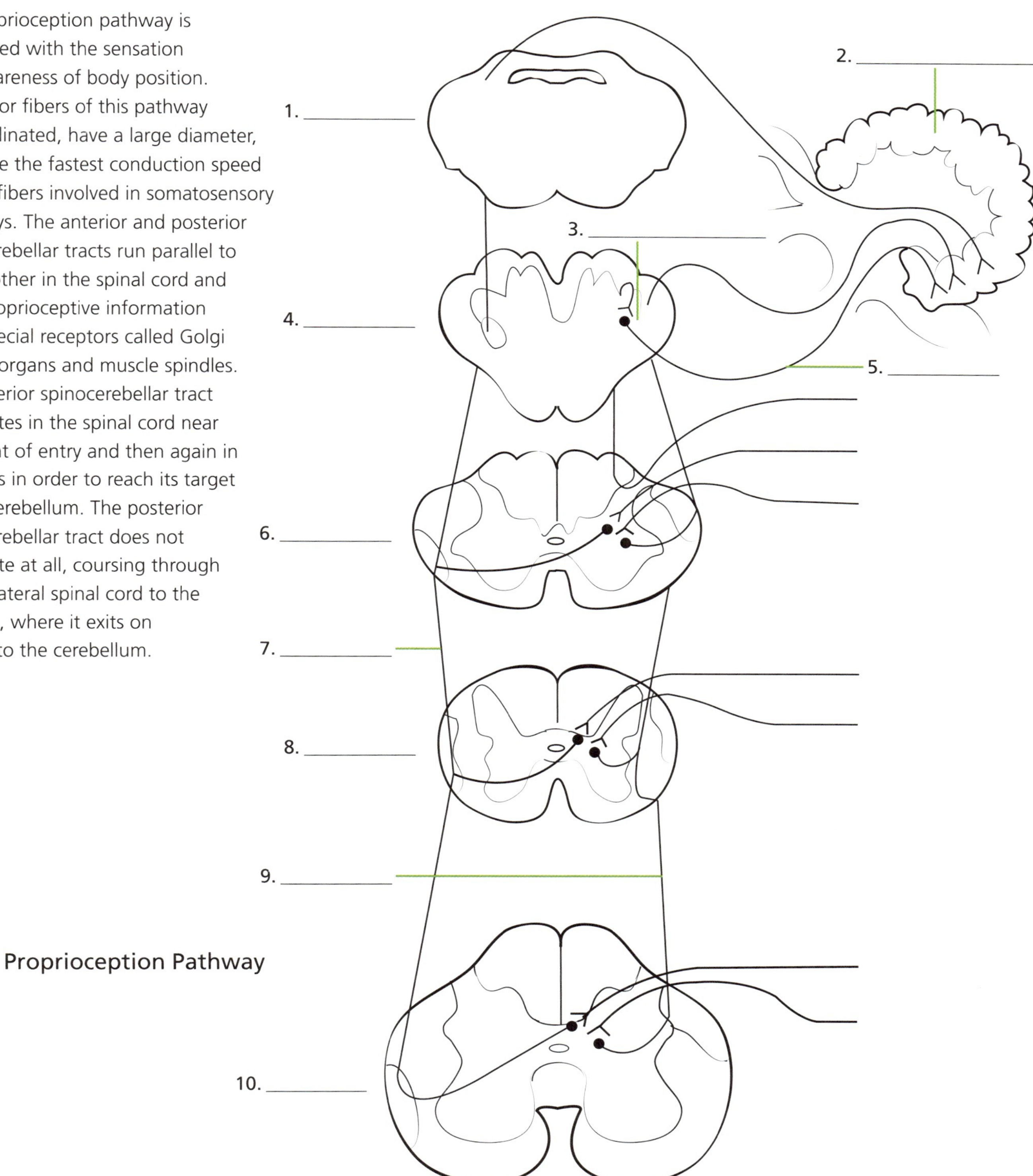

Proprioception Pathway

Answers

1. upper pons, 2. ipsilateral cerebellum, 3. accessory cuneate nucleus, 4. lower medulla, 5. cuneocerebellar tract, 6. cervical spinal cord, 7. anterior spinocerebellar tract, 8. thoracic spinal cord, 9. posterior spinocerebellar tract, 10. lumbar spinal cord

Visual Pathways

The visual cortices are found primarily in the occipital, temporal, and parietal lobes. They contain different function areas, denoted V1, V2, V3, and so on. V1, the primary visual cortex or striate cortex, contains neurons that are sensitive to a wide variety of visual features. Information processed in V1 is sent to multiple higher-order visual centers, such as V2, V3, V4, and V5, which process increasingly specific aspects of the visual world. Beyond V2, visual information diverges into two groups of pathways—the dorsal stream and the ventral stream. The dorsal stream is commonly called the "where" pathway and projects into the posterior parietal lobe. It processes information about motion and location of objects, as well as coordination of movements and other spatial relationships. The ventral stream projects into the inferior temporal lobe and is commonly called the "what" stream. It processes information about object shape and identity, as well as recognition of features and visual memory. Finally, a region called the frontal eye field (FEF) is located in the frontal cortex and exchanges information bidirectionally with other areas of visual cortex. The FEF is responsible for initiating voluntary eye movements and directing visual attention.

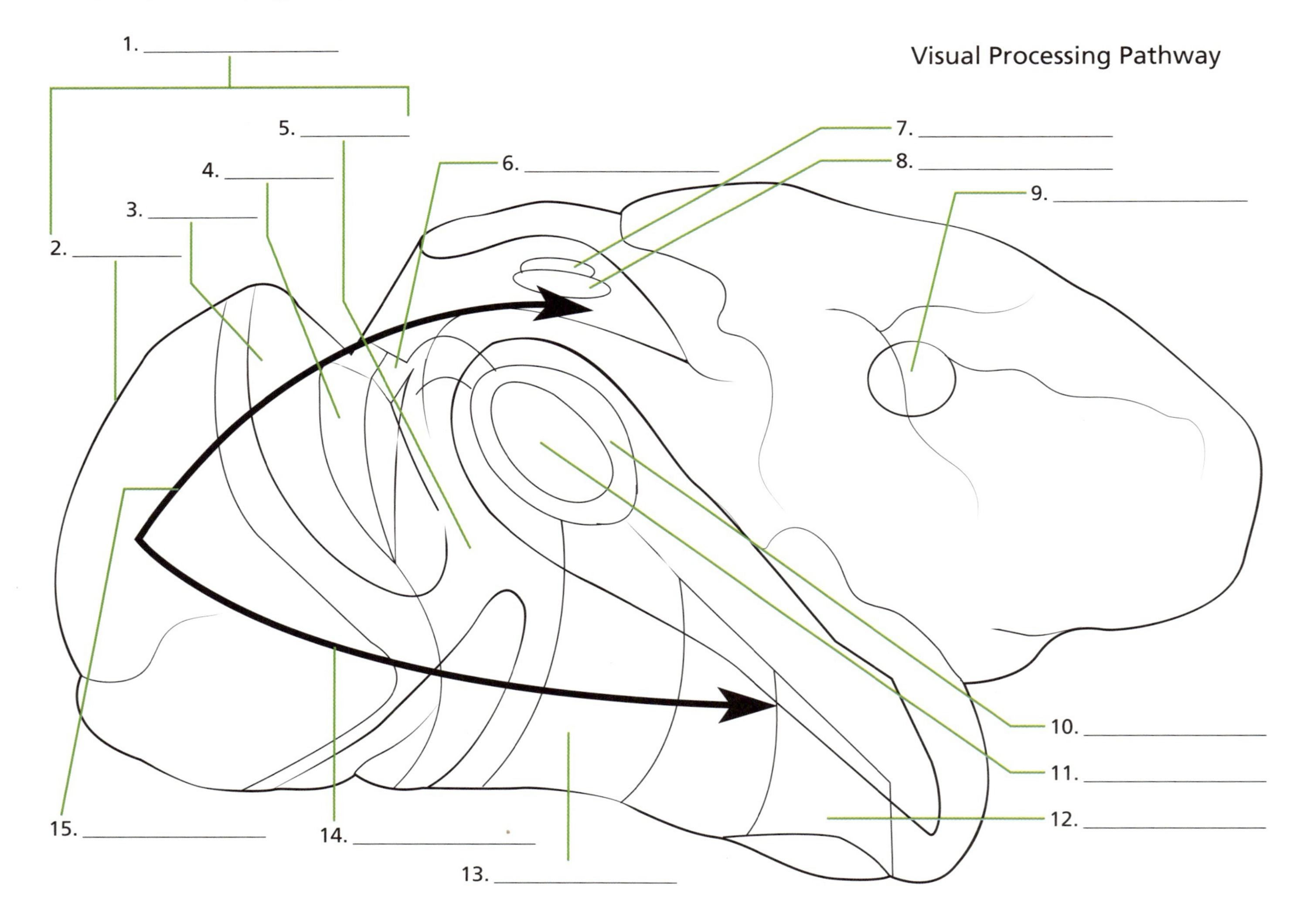

Answers

1. visual cortices, 2. V1, 3. V2, 4. V3, 5. V4, 6. posterior intraparietal area, 7. ventral intraparietal area, 8. lateral intraparietal area, 9. frontal eye field, 10. medial superior temporal area, 11. middle temporal area/V5, 12. anterior inferior temporal cortex, 13. posterior inferior temporal cortex, 14. "what" pathway, 15. "where" pathway

Auditory Pathways

The auditory pathway begins with the transduction of vibrations (sound) into electrical impulses in the cochlea in the inner ear. These impulses are transmitted to the brain via the vestibulocochlear nerve (CN VIII), terminating in the cochlear nuclei. From there, the information is sent to the superior olivary nucleus, which detects interaural differences in stimulus intensity and timing. The superior olive projects to the inferior colliculus, which integrates information about sound and source localization, and then sends projections to the medial geniculate nucleus (MGN) of the thalamus. Thalamocortical projections terminate in the primary auditory cortex in the temporal lobe of the brain, which is responsible for higher-order processing of auditory information and conscious perception of sounds.

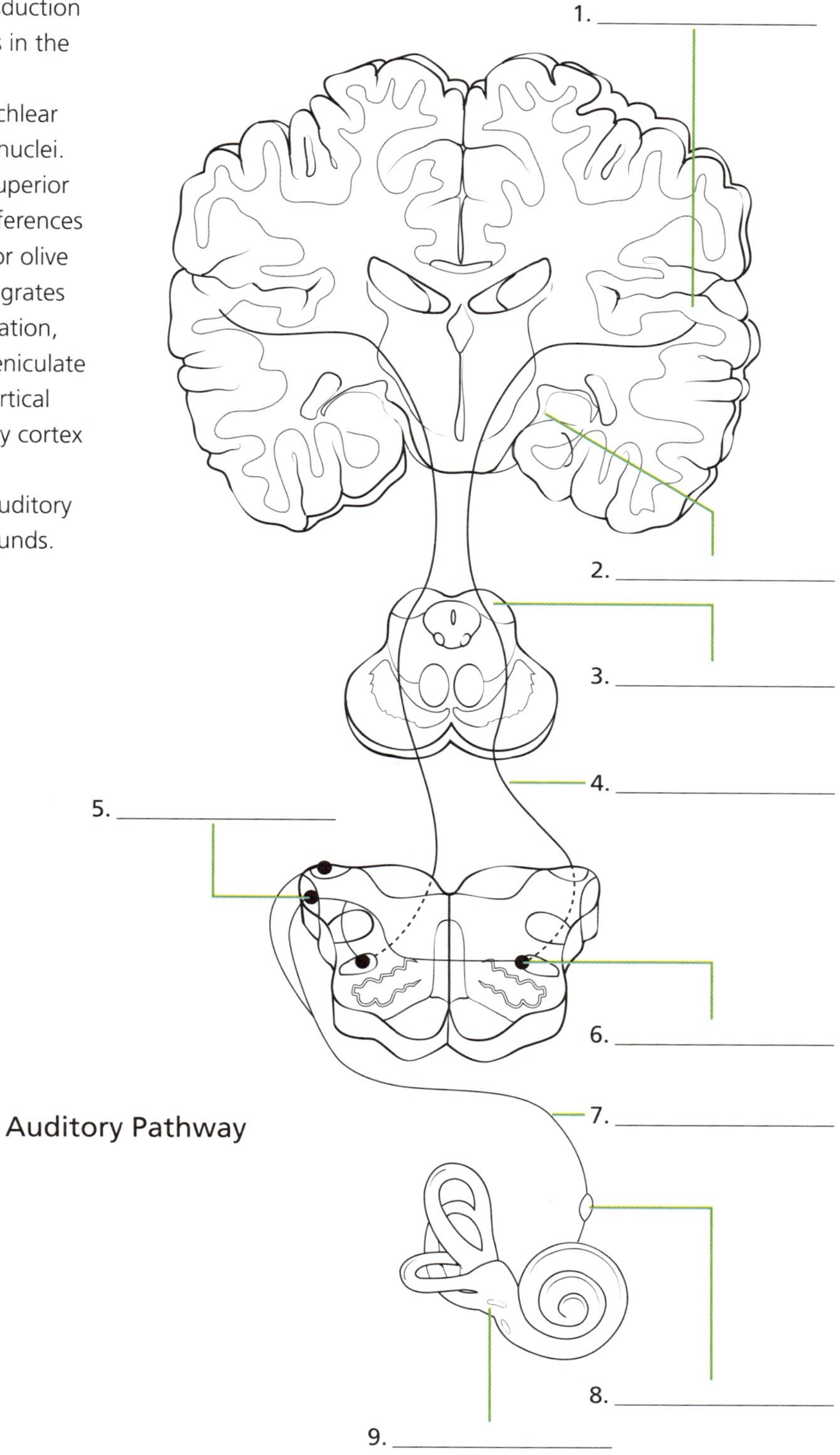

Auditory Pathway

Answers

1. auditory cortex, 2. medial geniculate nucleus, 3. inferior colliculus, 4. lateral lemniscus, 5. ventral cochlear nucleus, 6. superior olive, 7. auditory nerve, 8. spinal ganglion, 9. cochlea

Substantia Nigra and Dopamine

Dopamine is at the base of much of our motivated behavior. This can mean both motivation to initiate movement or motivation to pursue food, sex, or drugs of abuse. Dopamine is involved in salience, or how prominent something is to our attention. It is also involved in movement control.

With such an important role in our behavior, it makes sense that dopamine projections should be very wide. Dopaminergic neurons are concentrated in two separate areas: the compact part of the substantia nigra (Latin for "black substance," and so named because the synthesis pathway for dopamine leaves a dark pigment in the area), which merges into the medially adjacent ventral tegmental area on the midline.

Dopaminergic neurons project in three groups of fibers. Nigrostriatal fibers project from the substantia nigra toward the striatum. Mesolimbic fibers and mesocortical fibers come from the ventral tegmental area, with mesolimbic projecting to the hippocampus and amygdala and mesocortical projecting toward the cortex.

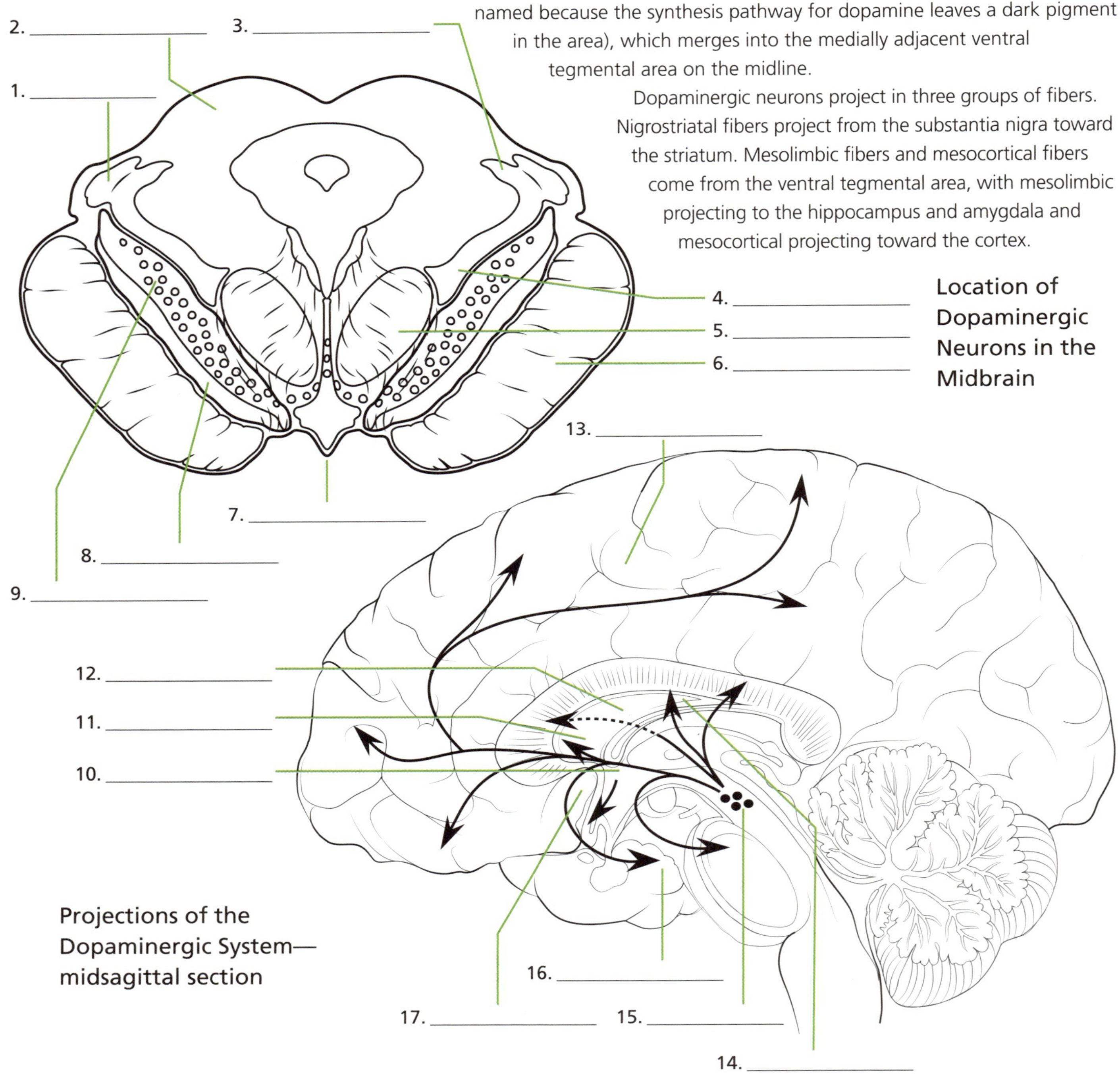

Location of Dopaminergic Neurons in the Midbrain

Projections of the Dopaminergic System—midsagittal section

Answers

1. medial geniculate nucleus, 2. superior colliculus, 3. spinothalamic tract, 4. medial lemniscus, 5. red nucleus, 6. cerebral peduncle, 7. ventral tegmental area, 8. substantia nigra, reticular part, 9. substantia nigra, compact part, 10. ventral striatum, 11. caudate, 12. putamen, 13. cortex, 14. tail of caudate, 15. dopamine cell bodies, 16. amygdala, 17. hippocampus

Locus Coeruleus and Norepinephrine

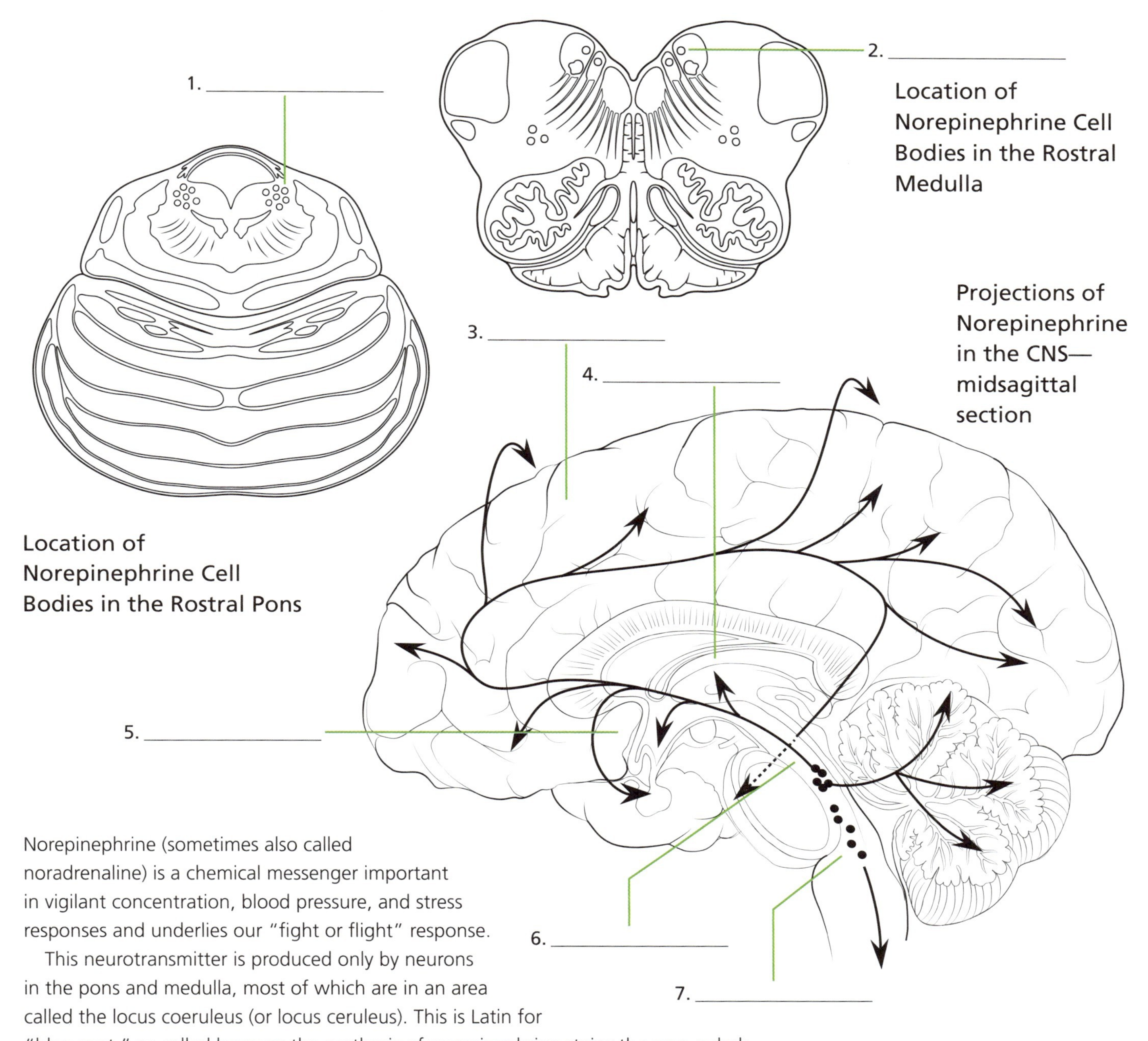

Norepinephrine (sometimes also called noradrenaline) is a chemical messenger important in vigilant concentration, blood pressure, and stress responses and underlies our "fight or flight" response.

This neurotransmitter is produced only by neurons in the pons and medulla, most of which are in an area called the locus coeruleus (or locus ceruleus). This is Latin for "blue spot," so called because the synthesis of norepinephrine stains the area a dark color. There are also some norepinephrine-producing cells in the medullary reticular formation.

From their basis, norepinephrine neurons extend widely over most of the CNS. There are projections upward the thalamus, hypothalamus, amygdala, and hippocampus, and wide projections extend throughout the cortex. There are even projections toward the cerebellum. The projection to the somatosensory cortex is especially strong. The neurons from the locus coeruleus extend primarily upward toward the cortex, while those from the reticular formation extend toward the spinal cord.

Answers

1. locus coeruleus, 2. nucleus of the solitary tract, 3. cortex, 4. thalamus, 5. hippocampus, 6. locus coeruleus, 7. reticular formation

Raphe Nuclei and Serotonin

Serotonin is synthesized and used throughout the body, from the brain to the gut and platelets in the blood. It plays roles in, among other things, arousal, mood, appetite, digestion, memory, and vasoconstriction. The system has between 14 and 17 different receptor subtypes (in contrast, dopamine and norepinephrine have only five receptors each), which speaks to the complexity of function. The neurotransmitter tends to modulate other systems, including those of dopamine and norepinephrine, rather than being directly responsible for any one state or function.

Serotonin is synthesized in a diffuse column of neurons called the raphe nuclei, including the dorsal raphe nucleus, the superior central nucleus, the nucleus raphe magnus, and the nucleus raphe obscurus. Projections from these neurons extend throughout the CNS, with a lot of density in the sensory cortex and in limbic regions such as the amygdala and hippocampus. There are also extensive projections to the brainstem and spinal cord.

Location of Serotonin Cell Bodies in the Caudal Midbrain

Projections of Serotonin in the CNS—midsagittal section

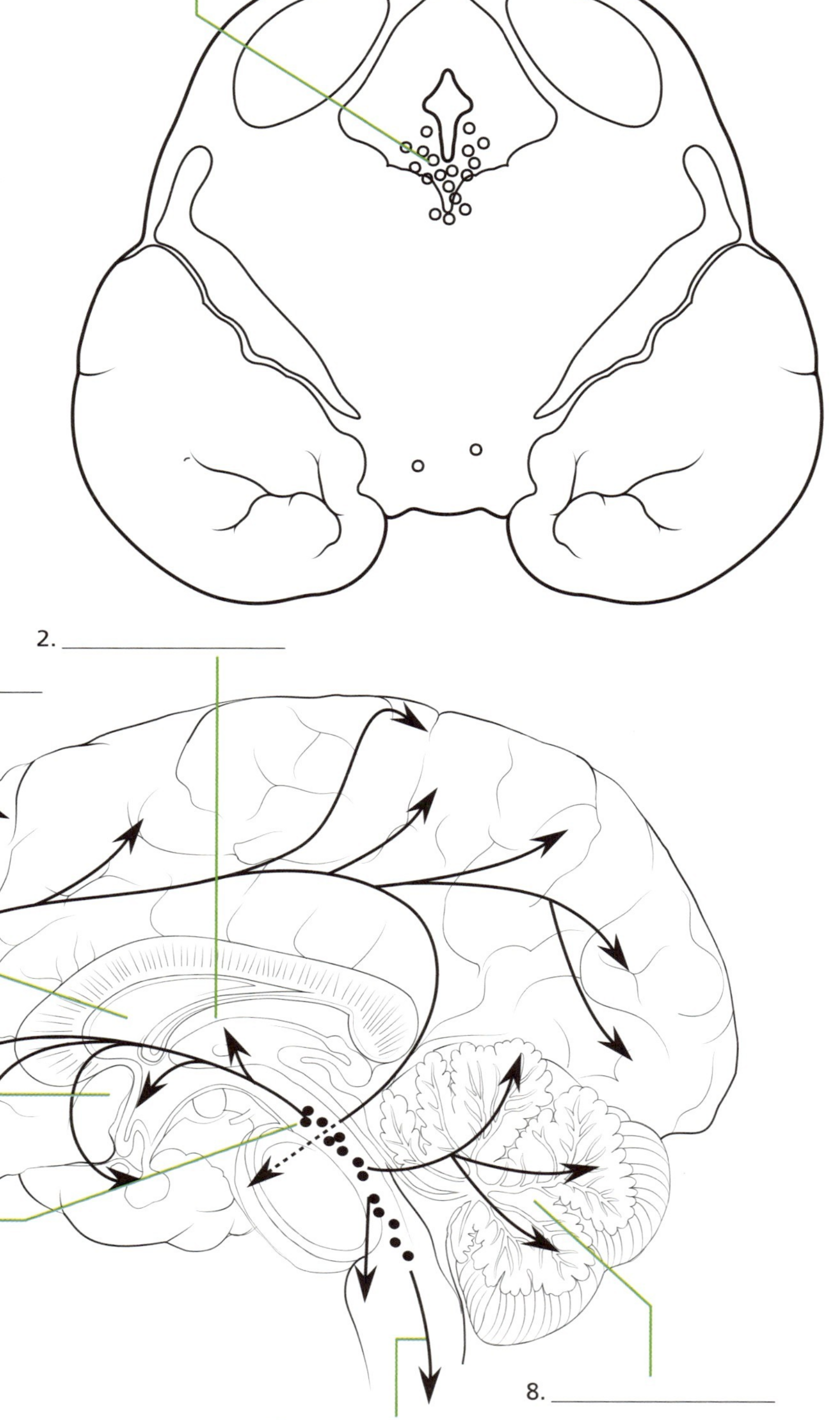

Answers

1. dorsal raphe nucleus, 2. thalamus, 3. cortex, 4. striatum, 5. hippocampus, 6. raphe neurons, 7. toward the spinal cord, 8. cerebellum

Acetylcholine

Acetylcholine is one of the most prominent chemical messengers in the PNS, where it serves as the main neurotransmitter for motor neurons. In the CNS, however, it plays roles in arousal, reward, attention, and sleep-wake cycles. Receptors for the PNS are primarily nicotinic, while those for the CNS are muscarinic, with slower responses than those in the PNS.

Acetylcholine is synthesized in cholinergic neurons in the reticular formation and in the basal nucleus of the forebrain (otherwise known as the basal nucleus of Maynert). These neurons project widely over the cortex and amygdala. There are also cholinergic neurons in the oculomotor nucleus, which project to the muscles controlled by the third cranial nerve.

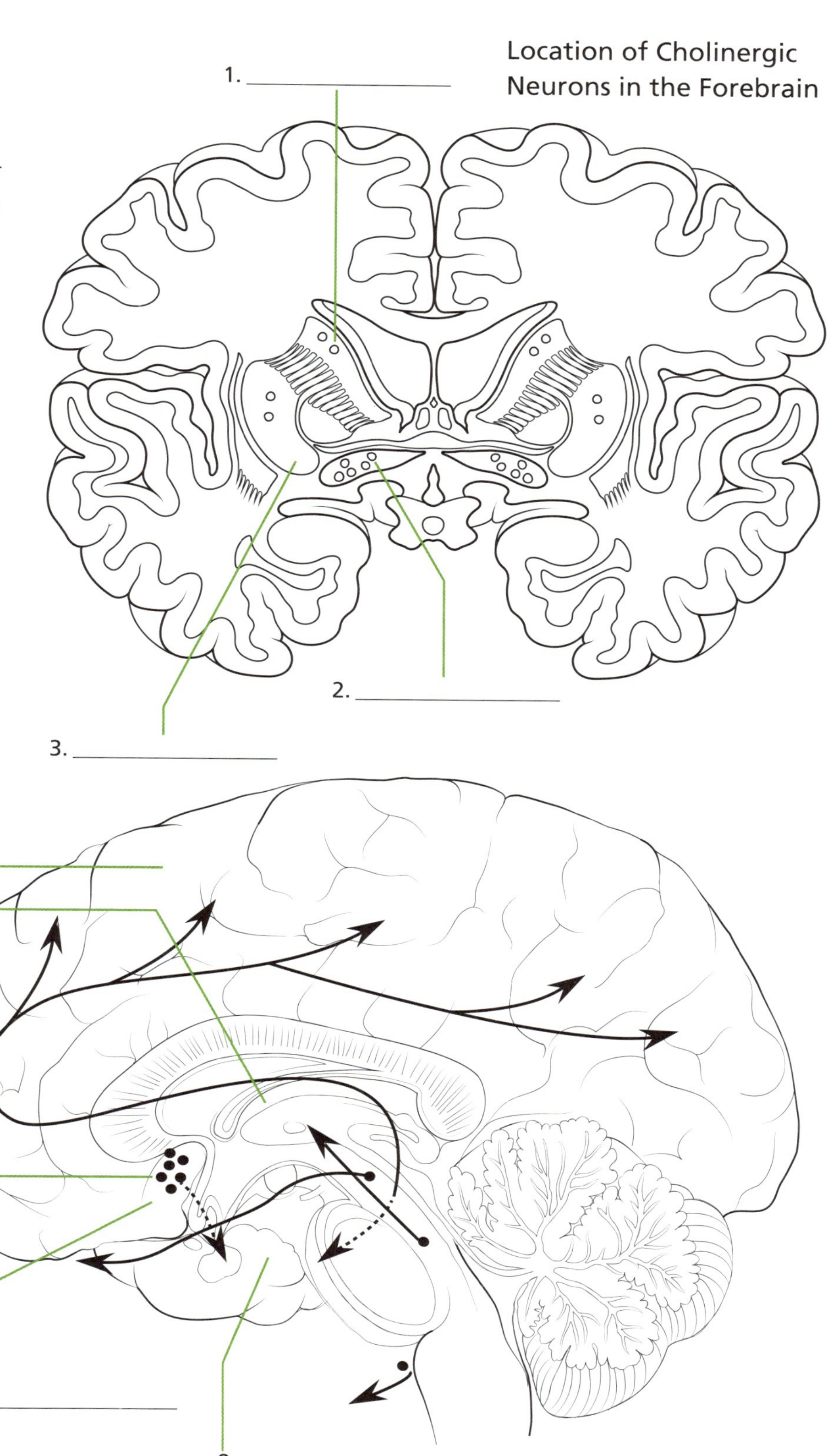

Location of Cholinergic Neurons in the Forebrain

Projections of the Cholinergic System in the CNS—midsagittal section

Answers

1. caudate, 2. basal nucleus, 3. putamen, 4. cortex, 5. thalamus, 6. basal nucleus, 7. hippocampus, 8. amygdala

Overview of the Arterial Blood Supply of the Brain

The blood supply of the brain is dependent on two arterial branches from the dorsal aorta. These include the vertebral arteries and the internal carotid arteries.

The internal carotid arteries branch to form the anterior and middle cerebral arteries. The anterior cerebral artery provides the anterior cerebral circulation that supplies blood to the anterior part of the brain. The middle cerebral artery supplies blood to the majority of the brain, including the frontal, temporal, and parietal lobes of the cerebral cortex.

The right and left vertebral arteries join together at the level of the pons (brainstem) to form the basilar artery. The basilar artery joins the blood supply from the internal carotid arteries, from which the posterior cerebral artery arises. The blood supply to the posterior part of the brain (including the midbrain, brainstem, and cerebellum) is via the posterior cerebral artery.

Arterial Blood Supply of Brain—right lateral view

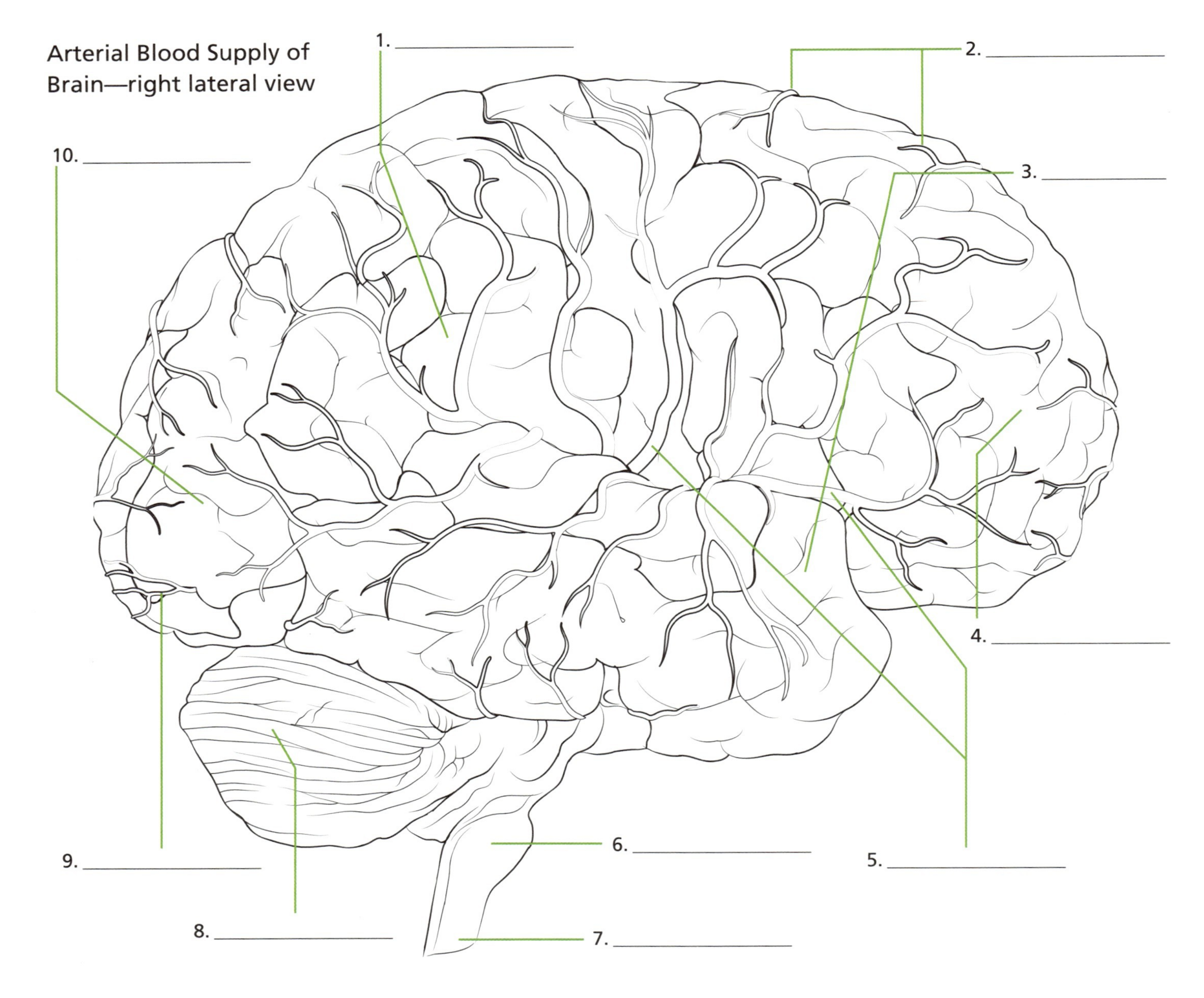

Answers

1. parietal lobe of cerebrum, 2. anterior cerebral artery, 3. temporal lobe of cerebrum, 4. frontal lobe of cerebrum, 5. middle cerebral artery, 6. brainstem, 7. spinal cord, 8. cerebellum, 9. posterior cerebral artery, 10. occipital lobe of cerebrum

Internal Carotid Branches

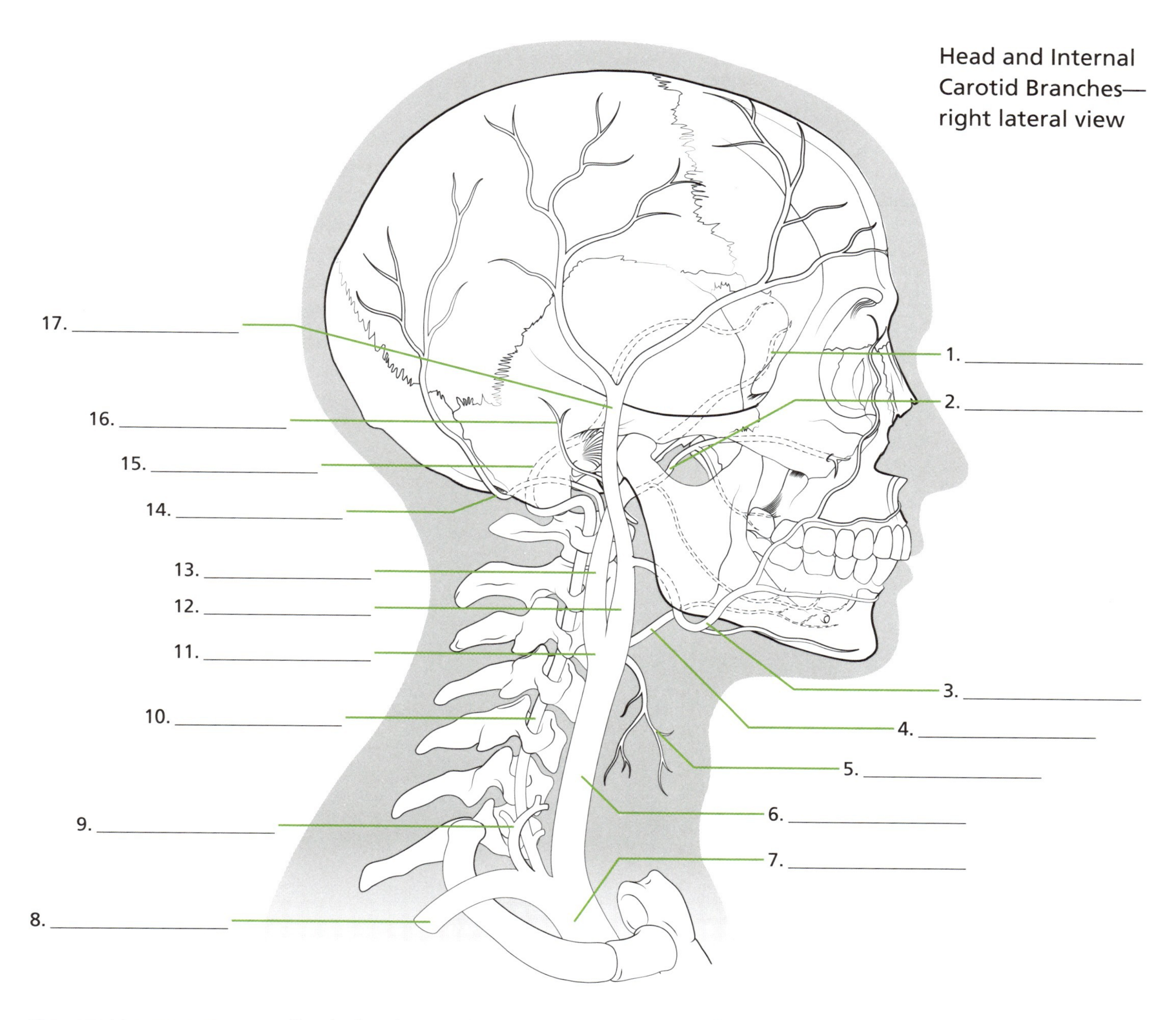

The carotid artery system supplies the head and neck with oxygenated blood. It bifurcates into the internal and external carotid arteries at the level of the third cervical vertebra. The internal carotid artery ascends into the carotid sinus (dilated section of the internal carotid artery), before entering the carotid canal in the skull. At the terminal end of the internal carotid artery, the ophthalmic artery and anterior choroidal artery branch off it before it divides into the anterior and middle cerebral arteries. The external carotid artery has many more branches than the internal carotid artery; in the neck it branches out to the superior thyroid artery, lingual artery, facial artery, occipital artery (opposite the facial artery), and maxillary artery. The vertebral artery also enters the skull through the neck and becomes the basilar artery prior to bifurcating into the posterior cerebral artery and the posterior communicating artery.

Answers

1. anterior choroidal artery, 2. maxillary artery, 3. facial artery, 4. lingual artery, 5. superior thyroid artery, 6. common carotid artery, 7. brachiocephalic artery, 8. subclavian artery, 9. thyrocervical axis, 10. vertebral artery, 11. carotid sinus, 12. external carotid artery, 13. internal carotid artery, 14. occipital artery, 15. basilar artery, 16. posterior auricular artery, 17. superficial temporal artery

Vertebral Artery Branches

The vertebral arteries provide the primary source of blood for the spinal cord of the neck and arise from the subclavian arteries. They travel in the transverse foramen of each cervical vertebra (one on each side) until they enter the cranium and join again to form the basilar artery.

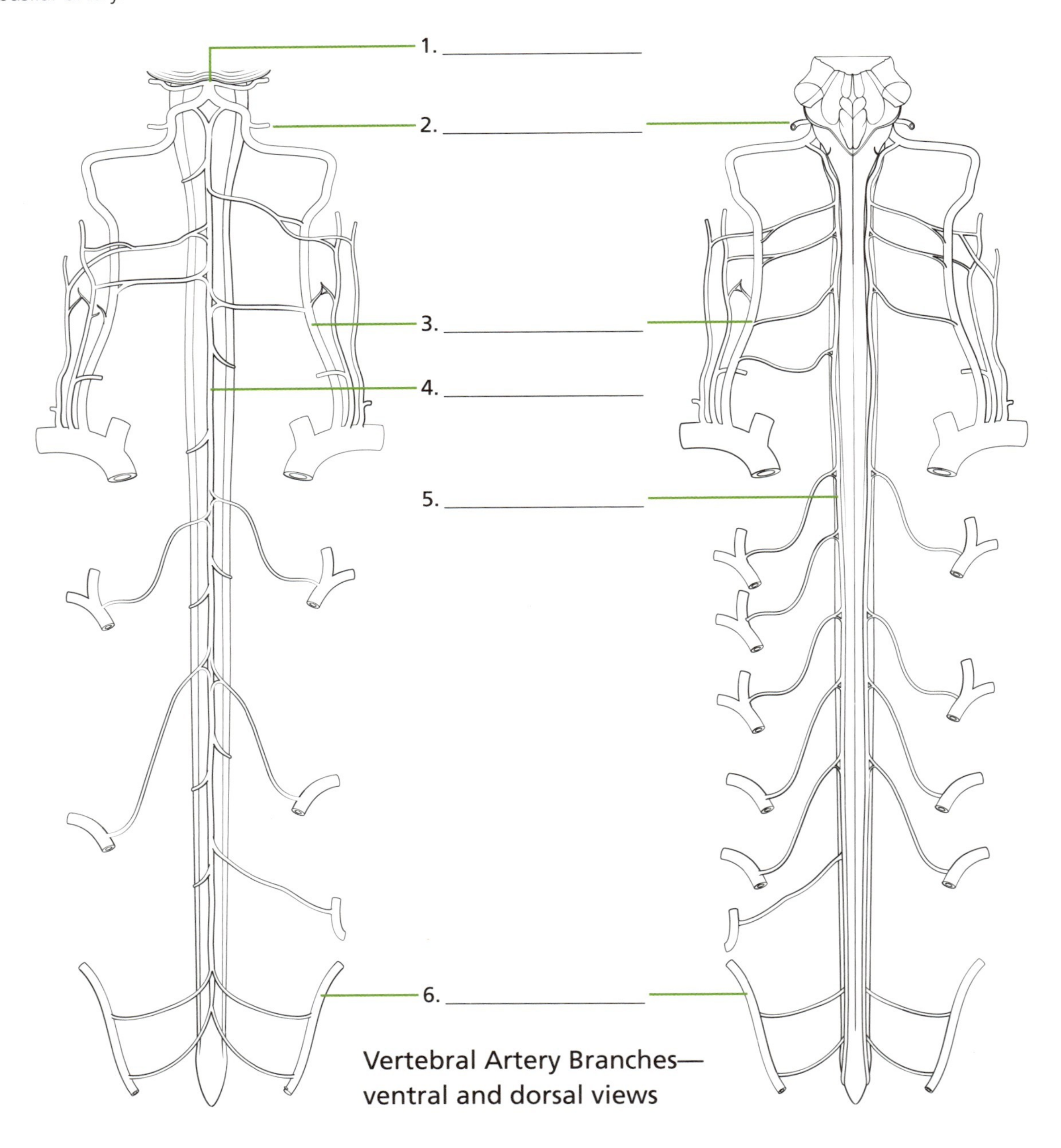

Vertebral Artery Branches—ventral and dorsal views

The ventral aspect of the spinal cord shows the formation of the anterior spinal artery from branches of the vertebral artery. The dorsal aspect of the spinal cord shows the posterior inferior cerebellar artery forming a pair of posterior spinal arteries. Approximately ten to twelve medullary arteries join to form the anterior and posterior spinal arteries at the level of the medulla.

Answers

1. basilar artery, 2. posterior inferior cerebellar artery, 3. vertebral artery, 4. anterior spinal artery, 5. posterior spinal artery, 6. medullary arteries

Circle of Willis

The circle of Willis is an anastomotic system of arteries at the base of the brain that encircles the pituitary gland. An anastomotic system of arteries is a network of separate blood vessels that come together to form a structure, in this case the circle of Willis. The anterior portion of the circle of Willis is formed from the division of the internal carotid artery into the anterior cerebral artery and middle cerebral artery. The left and right anterior cerebral arteries are connected by the anterior communicating artery. In the posterior section of the circle of Willis, the left and right vertebral arteries form the basilar artery, which branches into the left and right posterior cerebral arteries. The circle of Willis is completed by the posterior cerebral arteries joining the internal carotid system via the posterior communicating arteries. The circle of Willis is an important connection between the blood supply of the forebrain and hindbrain. However, a complete and well-developed circle of Willis is present in less than half the population; for the majority, there are anatomical differences in the diameter and presence of arteries that make up the system.

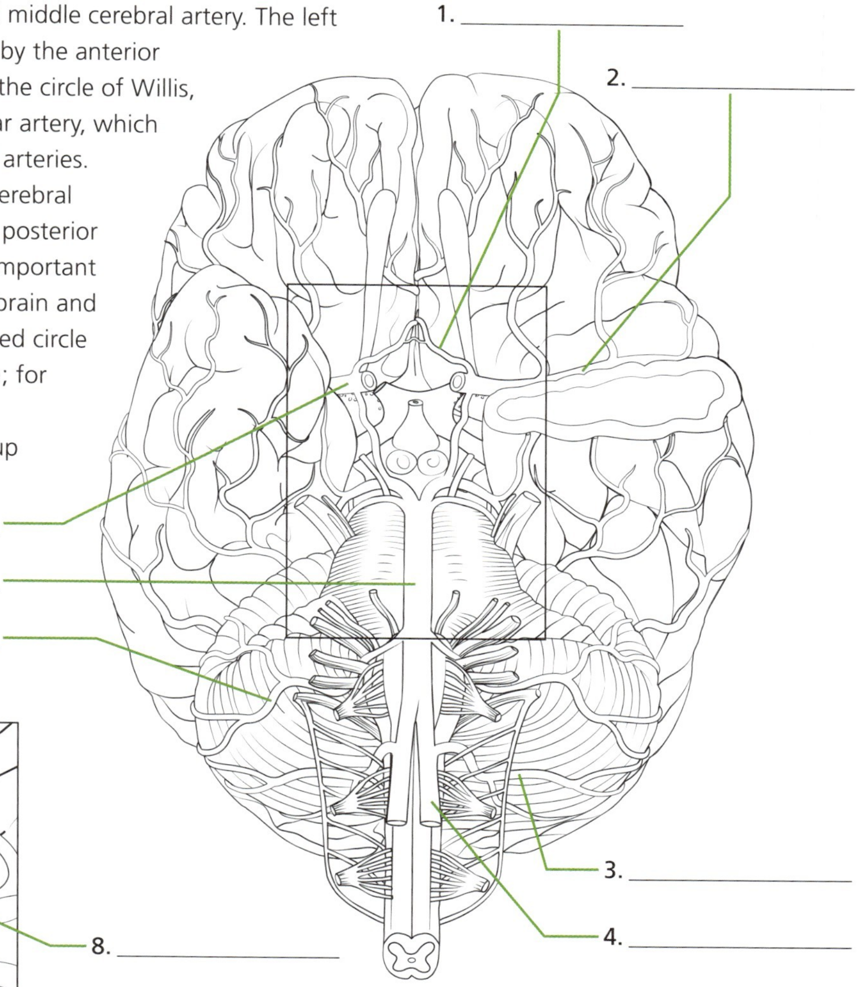

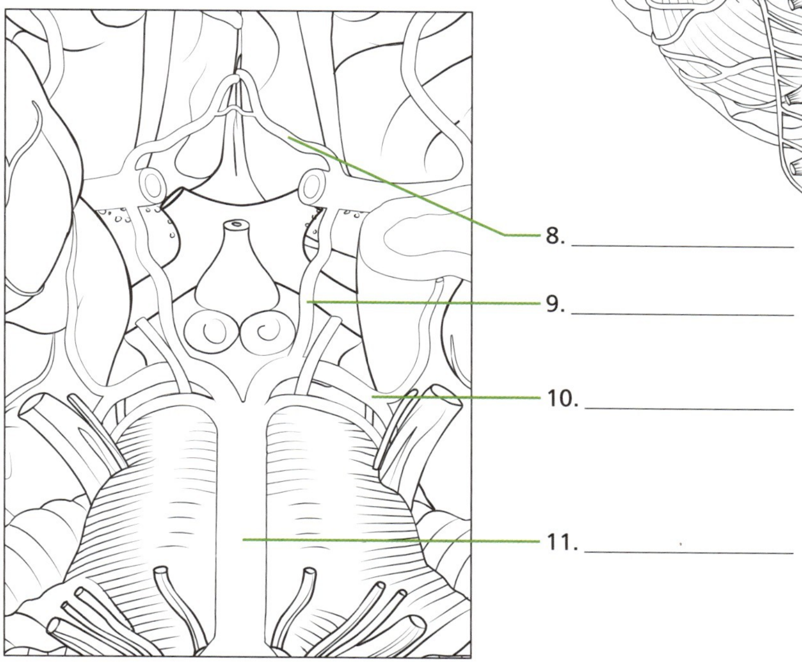

Brain and Circle of Willis—ventral view

Answers

1. anterior cerebral artery, 2. middle cerebral artery, 3. posterior inferior cerebellar artery, 4. vertebral artery, 5. anterior inferior cerebellar artery, 6. basilar artery, 7. internal carotid artery, 8. anterior communicating artery, 9. posterior communicating artery, 10. posterior cerebral artery (to midbrain), 11. basilar artery (to pons)

Blood Supply to the Cerebral Cortex

The blood supply of the cerebral cortex is dependent primarily on the main cortical arteries, including the anterior, middle, and cerebral arteries. These cortical cerebral arteries supply a network of blood vessels that vascularize both cerebral hemispheres.

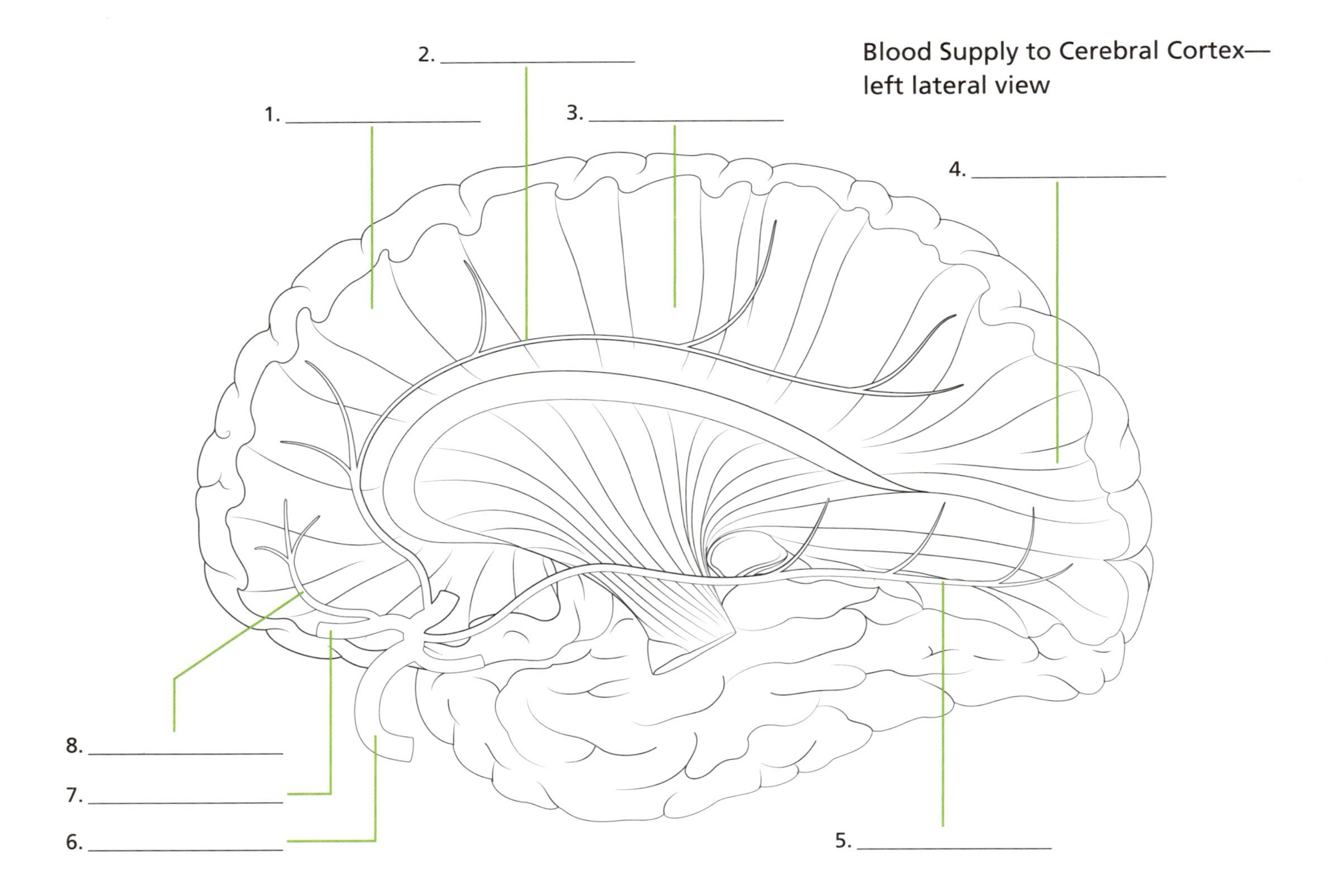

From the internal carotid artery at the circle of Willis, the anterior cerebral artery and the middle cerebral artery arise and project to all parts of the brain, including the cortex, basal ganglia, thalamus, and internal capsule. The anterior cerebral artery supplies the frontal lobes of the cortex, whereas the middle cerebral artery (largest branch) supplies part of the frontal and the majority of the lateral surface of the temporal and parietal lobes. The lateral striate arteries branch out from the middle cerebral artery; they are small but deeply penetrating arteries of the middle cerebral cortex. The precise location of the cortical cerebral arteries is difficult to describe because they are prone to significant variations in the territories to which they supply blood.

Answers

1. anterior cerebral cortex, 2. lateral striates, 3. middle cerebral cortex, 4. posterior cerebral cortex, 5. anterior choroidal artery, 6. middle cerebral artery, 7. anterior cerebral artery, 8. medial striates

Blood Supply to the Deep Forebrain

The forebrain is also known as the cerebrum and is divided into left and right hemispheres. The outer tissue of the cerebrum is gray matter, and the deeper inner tissue is called white matter. Deep in the forebrain are internal structures such as the thalamus, hypothalamus, and limbic system.

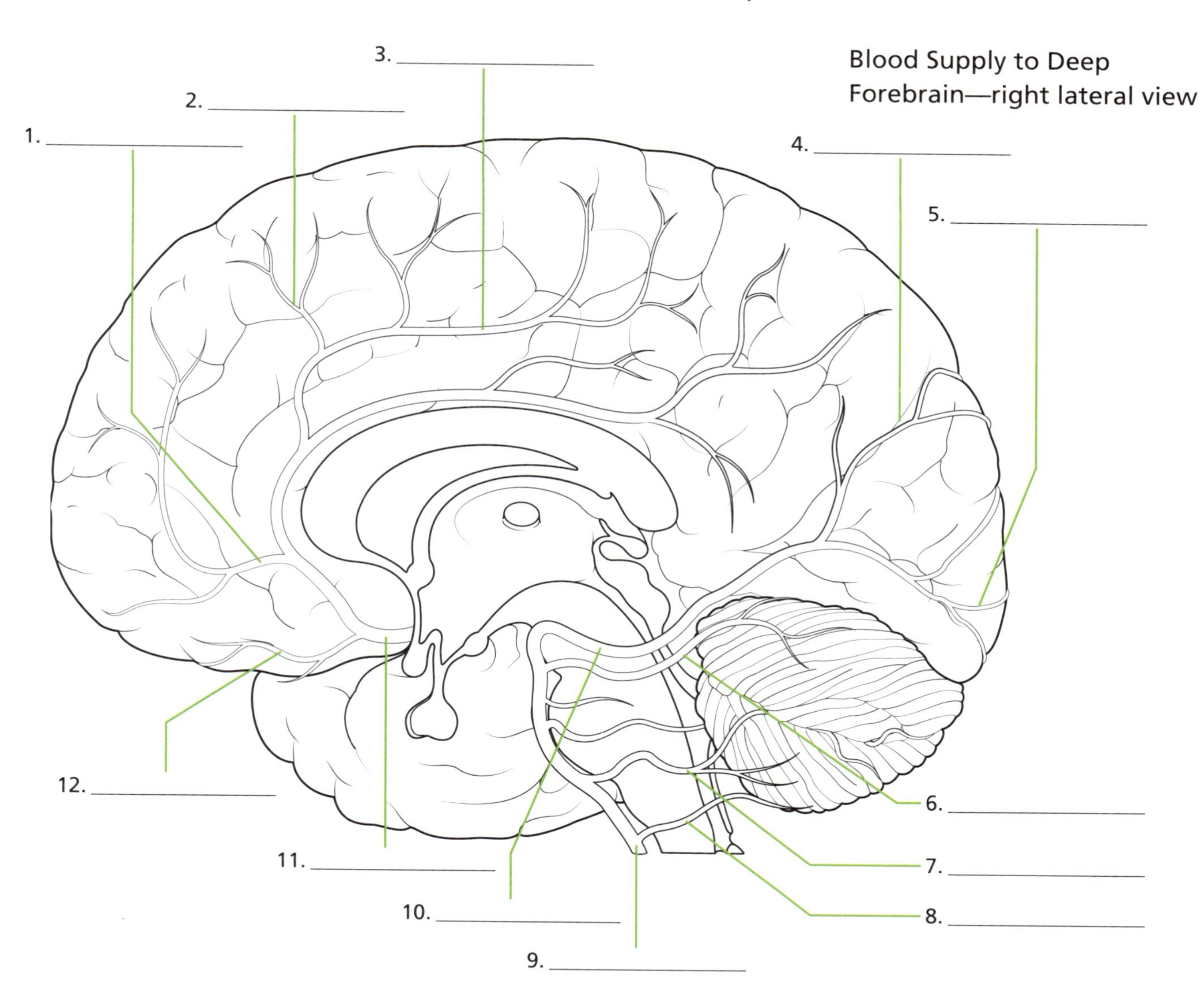

The basilar artery is the main blood supply to the brainstem and connects to the circle of Willis. It branches out to three pairs of blood vessels to supply blood to the cerebellum, including the anterior inferior cerebellar artery and superior cerebellar artery. At the circle of Willis, the anterior cerebral artery supplies oxygenated blood to the medial surface of the cerebral cortex. Its branches include the frontopolar artery and the orbital artery. The anterior cerebral artery then bifurcates into the pericallosal artery and the callosomarginal artery, which extend posteriorly.

Answers

1. frontopolar artery, 2. callosomarginal artery, 3. pericallosal artery, 4. parieto-occipital artery, 5. calcarine artery, 6. superior cerebellar artery, 7. anterior inferior cerebellar artery, 8. posterior inferior cerebellar artery, 9. basilar artery, 10. posterior cerebral artery, 11. anterior cerebral artery, 12. orbital artery

Blood Supply to the Brainstem

Circulation of blood to the brainstem comprises arterial branches from the posterior cerebral artery, basilar artery, and vertebral artery. The distribution of blood supply through each division of the brainstem—from midbrain to caudal medulla—is of a similar pattern. The medial region of the brainstem is supplied by paramedian branches of the basilar artery. The dorsolateral regions of the brainstem, however, are directly perfused by circumferential branches of the cerebral, basilar, and vertebral arteries.

The posterior inferior cerebellar artery is the largest branch of the vertebral artery and supplies blood to distinct lateral areas of the upper and caudal medulla. The anterior inferior cerebellar artery supplies blood to distinct lateral areas of the middle pons. Branches of the basilar artery supply blood to the lateral territories of the pons and midbrain.

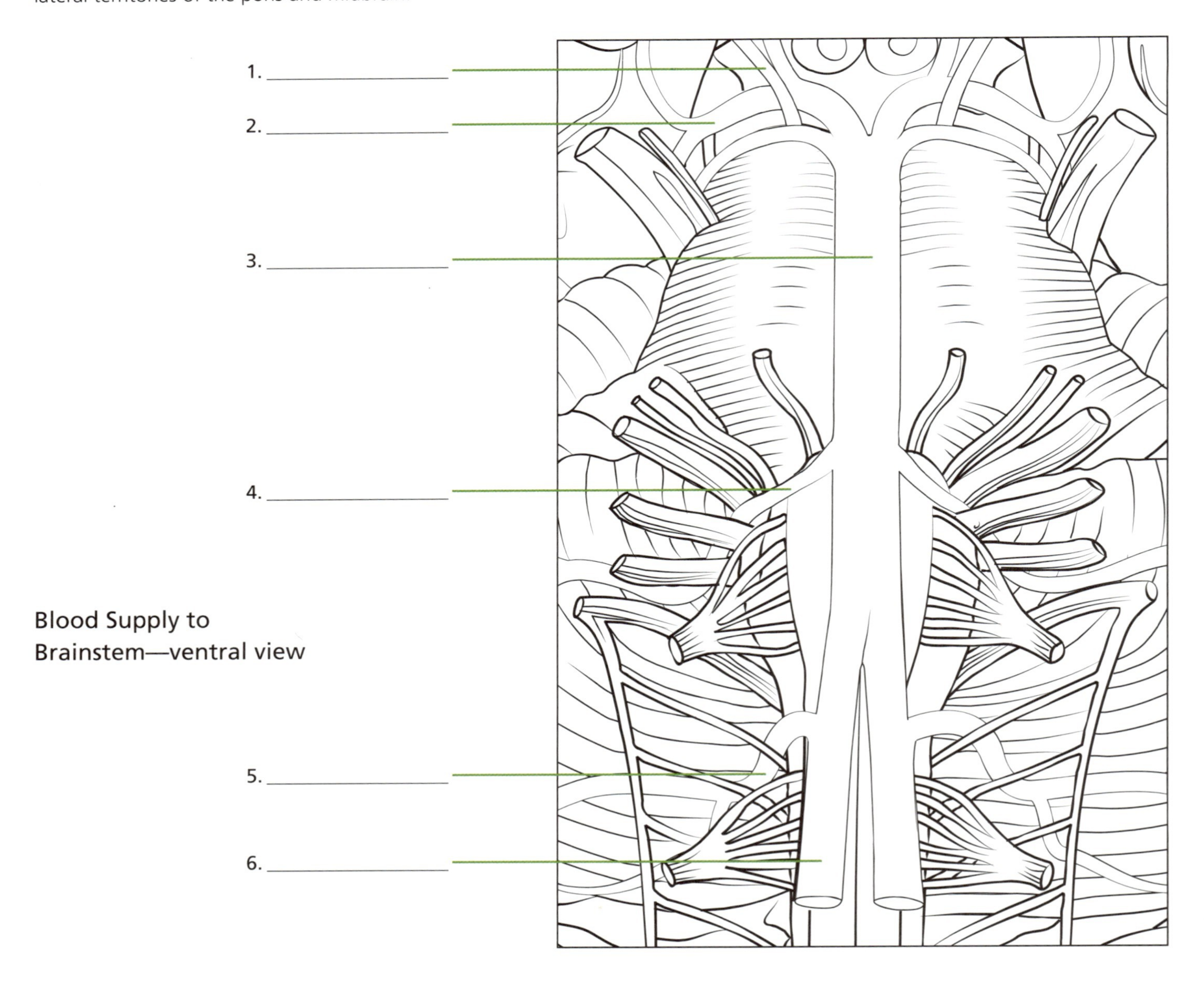

Blood Supply to Brainstem—ventral view

Answers

1. posterior communicating artery, 2. posterior cerebral artery (to midbrain), 3. basilar artery (to pons), 4. anterior inferior cerebellar artery, 5. posterior inferior cerebellar artery, 6. vertebral artery (to medulla)

Label which arteries supply each region.

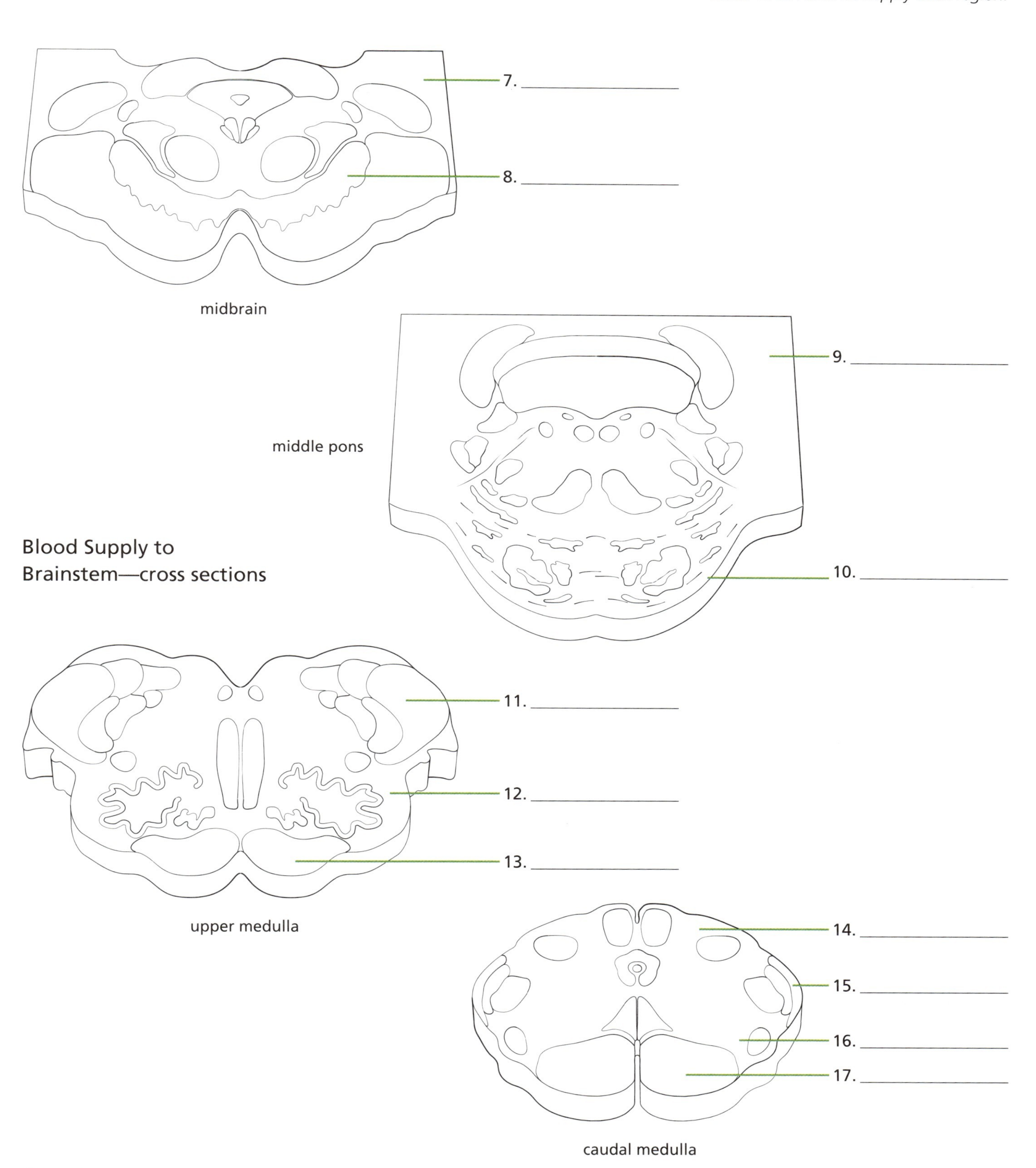

Answers

7. posterior cerebral artery, 8. basilar artery, 9. anterior inferior cerebellar artery, 10. basilar artery, 11. posterior inferior cerebellar artery, 12. vertebral artery, 13. anterior spinal artery, 14. posterior spinal artery, 15. posterior inferior cerebellar artery, 16. vertebral artery, 17. anterior spinal artery

Blood Supply to the Spinal Cord

The spinal cord blood supply comes from the anterior spinal artery and the paired posterior spinal arteries. There are also connecting arteries between these longitudinal spinal arteries, called the arterial vasocorona.

A cross section through the spinal cord illustrates that its blood supply is dependent on the anterior and posterior spinal cord arteries. The posterior inferior cerebellar artery divides into a pair of posterior spinal arteries that supply blood to the dorsal territory of the spinal cord. The anterior spinal artery supplies blood to the anterior two-thirds of the spinal cord and connects with the posterior spinal arteries through a network of blood vessels known as the vasocorona.

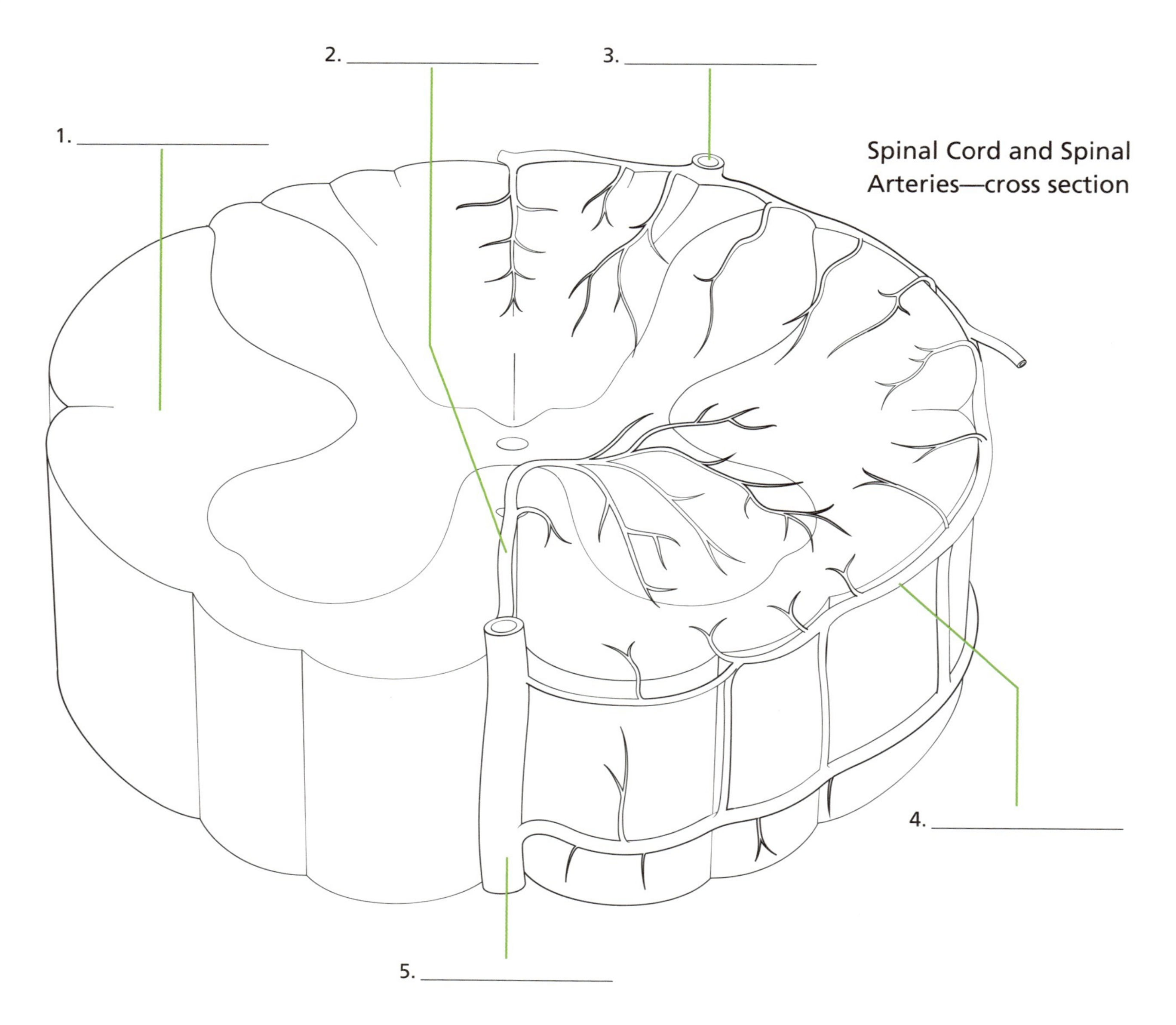

Spinal Cord and Spinal Arteries—cross section

Answers

1. spinal cord, 2. sulcal artery, 3. posterior spinal artery, 4. vasocorona, 5. anterior spinal artery

Venous Drainage of the Cerebral Hemispheres

The drainage of blood from the cerebral hemispheres begins in small veins that form larger venous structures in the pia mater. These veins then form into cerebral veins that bridge the subarachnoid space and enter the dura mater. The majority of the cerebral venous channels drain into the superior sagittal sinus, which flows posteriorly through the midline structure. The straight sinus also drains into the superior sagittal sinus at the confluence of sinuses, where it divides into the right and left transverse sinuses. Each transverse sinus drains blood into the internal jugular vein. The cavernous sinus is a network of channels that are drained via the inferior petrosal sinus (IPS) and superior petrosal sinus (SPS) into the internal jugular vein.

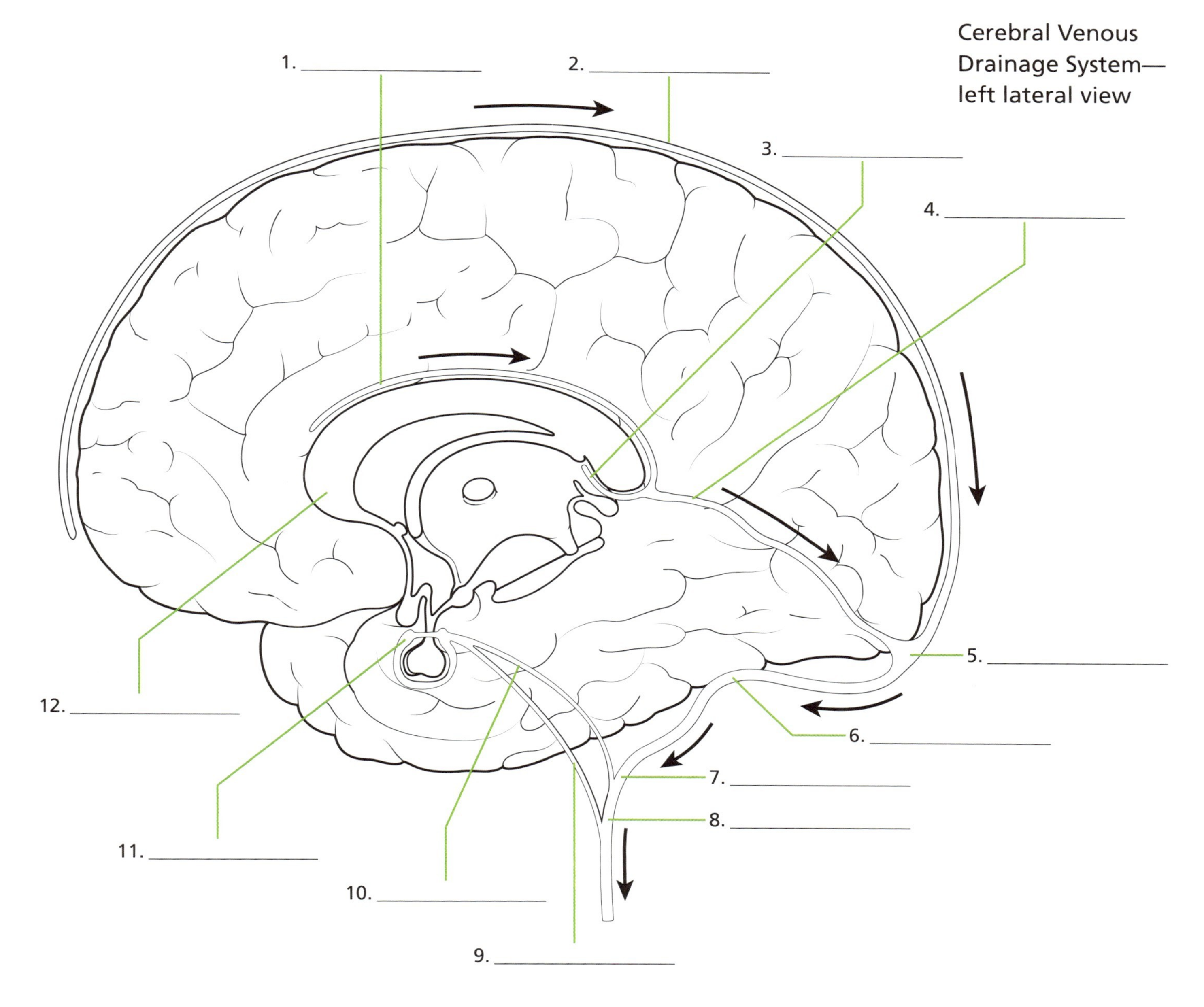

Cerebral Venous Drainage System—left lateral view

Answers

1. inferior sagittal sinus, 2. superior sagittal sinus, 3. great cerebral vein, 4. straight sinus, 5. confluence of sinuses, 6. transverse sinus, 7. sigmoid sinus, 8. internal jugular vein, 9. inferior petrosal sinus, 10. superior petrosal sinus, 11. cavernous sinus, 12. corpus callosum

Meningeal Layers

The meningeal layers of the skull provide protection and supply blood flow and a space for the cerebrospinal fluid. The meninges are composed of three layers: the dura mater, the arachnoid mater, and the pia mater. The dura mater consists of dense and inflexible fibrous connective tissue that is divided into two parts, with the outermost layer—the periosteal layer—lining the bone of the skull. The meningeal layer lines the periosteal layer, except at areas such as the midline, where the superior sagittal sinus can be found. The arachnoid mater is a delicate membrane directly beneath the dura mater and is filled with an intricate network of collagen. Between the arachnoid mater and the pia mater is the subarachnoid space, which is where the cerebral arteries and veins are found. The innermost layer is the pia mater, a thin, smooth vascular membrane that lines every contour and fold of the brain.

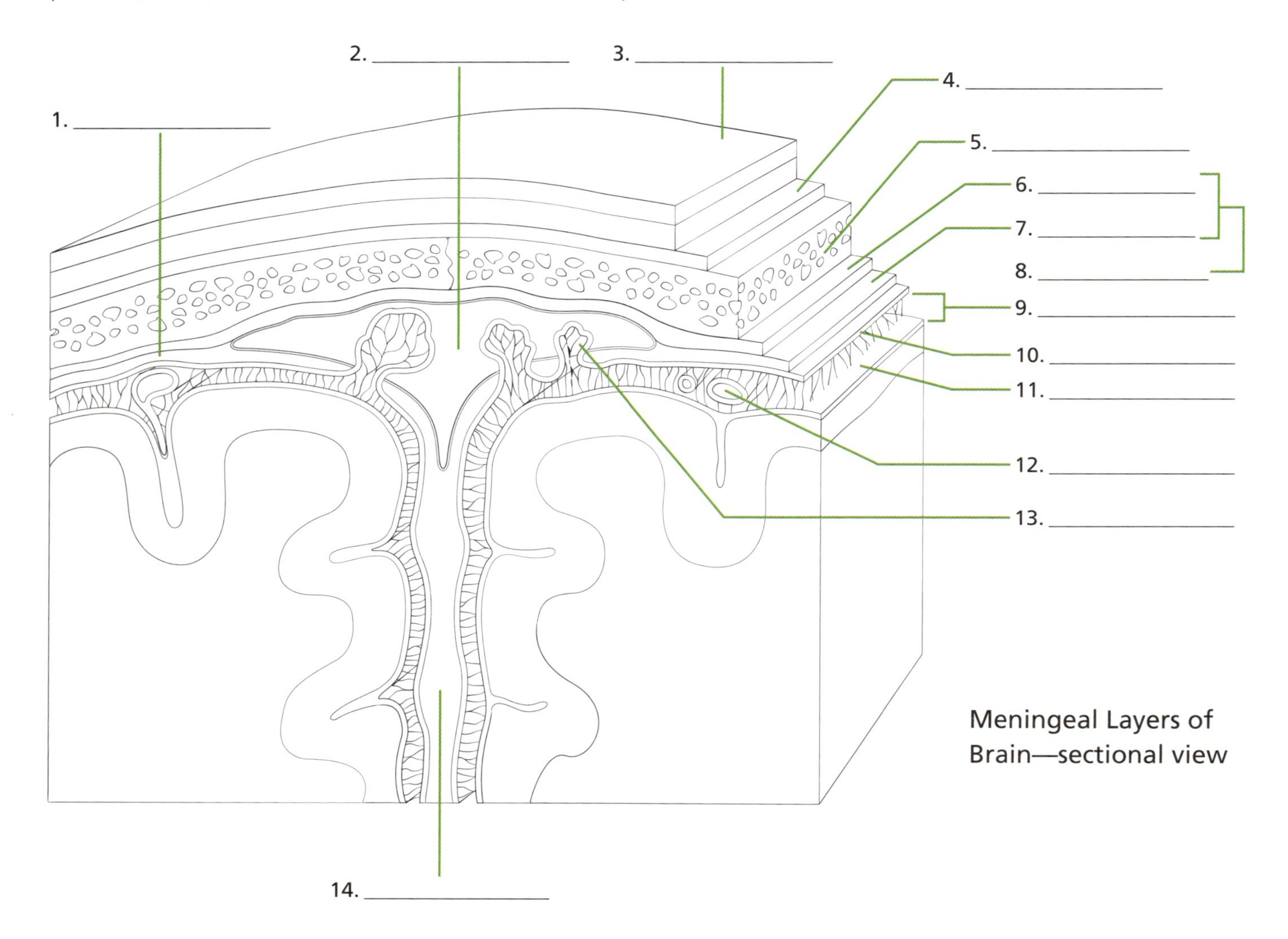

Meningeal Layers of Brain—sectional view

Answers

1. subdural space, 2. superior sagittal sinus, 3. skin of scalp, 4. periosteum, 5. bone of skull, 6. periosteal layer, 7. meningeal layer, 8. dura mater, 9. subarachnoid space, 10. arachnoid mater, 11. pia mater, 12. blood vessel, 13. arachnoid villus, 14. falx cerebri

Dural Venous Sinuses

The venous drainage system does not follow the arteries of the brain but instead drains into dural sinuses that run intracranially between the two layers of the dura mater. The intracranial dural venous sinuses are endothelium-lined spaces located between the meningeal layers of the dura mater.

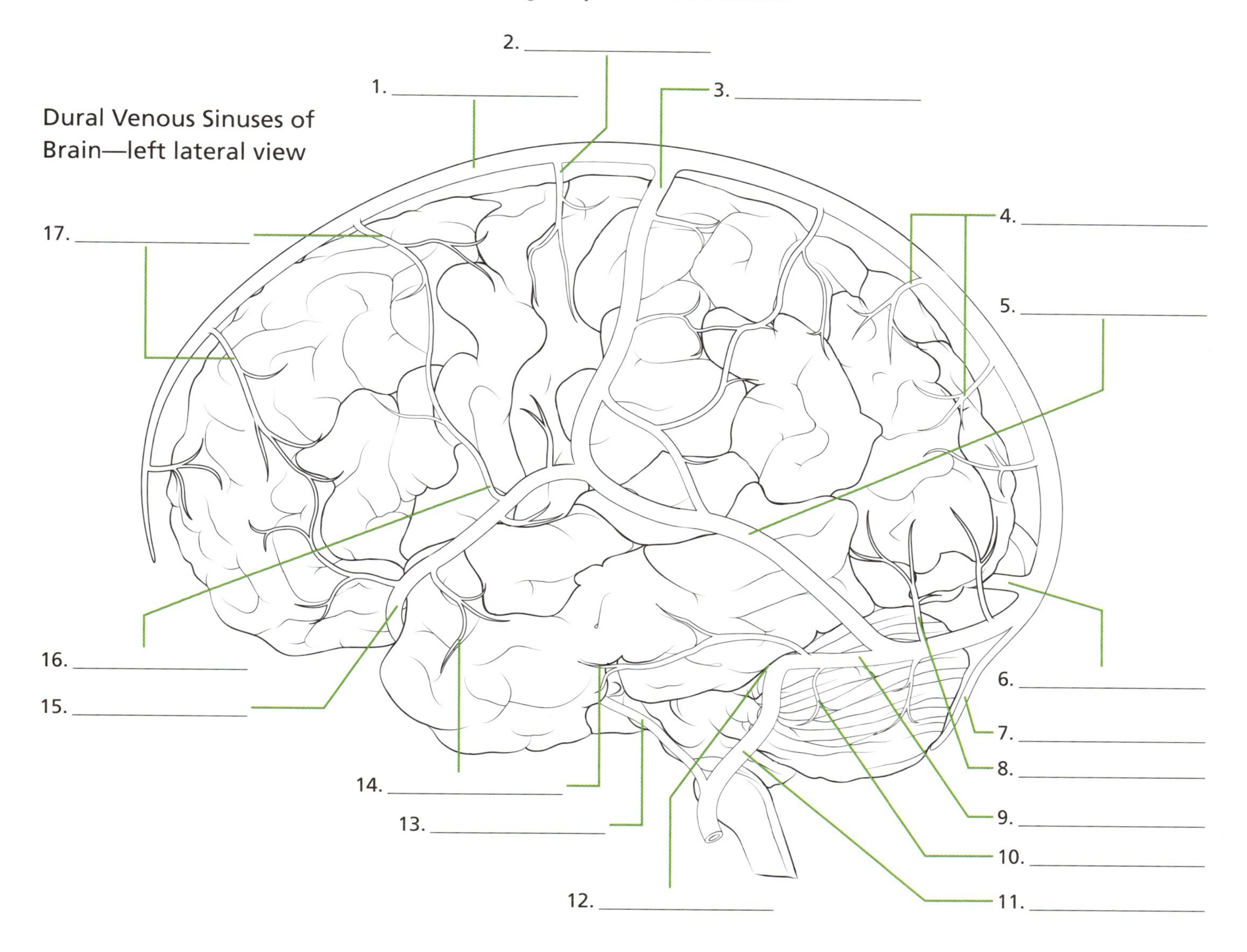

Many of the dural venous sinuses are considered paired, as they are positioned on the left and right of the head. In total, there are six paired sinuses: the superior petrosal sinus (SPS), inferior petrosal sinus (IPS), sphenoparietal sinus, transverse sinus, cavernous sinus, and sigmoid sinus. There are also four unpaired sinuses, which lie in the median plane: the superior sagittal sinus, inferior sagittal sinus, straight sinus, and occipital sinus. On the lateral surface of the hemisphere, the superficial middle cerebral vein is connected to the transverse sinus by the inferior anastomotic vein (vein of Labbé). It also connects with the superior sagittal sinus by the superior anastomotic vein (vein of Trolard), which drains into the superior cerebral veins.

Answers

1. superior sagittal sinus, 2. vein of the central sulcus (Rolandic vein), 3. superior anastomotic vein (vein of Trolard), 4. superior cerebral veins, 5. inferior anastomotic vein (vein of Labbé), 6. straight sinus, 7. occipital sinus, 8. inferior cerebral veins, 9. transverse sinus, 10. cerebellar vein, 11. sigmoid sinus, 12. superior petrosal sinus, 13. inferior petrosal sinus, 14. temporal cerebral veins, 15. superficial middle cerebral vein, 16. inferior cerebral veins, 17. superior cerebral veins

Ventricular System

The ventricular system is a series of four interconnected ventricles in the brain that connect with the central canal of the spinal cord. Within the ventricles are regions called choroid plexuses, which produce the cerebrospinal fluid (CSF). The function of the CSF is to provide a suitable environment for the brain to function by circulating key molecules as well as removing toxic byproducts. The CSF also acts as a shock absorber for the CNS to limit damage from cranial injuries.

The largest ventricles are the left and right lateral ventricles, which are characterized by an anterior and posterior horn. The system is continuous with interventricular foramina connecting each ventricle. From the lateral ventricles is an interventricular foramen that connects to the third ventricle. From the third ventricle is the cerebral aqueduct, which connects to the fourth ventricle. The fourth ventricle has a median aperture and a lateral aperture, which connect the fourth ventricle to the subarachnoid space. From here, the CSF also drains into the central spinal canal.

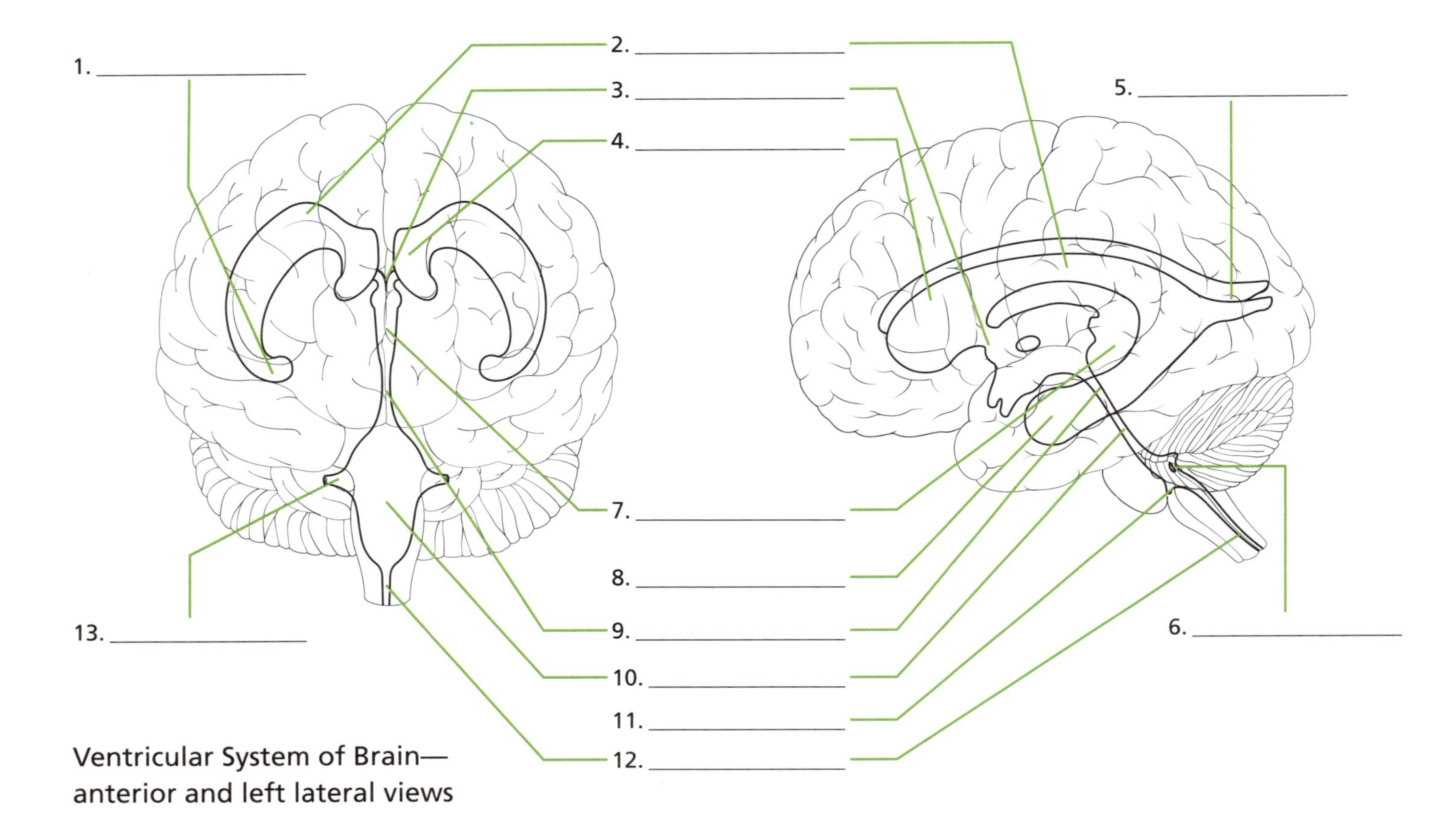

Ventricular System of Brain—anterior and left lateral views

Answers

1. inferior horn, 2. lateral ventricles, 3. interventricular foramen, 4. anterior horn, 5. posterior horn, 6. median aperture, 7. third ventricle, 8. inferior horn, 9. cerebral aqueduct, 10. fourth ventricle, 11. lateral aperture, 12. central canal of spinal cord, 13. lateral aperture

Flow of Cerebrospinal Fluid

CSF is formed within the ventricles by specialized tissue called the choroid plexus. The CSF is ultrafiltered plasma that provides the correct environment for the brain to function properly. It also protects the brain from cranial injuries and reduces its weight (by providing buoyancy), thereby limiting pressure on the base of the brain.

Brain and Flow of Cerebrospinal Fluid—left lateral view

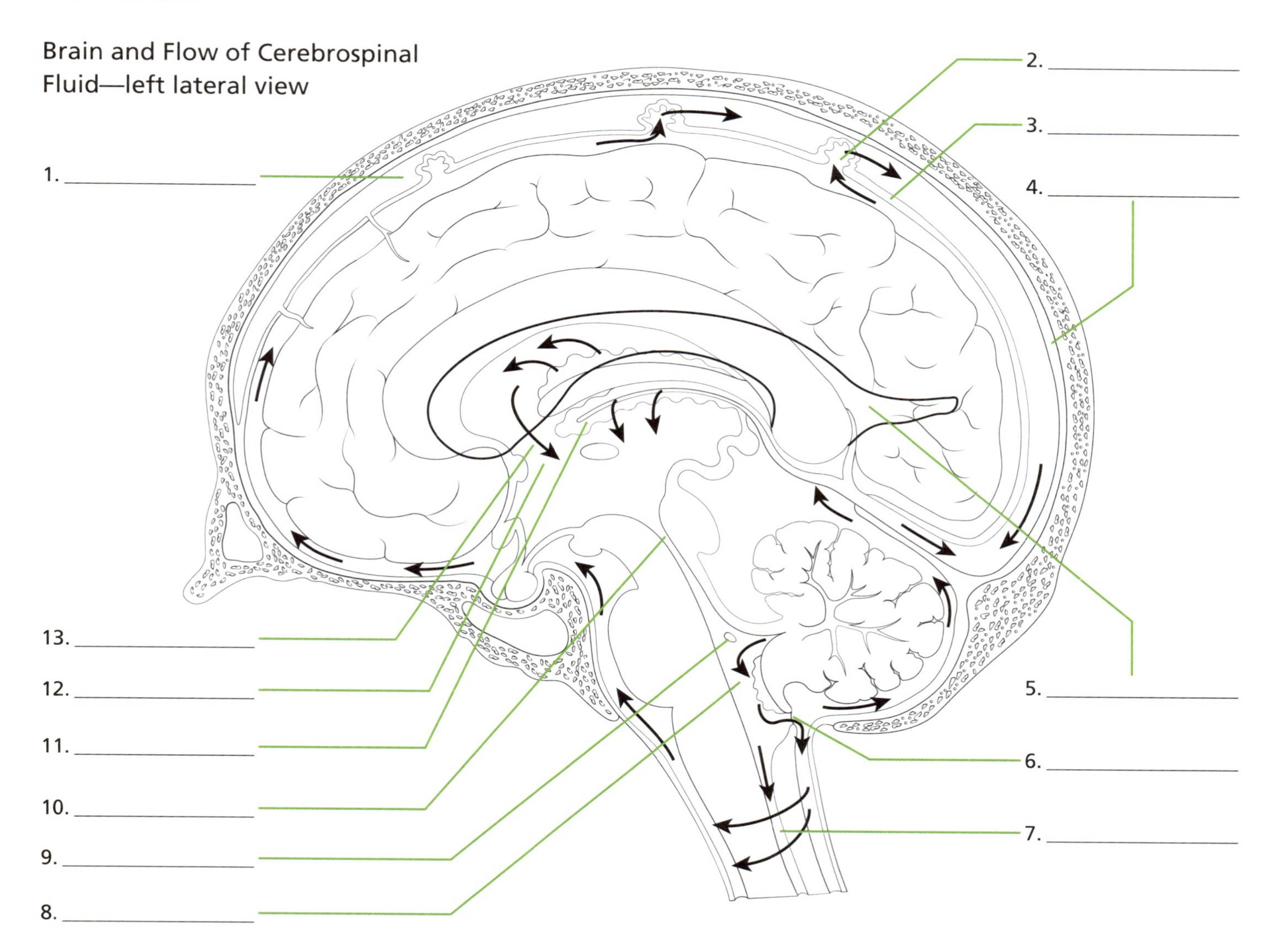

The CSF flow begins in the left and right lateral ventricles of the cerebrum and continues into the third ventricle through the foramen of Monro. The third ventricle is situated between the left and right hemispheres of the thalamus, and CSF flows through the cerebral aqueduct into the fourth ventricle of the brainstem. From the fourth ventricle, the CSF passes into the subarachnoid space around the brainstem and into the central canal of the spinal cord. The CSF is then partially absorbed into the bloodstream and passes through the superior sagittal sinus for filtering in the kidneys.

Answers

1. superior sagittal sinus, 2. arachnoid granulation, 3. subarachnoid space, 4. meningeal dura mater, 5. right lateral ventricle, 6. median aperture, 7. central canal, 8. fourth ventricle, 9. lateral aperture, 10. cerebral aqueduct, 11. choroid plexus, 12. third ventricle, 13. interventricular foramen

Index

Major topics are indicated with **bold** page numbers.

A

α-adrenergic receptor 66
abdomen 8
abdominal cavity 9
abdominopelvic cavity 9
abducens nerve (VI) 70, 72, 78, **86**
abducens nucleus 75, 79, 93
abduction of the hip L5, S1, 55
abduction of the shoulder C5 55
absolute refractory period 24
accessory basal nucleus 147
accessory cuneate nucleus 47, 74, 167
accessory nerve (CN XI) 78, **96**
acetylcholine 61, 65, 66, 67, **173**
acetylcholine effectors 65
acetylcholine receptor site 61
acetylcholinesterase 61
acoustic radiation 135
action potentials **24**
active zone 29
Aδ fibers 165
adduction of the hip L1, L2, L3 55
adduction of the shoulder C6, C7, C8 55
adipocytes 34
adipose tissue 116
adrenal cortex 120
adrenal glands 119
adrenal medulla 65, 67
adrenocorticotropic hormone 119, 120
afferent neurons 36, 62
after-hyperpolarization 24
aging **160–161**
agranular layer 138
ala 99
alar plate 33, 36
allocortex 39
alpha efferent motor fiber 58
Alzheimer's disease 115, 130, 156, 161
amygdala 83, 109, 123, 132, 146, 148, 149, 150, 154, 155, 170, 173
amygdalar region 133
amyloid beta 156
anatomical planes **10–11**, **15**
anatomical view orientation 10
angular gyrus 136, 151
anorexigenic neuron 116
anosmia **83**
ansa lenticularis 111
antagonistic flexor muscle 62
antebrachium 8
anterior belly of the digastric muscle 89
anterior cerebral artery 132, 174, 177, 178, 179
anterior cerebral cortex 178
anterior choroidal artery 175, 178
anterior cingulate cortex 155
anterior cingulate gyrus 154
anterior commissure 38, 39, **131**, 132, **132**, 146, 150
anterior communicating artery 177
anterior corticospinal tract 48, 162
anterior forceps 130
anterior horn 52
anterior horn cell neuron III 49
anterior horn (lateral ventricle) 186
anterior hypothalamus 114
anterior inferior cerebellar artery 177, 179, 180, 181
anterior inferior temporal cortex 168
anterior limb of internal capsule 131, 132
anterior lobe of cerebellum **99**
anterior median fissure 41, 52
anterior nucleus of thalamus 108, 110, **131**, 133, 146, 148, 150
anterior pituitary gland 119, 120
anterior pretectal nucleus 113
anterior spinal artery 176, 181, 182
anterior spinocerebellar tract 46, 74, 167
anterior temporal lobe 38
anterior thalamic radiation 135
anterior trunk of the mandibular nerve 89
anterior white commissure 52
anterolateral quadrant 44
antidiuretic hormone 119
aphasia **152**
arachnoid granulation 187
arachnoid mater 184
arachnoid villus 184
arcuate fasciculi 151, 160
arrector pili muscle 57
arterial blood supply **174**
arteriole 57
ascending nerve tracts **43–47**
ascending sensory tracts 77
association cortices 39, 138
association fibers 13
astrocyte end-feet 25, 30
astrocytes **25**, 30, 32, 156
atlas (C1) 71, 72
auditory association area 126
auditory cortex 110, 126, 142, **142**, 151, 169
auditory impulses 77
auditory nerve 169
auditory pathways **169**
auditory tube 92
autogenic inhibition reflex **63**
autonomic ganglion 65
autonomic motor neurons 164
autonomic nervous system **65–66**
axial plane 11, 15
axilla 8
axillary nerve 12
axon hillocks 19, 27
axon terminal 19, 29, 61
axonal degeneration/ regeneration **64**
axons 19, 20, 21, 22, **26**, **27**, 61

B

β-adrenergic receptor 66
back 8
basal ganglia **121**, **124–125**
basal lamina 30
basal nuclei 147, 173
basal plate 33, 36
basement membrane 30
basilar artery 175, 176, 177, 179, 180, 181
basilar pons 133, 134
basket cells 103
bed nucleus of stria terminalis (BNST) 147
Bell's palsy **91**
bipolar neurons 20
biventral lobule 100
bladder 67, 68, 118
blood–brain barrier (BBB) **30**
blood supply to brain **174–185**
body cavities **9**
body of caudate 123, 133, 134
body of corpus callosum **130**, 132, 133, 134
body of fornix 131, 133, 134
Bowman's gland (olfactory gland) 82
brachial plexus 12, 53
brachiocephalic artery 175
brachium 8
brainstem 15, 18, **36**, **70–73**, 86, 92, **98**, 174, **180–181**
branchial motor column 98
Broca's aphasia 152
Broca's area 17, 126, **151**
buccinator 90

C

C fibers 165
C7–C1 spinal cord levels 47
C8–L2 spinal cord levels 45
CA1 148, 149
Ca^{2+} channels 29
CA3 148, 149
calcaneus 8
calcarine artery 179
calcarine fissure 127
callosomarginal artery 179
capillaries 25, 57
capillary endothelial cells 30
cardiovascular centers 73
carotid artery system **175**
carotid body 95
carotid sinus 94, 95, 175
cartilage 60
cauda equina 40, 42, 53
caudal medulla 181
caudal view orientation 10, 14
caudate 13, 15, 39, 111, 121, 122, 123, 124, 125, 132, 170, 173
cavernosus sinus 183
celiac ganglia 69
central canal 41, 52, 74, 186, 187
central lobule 99
central nervous system (CNS) **12**
central nuclei 147
central sulcus 16, 17, 126, 129, 136, 139, 143, 144
central tegmental tract 75
central thalamic radiation 135
centromedian thalamic nucleus 108, 131
cephalic flexure 35
cerebral hemispheres **38**, **126–129**
cerebellar cortex **103**
cerebellar hemisphere 36
cerebellar nuclei neurons 37
cerebellar peduncles 71
cerebellar vein 185
cerebellum 15, 16, 18, 36, **37**, 45, 46, 47, **99–106**, 107, 112, 162, 172, 174
cerebral aqueduct 18, 76, 186, 187
cerebral cortex 13, 15, **39**, **136–138**, **145**, **178**
cerebral hemispheres 12, 18, 35
cerebral peduncle 71, 72, 76, 128, 162, 170
cerebral venous drainage **183**
cerebrocerebellum 102, **106**
cerebrospinal fluid 156, **187**
cerebrum 15, 16, 18, 162 *see also* forebrain
cervical branch of facial nerve 90
cervical enlargement 12, 40, 42
cervical flexure 35
cervical plexus 12, 53
cervical spinal cord 162, 167
cervical spinal nerves 40
cheek 8
chin 8
chloride (Cl^-) ions 23
cholinergic neurons 173
chondrocytes 34
choroid 84
choroid plexus 39, 71, 72, 74, 142, 187
chromatolysis 64
cilia of olfactory sensory neurons 82
ciliary body 84
ciliary ganglion 86
cingulate 132
cingulate gyri 127, 130, 132, 146, 150
cingulate sulcus 132
cingulum 148
circle of Willis **177**
circumolivary bundle 70
circumventricular organs 117
claustrum 133
climbing fibers 103
coccygeal ligament 40
coccygeal nerve 53
cochlea 92, 142, 169
cochlear branch of CN VIII 92
cochlear nuclei 74, 92
collagen 57, 59, 60, 184
collateral ganglion 164
collateral sulcus 128
colon 95
columns of fornix 131
commissure of fornix 38
common carotid artery 94, 175
common fibular nerve 12
conduction aphasia 152
confluence of sinuses 183
connective tissue capsule 58
contralateral view 14
conus medullaris 40, 53
cornea 84, 89
corneal reflex 89
corona radiata 13, 130
coronal plane 10, 11, 15
corpus callosum 13, 15, 17, 18, 38, 39, 112, 127, **130**, 146, 183
cortex 170, 171, 172, 173
corticobulbar fibers 76
corticobulbar tract 70, 75
corticonuclear tract 135
corticopontine tract 135
corticorubral tract 135
corticospinal fibers 76, 134
corticospinal tract 48, **48**, 70, 71, 75, 135
corticotropin-releasing factor 120
cortisol 120
cranial cavity 9
cranial nerve functional columns **98**
cranial nerve nuclei 36, **79**
cranial nerves **78–81**
cranial root of accessory nerve 96
cranial view orientation 10
cranium 8
cribriform plate 82
crus 8
crus cerebri 76, 134
crus of fornix 130, 134
culmen 99
cuneate nucleus 73, 74

cuneocerebellar tract 47, **47**, 167
cuneus 127
cytoplasm 28

D
declive 99
decussation of medial lemniscus 43, 74
decussation of pyramids 13, 70, 74, 162
deep fibular nerve 12
deep forebrain **179**
deep frontal white matter 155
deep pontine nuclei 75
dementia 160
dendrites 19, 20, 22, 27, 29
dendritic spine 29
dendritic tree 21
dentate gyrus 148, 149
dentate nucleus 109
depolarization 23, 24
depressor anguli oris 90
depressor labii inferioris 90
dermatomes **54**
dermis 56, 57
descending motor projections 77
descending nerve tracts **48–51**
diaphragm 9
diencephalon 18, 35, 38, 107, 127
digital nerve 12
digits 8
distal view orientation 10
dopamine **170**
dopamine cell bodies 170
dorsal cavity 9
dorsal cochlear nucleus 79
dorsal column 41
dorsal column nuclei 43
dorsal column pathway **43**
dorsal horn 33, 41
dorsal median sulcus 71
dorsal motor nucleus of vagus 74, 79
dorsal plane 10
dorsal (posterior) view 14
dorsal raphe nucleus 172
dorsal root 33, 41, 52, 164
dorsal root ganglion 33, 41, 43, 52, 164
dorsal spinocerebellar tract **45**
dorsal striatum **123**, 133
dorsal thalamus 135
dorsiflexion of the ankle L4, L5 55
dorsolateral prefrontal cortex 155
dorsomedial hypothalamus 114
dorsomedial prefrontal cortex 144
dorsomedial thalamic nucleus 131
dorsomedial thalamus 133
dura mater 184
dural venous sinuses **185**

E
ear 8
ectoderm 31, 34
Edinger-Westphal nucleus (CN III) 79, 113
efferent neurons 36, 62
elbow 8
electrical synapses 29
emotional expression **147**, 150
encapsulated Golgi-type endings 60
endocytosis 159
endplate potential 61
enteric nervous system (ENS) **69**
entorhinal cortex 39, 83, 148, 149, 150, 157
ependymal cells 25
epidermis 56, 57
epinephrine 67
epithalamus **112**
esophagus 69
excitatory neuron 62
exocytosis 22, 159
extension of the elbow C6, C7 55
extension of the hip L4, L5 55
extension of the knee L3, L4 55
extension of the shoulder C6, C7, C8 55
external acoustic meatus (CN VII, VIII) 81
external capsule 134
external carotid artery 175
external geminal zone 37
external granular layer 137
external pyramidal layer 137
extrafusal muscle fibers 58, 59
extrastriate visual areas 154, 155
extrastriate visual cortex 109
extreme capsule 145
eye glands 67, 68
eye movements **86–87**
eyes 8, 118

F
face 8, 158
facial artery 175
facial colliculus 71, 75
facial motor nucleus 89
facial nerve (CN VII) 70, 72, 78, **90–91**, 92
facial nucleus (CN VII) 75, 79
falx cerebri 184
fasciculi **145**
fasciculus cuneatus 47, 71, 74
fasciculus gracilis 74
femoral nerve 12, 53
femur 8
fiber bundles of internal capsule **135**
fibers of vagus nerve 74
fibroblasts 57
fimbria of fornix 131
finger abduction and adduction T1 55
finger extension C7, C8 55
finger flexion C7, C8 55
first cervical nerve 71
first-order neuron 43, 44
flexion of the elbow C5, C6 55
flexion of the hip L2, L3 55
flexion of the knee L5, S1 55
flexion of the shoulder C5 55
flocculonodular lobe **101**
flocculus 101, 102
floor plate 33
folium 100
foot 158
foramen lacerum 80, 81
foramen magnum 53, 80, 81, 96
foramen ovale 80, 81, 88
foramen rotundum 80, 88
foramen spinosum 80, 81
foramina **80–81**
forceps major of corpus callosum 131
forceps minor of corpus callosum 131
forearm pronation C7, C8 55
forearm supination C6 55
forebrain 18, 48, 165, **179**
see also cerebrum
forehead 8
fornix 13, 15, 38, 109, 115, 127, 146, 148, 149
fourth ventricle 36, 73, 74, 75, 134, 186, 187
fovea centralis 84
free nerve endings 60
frontal (coronal) plane 10, 11, 15
frontal cortex **144**, 154
frontal eye field 136, 168
frontal gyrus 132
frontal lobe 16, 17, 38, 82, 126, 129, 174
frontopolar artery 179
frontopontine fibers 76
frontopontine tract 135
fusiform gyrus 127

G
gallbladder 69, 95
gamma efferent motor fiber 58
ganglia 13, 65
gap junctions 28
gastrointestinal tract 116
general sensory afferent neurons 36
general sensory efferent neurons 36
general somatic sensory column 98
general visceral afferent neurons 36
general visceral efferent neurons 36
genioglossus 97
genitals 158
genu of corpus callosum 130, 132, **132**
genu of facial nerve 75
genu of internal capsule 131
ghrelin 116
glia **25**
glia cells **32**
glioblast 32
global aphasia 152
globus pallidus 13, 15, 39, 109, 111, 123, 131, 132, 133
globus pallidus external 121, 122, 123, 124
globus pallidus internal 121, 122, 123, 124
glomeruli 82
glossopharyngeal nerve (CN IX) 70, 72, 78, **94**
gluteus 8
glymphatic system 156
Golgi apparatus 22
Golgi cells 103
Golgi stain 137
Golgi tendon organs 45, 46, 47, 59, 63, 167
gonadotropins 119
gonads 119
gracilis nucleus 73, 74
graded potentials **23**
granular layer 37, 138
granule cells 103
gray commissure 41, 52
gray matter 13, 33
gray matter—anterior horn 52
gray matter—lateral horn 52
gray matter—posterior horn 52
gray ramus 164
great cerebral vein 183
growth cone sprouting 64
growth hormone 119
gustatory cortex **139**
gyri 13, 160
gyrus rectus 128, 132

H
habenula 71, 112, 131
hair follicles **57**
hallux 8
hand 8, 158
head 8
head of caudate 122, 123, **130**, 131, 135
heart 67, 68, 95, 118
hemispheres **17**
hindbrain 18, 35
hippocampal formation 38, 150
hippocampus 15, 32, 133, 134, 146, 147, 148, 149, 154, 155, 157, 170, 171, 172, 173
horizontal fissure 100
Huntington's disease 125, 130
hyoglossus 97
hyperpolarization 23
hypoglossal canal 80, 81, 97
hypoglossal motor nucleus 74
hypoglossal nerve (CN XII) 70, 72, 78, 97, **97**
hypoglossal nucleus (CN XII) 79
hypophyseal portal system 119
hypothalamic-pituitary-adrenal axis **120**
hypothalamus 15, 18, 107, 112, 115, **116–118**, 119, 132, 133, 146, 154

I
inferior alveolar nerve 88
inferior anastomotic vein 185
inferior cerebellar peduncle 45, 47, 74, 75, 104, 105
inferior cerebral veins 185
inferior colliculus 51, 71, 72, 76, 109, 134, 142, 169
inferior frontal gyrus 136
inferior frontal sulcus 132, 136
inferior fronto-occipital fasciculus 145
inferior ganglion 94, 95
inferior horn 133, 134, 186
inferior longitudinal fasciculus 145
inferior oblique 86, 87
inferior olivary nucleus 74, 128
inferior orbital fissure 81
inferior parietal lobule 109
inferior petrosal sinus (IPS) 183, 185
inferior rectus 86, 87
inferior sagittal sinus 183
inferior salivatory nucleus (CN IX) 79
inferior semilunar lobule 100
inferior temporal gyrus 128, 136, 143
inferior temporal sulcus 128
inferior view orientation 10
infraorbital foramen 81, 88
infraorbital nerve 88
infundibular stalk 128
infundibulum 70, 133, **133**
inguen 8
inhibited motor neuron 63
inhibitory interneuron 62, 63
inner plexiform layer 84
insula 16, 132
insulin 116
intercostal nerve 12
internal acoustic meatus 80, 92
internal capsule 13, 39, 111, 122, 123, 131, **135**, 162
internal carotid artery 175, 177
internal carotid branches **175**
internal granular layer 137, 138
internal jugular vein 183
internal medullary lamina 107
internal pyramidal layer 137
interneurons 21, 163
interparietal sulcus 136
interpeduncular nucleus 113
interstitial nucleus of Cajal 113
interthalamic adhesion 108, 112
interventricular foramen 39, 88, 186, 187

intestines 118
intrafusal muscle fiber 58
intralaminar nuclei 108, 110
intrinsic muscles of the tongue 97
involuntary muscle contraction pathway **164**
ions 27, 28
ipsilateral cerebellum 167
ipsilateral fasciculus cuneatus 47
ipsilateral view 14
iris 84

J
joint capsule 60
joint receptors **60**
jugular foramen 80, 81, 94, 95, 96

K
kidneys 95, 119
kinesthesia 60

L
lacrimal fossa 81
lacrimal glands 67, 68
large intestine 67
laryngeal branches 95
lateral aperture 186, 187
lateral column 41
lateral corticospinal tract 48, 162
lateral descending tracts 105
lateral diffusion 159
lateral dorsal nucleus 110
lateral femoral cutaneous nerve 12
lateral fissure 16
lateral geniculate nucleus 70, 71, 72, 85, 108, 110, 113, 135, 140, 141
lateral horn 41, 52
lateral hypothalamus 114, 116, 147
lateral intraparietal area 168
lateral lemniscus 76, 109, 169
lateral nuclei 147
lateral occipital area of visual cortex 140, 141
lateral olfactory stria 128
lateral olfactory tract 83
lateral posterior nucleus 108
lateral pterygoid muscle 89
lateral rectus 86, 87, 93
lateral reticular nucleus 51
lateral rotation of the hip L5, S1 55
lateral rotation of the shoulder C5 55
lateral striates 178
lateral sulcus 136, 143, 144
lateral ventricles 13, 15, 39, 107, 130, 131, 133, 186
lateral vestibular nucleus 49
lateral vestibulospinal tract 49
lateral view orientation 10, 14
left hemisphere 129
left medial rectus 93
left visual field 85
lens 84
lentiform nucleus 135
leptin 116
levator labii superioris 90
levator palpebrae superioris 86
limbic system 117, **146–150**
lingual artery 175
lingual gyrus 127
lingual nerve 88
lingula 99
lips 158
Lissauer's tract 44
liver 67, 95, 118
lobes (brain) **16**, 126
lobulus simplex 99
locus coeruleus **171**
longitudinal fissure 13, 129, 144
lower back 8
lower buccal branch of facial nerve 90
lower limb 8
lower medulla 167
lower motor neurons 162
lumbar plexus 12
lumbar spinal cord 162, 167
lumbar spinal nerves 40
lumbosacral enlargement 12, 40, 42
lumbosacral plexus 53
lunate sulcus 17
lungs 67, 68, 95, 118

M
macula lutea 84
mammillary bodies 15, 70, 114, **115**, 128, 133, **133**, 146, 148, 149, 150
mammillothalamic tract 109, 133, 146, 150
mandibular foramen 88
mandibular nerve (V_3) 88
mantle layer 33
marginal layer 33
massa intermedia 39
masseter muscle 89, 90
mastication 89
maxillary artery 175
maxillary nerve (V_2) 88
mechanoreceptors 43, 56, 57
medial descending tracts 105
medial dorsal nucleus of the thalamus 146
medial frontal gyrus 132
medial geniculate nucleus 70, 71, 108, 110, 135, 142, 169, 170
medial lemniscal pathway 166
medial lemniscus 49, 51, 71, 74, 75, 76, 109, 170
medial longitudinal fasciculus 74, 93
medial longitudinal fasciculus fibers 75
medial longitudinal fissure 17
medial olfactory stria 128
medial olfactory tract 83
medial prefrontal cortex 157
medial pretectal nucleus 113
medial pterygoid muscle 89
medial rectus 86, 87, 93
medial rotation of the hip L1, L2, L3 55
medial rotation of the shoulder C6, C7, C8 55
medial striates 178
medial superior temporal area 168
medial thalamic nuclei 148
medial vestibular nucleus 49
medial vestibulospinal tract 49
medial view orientation 10, 14
median aperture 186, 187
median nerve 12, 53
median plane 10
mediodorsal nucleus 108, 110
medulla oblongata 12, 13, 18, 36, 39, 43, 44, 46, 48, 50, 51, 70, 72, **74**, 92, 95, 97, 162, 165
medullary arteries 176
medullary center 132
medullary reticular formation 50, 171
medullary reticulospinal tract 50
Meissner's corpuscles **56**
melanocytes 34
memory formation circuitry **148–149**
meningeal dura mater 187
meningeal layers **184**
meninges 40
mental foramen 88
mental nerve 88
mentalis 90
Merkel disks **56**
mesencephalic nucleus of trigeminal nerve (CN V) 79
mesencephalon 35, 37
see also midbrain
mesocortical fibers 170
mesolimbic fibers 170
metencephalon 18, 35
microglia **25**, 30
microtubule 22, 28
midbrain 18, 35, 39, 48, 51, **76**, 162, 165
see also mesencephalon
middle cerebellar peduncle 70, 72, 75, 106, 134
middle cerebral artery 174, 177, 178
middle cerebral cortex 178
middle frontal gyrus 17, 136
middle longitudinal fasciculus 145
middle pons 181
middle temporal area 168
middle temporal gyrus 128, 136, 143
middle temporal sulcus 136
midline view 14
midsagittal plane 10, 11
mitochondria 22, 28
molecular layer 37, 137
mossy fibers 103, 148, 149
motor cortex 48, 110, **143**, 151, 162
motor endplates 49, 61
motor neurons 12, 21, 61, 63, 163
motor nucleus of trigeminal nerve 75
motor pathways 33, **162–164**
motor speech (Broca's) area 126
mouth 8
multiform layer 137
multipolar neurons 20, 21
muscarinic receptor 66
muscle fiber 61, 64
muscle spindle fiber 62
muscle spindles 45, 47, 58
muscle stretch receptors **58**
musculocutaneous nerve 12
myelencephalon 18, 35
myelin 13, 19
myelin sheath 19, 21, 22, 26, 27, 61
myelination **26**
myotomes **55**

N
nasal cavity 82
neck 8
neocortex 39
neospinothalamic tract 165
nerve fibers 26, 56
nerve plexuses **53**
nerve rootlets 52
nervous system overview **12–18**
neural crest **34**
neural crest cells 31, 34
neural fold 31
neural groove 31
neural plasticity **158–159**
neural plate **31**
neural tube **31**, 34
neuroblast 32
neurocytes 33
neurodegeneration 161
neuroectoderm 31
neuroendocrine hormones 147
neurogenesis 32
neuromuscular junctions **61**
neuron development 161
neuronal membrane 22
neurons **19–22**, 25, 30, **32**, 34
neurotransmitter transporter 29
neurotransmitters 29, **66**
nicotinic receptor 66
nigrostriatal fibers 170
Nissl stain 137
nociceptors 44, 57, **165**
nodes of Ranvier 19, 26, 27
nonrapid eye movement (NREM) sleep 153
norepinephrine 65, 66, 67, **171**
norepinephrine effectors 65
nose 8
notochord 31, 34
nuclear transitory zone 37
nuclei 13, 20, 21, 22, 25, 26, 27, 30
nuclei of hypothalamus **114**
nucleolus 22
nucleus accumbens **122**, 123
nucleus ambiguus 74, 79
nucleus dorsalis of Clarke 45
nucleus of posterior commissure 113
nucleus of the optic tract 113
nucleus of the solitary tract 116, 117, 171
nucleus raphe magnus 172
nucleus raphe obscurus 172
nucleus reticularis gigantocellularis 50
nucleus solitarius, caudal portion (CN IX, X) 79
nucleus solitarius, rostral portion (CN VII, IX, X), 79

O
obturator nerve 12
occipital artery 175
occipital lobe 16, 17, 38, 126, 129, 174
occipital notch 16
occipital sinus 185
occipito fibers 76
occipitotemporal gyrus 128
oculomotor nerve (CN III) 70, 72, 76, 78, **86**
oculomotor nucleus (CN III) 79, 93
olecranon 8
olfactory bulb 38, 39, 78, 82, 83, 117, 128
olfactory centers 82
olfactory cortex **139**
olfactory epithelium **82**
olfactory foramina 80, 82
olfactory mucosa 82
olfactory receptor neuron 82
olfactory sulcus 132
olfactory tract 82, **83**, 128
olfactory tubercle 39, 83, **122**
oligodendrocytes 19, 25, 26, 32
olivary nucleus 73
olivary pretectal nucleus 113
olive 70, 73
ophthalmic nerve (V_1) 88, 89
optic canal 80, 81
optic chiasm 70, 83, 85, 86, 112, 128, 146
optic disc 84
optic nerve (CN II) 72, 78, **84–85**, 85, 86
optic nerve head 84
optic radiation 85, 131, 135

optic tract 70, 72, 85, 109, 113, 132
optic vesicle 35
orbicularis oculi 89, 90
orbicularis oris 90
orbital artery 179
orbital gyrus 132, 136
orbitofrontal cortex 128, 144, 148
orexigenic neuron 116
osteoblasts 34
osteoclasts 34
otic ganglion 94
outer plexiform layer 84
oxytocin 119

P

Pacinian corpuscles **56**, 57
Paciniform receptors 60
pain pathway **165**
pain receptors **57**
palatine tonsil 94
paleospinothalamic tract 165
palm 8
palmar view orientation 10
pancreas 67, 68, 116
Papez circuit **150**
papillary 56
paracentral lobule 127
parahippocampal cortex 155, 157
parahippocampal gyrus 127, 128, 146
parallel fibers 103
parasagittal plane 11
parasympathetic activation 147
parasympathetic column 98
parasympathetic fibers 94
parasympathetic nervous system **68**
parasympathetic postganglionic neuron (unmyelinated) 65
parasympathetic preganglionic neuron (myelinated) 65
paraventricular nucleus 114, 116, 120, 147
parietal lobe 16, 17, 38, 126, 129, 148, 174
parieto fibers 76
parieto-occipital artery 179
parieto-occipital fissure 127, 129
parieto-occipital sulcus 16, 130
Parkinson's disease 111, 112, 125, 130, 134
parotid gland 94
pars compacta 76
patella 8
pelvic cavity 9
pelvic pathways 69
pelvis 8
peptide YY 116
perforant path 148, 149
periaqueductal gray 76, 147, 165
pericallosal artery 179
pericardial cavity 9
pericytes 30
periosteal layer 184
periosteum 184
peripheral nervous system **12**
peripheral retina 84
perirhinal cortex 157
pes 8
phagocytosis 25
phantom limb 158
pharyngeal muscles 94
pharyngeal nerve branches 95
pharynx 158
photoreceptors 84
phrenic nerve 53
pia mater 184
pineal body 71, 112
pineal gland 15, 134
piriform cortex 83
pituitary gland 112, 118, **119**
plantar nerve 12
plantar view orientation 10
plantarflexion of the ankle S1, S2 55
plasma membrane of muscle fiber 61
pollex 8
pons 13, 18, 35, 36, 39, 46, 48, 50, 51, 70, 71, **75**, 88, 92, 94, 95, 142, 154, 162, 165
pons–medulla junction 49
pontine flexure 35
pontine nuclei 106
pontine parabrachial nucleus (PBN) 147
pontine reticular formation 50
pontine reticulospinal tract 50
pontine tegmentum 155
popliteus 8
postcentral gyrus 17, 129, 136
postcentral sulcus 17
postcommissural fornix 150
posterior auricular artery 175
posterior auricular nerve 90
posterior cerebral artery 174, 177, 179, 180, 181
posterior cerebral cortex 178
posterior cingulate cortex 155
posterior cingulate gyrus 154
posterior commissure 39
posterior communicating artery 177, 180
posterior dorsal nucleus 110
posterior forceps 130, 131
posterior horn 52, 134, 186
posterior hypothalamus 114
posterior inferior cerebellar artery 176, 177, 179, 180, 181
posterior inferior temporal cortex 168
posterior intraparietal area 168
posterior limb of internal capsule 131, 133, 134
posterior lobe of cerebellum **100**
posterior median sulcus 41, 52, 73
posterior parietal cortex 157
posterior pituitary gland 119
posterior pretectal nucleus 113
posterior spinal artery 176, 181, 182
posterior spinocerebellar tract 45, 74, 167
posterior tibial nerve 12
posterior view orientation 10
postganglionic fibers 65, 66
postsynaptic density 29
postsynaptic membrane 28, 29
potassium (K⁺) ions 23, 24
precentral gyrus 129, 136
precentral sulcus 17, 136
precuneus 127
prefrontal cortex 109, 110, 128, 146, 147, 148
preganglionic fibers 65, 66
preganglionic neurons 164
premotor cortex 143, 162
preoptic area 114
presynaptic terminal 28, 29
pretectal nuclei **113**
primary auditory cortex 142
primary cortices 39
primary fissure 99
primary motor cortex 110, 126, 143, 144, 162
primary sensory cortex 126, 138
primary somatosensory cortex 43, 44
primary visual cortex 85, 110, 151, 154, 155
principal sensory nucleus 166
progenitor cells 32
projection fibers 13
prolactin 119
proprioception 59, **167**
proprioceptors 43
proximal view orientation 10
pseudounipolar neurons 20
pubis 8
pudendal nerve 12
pulmonary plexus 95
pulvinar 71, 108, 109, 110, 113, 131, 134
Purkinje cell clusters 37
Purkinje cell layer 37
Purkinje cell precursors 37
Purkinje cells 103
putamen 13, 15, 39, 111, 121, 122, 123, 124, 125, 130, 131, 132, 133, 134, 170, 173
pyramidal decussation 48
pyramids 48, 49, 70, 74, 100, 128, 162
pyriform cortex 39

Q

quadrangular lobule 99
quadriceps muscle 62

R

radial nerve 12, 53
radiations to cerebral cortex 77
raphe nuclei 74, 75, **172**
rapid eye movement (REM) sleep 153
reading comprehension area 126
rectum 67, 68
red nucleus 51, 76, 105, 113, **134**, 170
relative refractory period 24
repolarization 24
reproductive organs 67, 68
respiratory rhythmicity center 73
resting potential **23**, 24
reticular formation 73, 74, **77**, 105, 117, 154, 165, 171
reticular nucleus 110
reticularis pontis caudalis (RPC) 147
reticulospinal tracts **50**
retina **84**, 85
retinal ganglion cells 84
retrosplenial cortex 148, 149
rhombic lip 36, 37
right hemisphere 129
right lateral rectus 93
right lateral ventricle 187
right visual field 85
risorius 90
Rolandic vein 185
roof plate 33, 36
root fibers of oculomotor nerve 76
rostral (anterior) view 14
rough endoplasmic reticulum 22
rubrospinal tract **51**
Ruffini-like receptors 60
Ruffini's endings **56**

S

saccule 92
sacral plexus 12
sacral spinal cord 69
sacral spinal nerves 40
sacrum 68
sagittal fissure 15
sagittal plane 10, 11, 15
salivary glands 67, 68, 118
saltatory conduction **27**
sarcoplasmic reticulum 61
satellite cells 25
Schaffer collaterals 148, 149
schizophrenia 134
Schwann cells 25, 26, 34, 64
sciatic nerve 12, 53
sclera 84
sebaceous gland 57
second-order neuron 43, 44
semicircular canals 92, 93
sensory nerve endings 57
sensory neurons 12, 21, 163
sensory nucleus of trigeminal nerve 75, 79
sensory pathway 33
septal nuclei 148
septum pellucidum 38, 39, 130, 132
serotonin **172**
severed axon 64
shingles 54
shoulder 8
sigmoid sinus 183, 185
skeletal muscle 162
skin sensory organs **57**
sleep **153–156**
slow-wave sleep 153
small intestine 67, 68, 95, 116
smooth endoplasmic reticulum 22
smooth muscle cells 34
sodium (Na⁺) ions 23, 24, 61
sole 8
solitary nucleus 73, 74, 75
solitary tract 75
soma 19, 21, 27
somatic motor column 98
somatic sensory association area 126
somatosensory cortex 110, 143, **143**, 144, 158, 165, 166
somatosensory pathways **165–167**
spatial perception **157**
special sensory afferent neurons 36
special somatic sensory column 98
special visceral afferent neurons 36
special visceral efferent neurons 36
sphenoid bone 82
sphincters 69
spinal accessory nerve 71, 72
spinal accessory nucleus (CN XI) 79
spinal arteries 182
spinal canal 9
spinal cord 18, **33**, 35, 39, **40–42**, 43, 44, 46, 53, 63, 65, 67, 68, 73, 162, 163, 164, 165, 174, **182**
spinal cord reflexes **62–63**
spinal ganglion 169
spinal nerves 33, 41, **52–53**, 164
spinal nucleus 74, 75, 166
spinal root of accessory nerve 96
spinal tract 74, 75, 166
spinal trigeminal sensory nucleus (CN V, VII, IX, X) 79, 89
spinocerebellum 102, **105**
spinothalamic fibers 76
spinothalamic tracts **44**, 71, 74, 75, 76, 109, 170
splanchnic nerve 164
spleen 95

splenium of corpus callosum 130, **134**
stellate cells 103
stem cells 32
sternocleidomastoid muscle 96
stomach 67, 68, 95, 118
straight sinus 183, 185
stratum basale 56
stratum corneum 56
stratum granulosum 56
stratum lucidum 56
stratum spinosum 56
stretch reflex **62**
stretching of the muscle via patella tendon 62
striatum 172
styloglossus 97
stylomastoid foramen 81
stylopharyngeus muscle 94
subarachnoid space 184, 187
subcallosal gyrus 132
subclavian artery 175
subcutis (hypodermis) 56, 57
subdural space 184
subiculum 147, 148, 149
substantia nigra 76, 109, 113, 121, 124, 125, 134, **170**
substantia nigra, compact part 170
substantia nigra, reticular part 170
subthalamic fasciculus 111
subthalamic nucleus 111, 121, 124
subthalamus **111**, 112
subventricular zone 32
sulcal artery 182
sulci 13, 160
sulcus limitans 33, 36, 71
superficial fibular nerve 12
superficial middle cerebral vein 185
superficial temporal artery 175
superior alveolar nerves 88
superior anastomotic vein 185
superior central nucleus 172
superior cerebellar artery 179
superior cerebellar peduncle 46, 75, 76, 105, 106, 134
superior cerebral veins 185
superior colliculus 71, 72, 76, 134, 170
superior frontal gyrus 127, 132, 136
superior frontal sulcus 17, 132, 136
superior ganglion 94, 95
superior longitudinal/arcuate fasciculus 145
superior oblique 86, 87
superior olivary nucleus 142
superior olive 169
superior orbital fissure 80, 81, 88
superior petrosal sinus (SPS) 183, 185
superior rectus 86, 87
superior sagittal sinus 183, 184, 185, 187
superior salivatory nucleus (CN VII) 79
superior semilunar lobule 100
superior temporal gyrus 136, 143, 144
superior temporal sulcus 136
superior thyroid artery 175
superior view orientation 10
supplementary motor area 110, 143, 162
suprachiasmatic nucleus 114, 117
supragranular layers 137
supramarginal gyrus 136
supraorbital foramen 88
supraorbital nerve 88
supraorbital notch 81
sura 8
sweat gland 56, 57
Sylvian fissure 136
sympathetic activation 147
sympathetic chain 164
sympathetic chain ganglion 164
sympathetic nervous system 65, **67**
sympathetic nervous trunk 69
sympathetic pathways 69
sympathetic postganglionic neuron (unmyelinated) 65
sympathetic preganglionic neuron (myelinated) 65
synapses 21, **28–29**
synaptic boutons 19
synaptic cleft 22, 29, 159
synaptic terminal 21
synaptic vesicle 22, 29

T
tail of caudate 123, 170
tail of caudate nucleus 15
tarsus 8
tectum 18, 76
tegmentum (reticular formation) 76
tela choroidea 38
telencephalon 18, 35, 38
temporal branch of facial nerve 90
temporal cerebral veins 185
temporal lobe 16, 38, 109, 126, 146, 148, 174
temporal pole 132
temporalis muscle 89
temporopontine fibers 76
tendon organ receptor capsule 59
tendon organ receptors **59**
tendons 58, 59
thalamic nuclei **109–110**, 165
thalamocortical fibers 166
thalamus 13, 15, 18, 43, 44, 71, **107–108**, 111, 112, 121, 123, 124, 131, **131**, 133, 134, 147, 154, 171, 172, 173
thermoceptors 44
third-order neuron 43, 44
third ventricle 13, 15, 39, 107, 132, 133, 186, 187
thoracic cavity 9
thoracic region 42
thoracic spinal cord **40–41**, 167
thoracic spinal nerves 40
thoracolumber spinal cord 69
thorax 8
thumb 158
thyrocervical axis 175
thyroid gland 119
thyroid-stimulating hormone 119
tight junction 30
tongue 94, 97
tonsil 100
touch pathway **166**
trachea 69
transverse (axial) plane 11, 15
transverse pontine fibers 75
transverse sinus 183, 185
trapezius muscle 96
trapezoid body 75
trigeminal ganglion 88
trigeminal motor nucleus (CN V) 79
trigeminal nerve (CN V) 70, 72, 78, 88, **88–89**, 134
trigeminal pathway 166
trigeminal thalamic fibers 76
trigeminothalamic tract 76, 109
trochlea 86
trochlear nerve (CN IV) 70, 71, 72, 76, 78, **86**
trochlear nucleus (CN IV) 79
trunk 8, 158
trunk of facial nerve 90
tuber 100
tuber cinereum 70
tympanic cavity 92
tympanic membrane 92
type Ia sensory fiber 58
type Ib afferent neuron 59
type Ib afferent sensory neuron 59
type II sensory fiber 58

U
ulnar nerve 12, 53
umbilicus 8
uncinate fasciculus 145
uncus 127, 128, 132
unipolar neuron 20
upper buccal branch of facial nerve 90
upper limb 8, 47
upper medulla 181
upper motor neurons 162
upper pons 167
utricle 92
uvula 100
uvulonodular fissure 101

V
V1 54, 140, 141, 168
V2 54, 140, 141, 168
V3 54, 140, 141, 168
V3a 140, 141
V4 54, 140, 141, 168
V5 140, 141, 168
V7 140, 141
V8 140, 141
vagal pathways 69
vagus nerve (CN X) 70, 72, 78, **95**, 96
vasocorona 182
vein of Labbé 185
vein of the central sulcus 185
vein of Trolard 185
ventral anterior nucleus 108, 110
ventral (anterior) view 14
ventral basal ganglia 146
ventral cavity 9
ventral cochlear nucleus 79, 169
ventral column 41
ventral horn 33, 41
ventral (inferior) view 14
ventral intraparietal area 168
ventral lateral nucleus 108, 110
ventral posterior nucleus 110
ventral posterolateral nucleus 108
ventral posteromedial nucleus 108, 166
ventral root 33, 41, 52, 164
ventral spinocerebellar tract **46**
ventral striatum **122**, 170
ventral tegmental area 76, 113, 170
ventral tegmental decussation 51
ventricle 125
ventricular layer 33
ventricular system **186**
ventricular zone 37
ventrolateral thalamus 106
ventromedial hypothalamus 114, 116
ventromedial prefrontal cortex 144
vermis 36, 102
vertebral artery 175, 176, 177, 180, 181
vertebral artery branches **176**
vesicles **35**
vestibular labyrinth 104
vestibular nuclear complex 74
vestibular nuclei 75, 79, 92, 93, 104, 105
vestibulo-ocular reflex (VOR) **93**
vestibulocerebellum 102, **104**
vestibulocochlear nerve (CN VIII) 70, 72, 78, **92–93**
vestibulospinal tracts **49**, 104
view orientation **10–11**
visceral effector 164
visceral sensory column 98
visual association area 126
visual cortex 126, **140–141**, 168
visual impulses 77
visual pathways **168**
visual radiation 140, 141
voltage-gated calcium channel 61
voluntary muscle contraction pathway **163**
VP 140, 141

W
Wallerian degeneration 64
water channels 156
Weigert stain 137
Wernicke's aphasia 152
Wernicke's area 17, 126, **151**
"what" pathway 168
"where" pathway 168
white matter 13, 33
white matter—dorsal column 52
white matter—lateral column 52
white matter—ventral column 52
white ramus 164
Whitnall's tubercle 81
wrist 8
wrist extension C6, C7 55
wrist flexion C6, C7 55

Z
zona incerta 111
zygomatic branch of facial nerve 90
zygomaticus 90